Springer Series on
ATOMIC, OPTICAL, AND PLASMA PHYSICS    44

Springer Series on
# ATOMIC, OPTICAL, AND PLASMA PHYSICS

The Springer Series on Atomic, Optical, and Plasma Physics covers in a comprehensive manner theory and experiment in the entire field of atoms and molecules and their interaction with electromagnetic radiation. Books in the series provide a rich source of new ideas and techniques with wide applications in fields such as chemistry, materials science, astrophysics, surface science, plasma technology, advanced optics, aeronomy, and engineering. Laser physics is a particular connecting theme that has provided much of the continuing impetus for new developments in the field. The purpose of the series is to cover the gap between standard undergraduate textbooks and the research literature with emphasis on the fundamental ideas, methods, techniques, and results in the field.

Takashi Fujimoto    Atsushi Iwamae
(Editors)

# Plasma Polarization Spectroscopy

With 180 Figures

 Springer

Professor Dr. Takashi Fujimoto
Dr. Atsushi Iwamae
Kyoto University, Graduate School of Engineering
Department of Mechanical Engineering and Science
Kyoto 606-8501, Japan
E-mail: t.fujimoto@zo4r2005.mbox.media.kyoto-u.ac.jp, iwamae@kues.kyoto-u.ac.jp

ISSN 1615-5653

ISBN 978-3-540-73586-1 Springer Berlin Heidelberg New York

Library of Congress Control Number:

Springer is a part of Springer Science+Business Media.

springer.com

© Springer-Verlag Berlin Heidelberg 2008

Typesetting and prodcution: SPI Publisher Services
Cover design: eStudio Calmar Steinen

Printed on acid-free paper     SPIN: 12038763     57/3180/SPI - 5 4 3 2 1 0

# Preface

Excited atoms and ions in a plasma produce line radiation, and continuum-state electrons emit continuum radiation. This radiation has been the subject of traditional plasma spectroscopy: the observed spectral distributions of the radiation or the intensities of the line and continuum radiation are reduced to the populations of electrons in the excited or continuum states, and these populations are interpreted in terms of the state of the plasma. As the word *intensity* suggests, it is implicitly assumed that this radiation is unpolarized. This is equivalent to assuming that the plasma is isotropic. If the plasma is under a magnetic and/or electric field, however, this assumption naturally breaks down: the atoms and ions are subjected to the Zeeman or Stark effect. A spectral line splits into several components and each component is polarized according to the field. This polarization is due to the anisotropy of the space in which the atoms or ions are present. Light or the radiation field, except for the case of an isotropic field like the blackbody radiation, is usually anisotropic. A polarized laser beam is an extreme example. Such a field can create anisotropy in atoms or ions when they are excited by absorbing photons from the field. Electrons having an anisotropic distribution in velocity space can create atomic anisotropy when these electrons excite the atoms or ions. Thus, it may be expected that we encounter many anisotropic plasmas, so that the radiation emitted by them is polarized. The polarization phenomena noted above have been recognized, of course, and the Zeeman or Stark effect is an important element of standard plasma diagnostic techniques. Other polarization phenomena have also been investigated to a certain extent, especially in the solar atmosphere research. In the laboratory plasma research, however, relatively little attention has been paid to the polarization of radiation. Given the fact that polarization is one of the important features of light, this situation may be regarded rather strange. This lack of interest in polarization may be ascribed to the fact that an electron velocity distribution is rather easily thermalized, especially in dense plasmas. Another factor may be experimental: if we want to detect polarization, and further, to measure it quantitatively, we have to do substantial preparations to perform such an experiment. Especially, if our plasma is time dependent or unstable or the wavelength of the radiation to be detected is outside the visible region, performing an experiment itself is extremely difficult.

In the past, there have been several plasma spectroscopy experiments in which an emphasis was placed on the polarization properties of the plasma radiation. Still these experiments are rather exceptional. Therefore, investigations in this direction may form a new research area; this new discipline may be named plasma polarization spectroscopy (PPS). As the brief account above suggests, PPS would provide us with information to which no other techniques have an access or information about finer details of the plasma, e.g., the anisotropic velocity distribution function of plasma electrons; this last aspect is important in plasma physics, e.g., the plasma instabilities.

There have been groups of workers who are interested in PPS, and, in the last decade, a series of international workshops have been held as the forum among them. It was agreed by the participants that PPS has now reached the point of some maturity and a book be published which summarizes the present status of PPS. These discussions have resulted in the present monograph.

As the editors of this book, we tried to make this book rather easy to understand for beginners, so that it should be useful for students and researchers who want to enter this new research area. We believe this book can be a step forward to establish PPS as a standard plasma diagnostic technique.

Kyoto, Japan                                                    *Takashi Fujimoto*
(August 2007)                                                   *Atsushi Iwamae*

# Contents

**11 Polarized Atomic Radiative Emission
in the Presence of Electric and Magnetic Fields**

# List of Contributors

**Moon Gu Baik**
Kyungwon University
Seongnam, Korea

**Elena O. Baronova**
Nuclear Fusion Institute
RRC Kurchatov Institute
Moscow 123182
Russia

**Igor Bray**
Curtin University of Technology
Perth, Western Australia
6845 Australia

**Roberto Casini**
High Altitude Observatory
National Center for Atmospheric
Research
P.O. Box 3000
Boulder, CO 80307-3000, USA

**George Csanak**
Theoretical Division
Los Alamos National Laboratory
Los Alamos, NM 87545, USA

**Takashi Fujimoto**
Department of Mechanical
Engineering and Science
Graduate School of Engineering
Kyoto University
Kyoto 606-8501, Japan
t.fujimoto@z04r2005.mbox.
media.kyoto-u.ac.jp

**Dmitry V. Fursa**
Curtin University of Technology
Perth, Western Australia
6845 Australia

**Motoshi Goto**
National Institute for Fusion
Science Toki 509-5292
Japan

**Peter Hakel**
Department of Physics
University of Nevada
Reno, NV 89557-0058, USA

**Atsushi Iwamae**
Department of Mechanical
Engineering and Science
Graduate School of Engineering
Kyoto University
Kyoto 606-8501
Japan
iwamae@kues.kyoto-u.ac.jp

**Verne L. Jacobs**
Materials Science and Technology
Division, Center for Computational
Materials Science
Naval Research Laboratory
Washington, DC 20375-5345, USA

**Lech Jakubowski**
Soltan Institute for Nuclear Studies
05-400 Swierk-Otwock, Poland

**Takako Kato**
National Institute for Fusion Science
Toki 509-5292, Japan

**Tetsuya Kawachi**
Japan Atomic Energy Agency
Kizu, Kyoto 619-0216, Japan

**David P. Kilcrease**
Theoretical Division
Los Alamos National Laboratory
Los Alamos, NM 87545, USA

**Young Soon Kim**
Myongji University
Yong-In, Korea

**Yong W. Kim**
Department of Physics
Lehigh University
Bethlehem, PA 18015, USA

**Egidio Landi Degl'Innocenti**
Dipartimento di Astronomia
e Scienze dello Spazio
Universitá di Firenze
Largo E. Fermi 2
I-50125 Firenze, Italy

**Richard More**
National Institute for Fusion Science
Toki 509-5292, Japan

**Mikhail M. Stepanenko**
Nuclear Fusion Institute
RRC Kurchatov Institute
Moscow 123182, Russia

# 1

## Introduction

T. Fujimoto

## 1.1 What is Plasma Polarization Spectroscopy?

Plasma spectroscopy is one of the disciplines in plasma physics: a spectrum of radiation emitted from a plasma is observed and its features are interpreted in terms of the properties of the plasma. In conventional plasma spectroscopy, line (and continuum radiation) intensities and broadening and shift of spectral lines have been the subject of observation. Attributes of the plasma, e.g., whether it is ionizing or recombining, what are its electron temperature and density, are deduced or estimated from the observation. We can expand the ability of plasma spectroscopy by incorporating in our framework the polarization characteristics of the radiation.

Figure 1.1 is an image of a plasma; a helium plasma is produced by a microwave discharge in a cusp-shaped magnetic field and this picture shows the intensity distribution of an emission line of neutral helium. The symmetry axis of the magnetic field and thus of the plasma lies horizontally below the bottom frame of the picture; this picture shows the upper one third of the plasma. The magnetic field is mirror symmetric with respect to the vertical plane (perpendicular to the axis) located at the center of this picture, and the magnetic field on this plane is purely radial. An interference filter placed in front of the camera lens selects the emission line of HeI $\lambda501.6\,$nm ($2^1S_0$–$3^1P_1$), and the intensity distribution of this line is recorded, as shown in this picture. Here, throughout this book, we adopt the convention for a transition that the lower level comes first and the upper level follows. A linear polarizer is also placed. From the comparison of the images for various directions of the transmission axis of the polarizer, the field view map of the directions and magnitudes of linear polarization is obtained; the result is shown with the direction and length of the bars. (The procedure to construct this picture is given in Chap. 14 later.) The meaning of the intensity distribution is rather straightforward; i.e., it shows the spatial distribution of the upper-level population of this line, i.e., He($3^1P$) in this case, or even the shape of the plasma. What does the polarization mean, especially in relation with the characteristics of

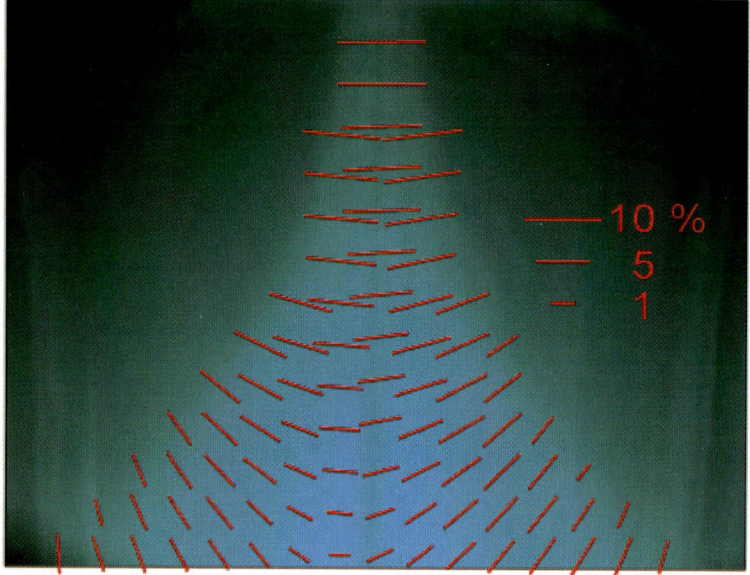

**Fig. 1.1.** The map of the intensity and polarization of the HeI $\lambda501.6\,\mathrm{nm}$ ($2\,^1\mathrm{S}_0 -$ $3\,^1\mathrm{P}_1$) line emitted from a microwave discharge plasma produced in a cusp-shaped magnetic field. The plasma axis lies horizontally below the bottom frame of the picture. The *short lines* indicate the magnitude and direction of linear polarization of this emission line

the plasma? This is the question to which *plasma polarization spectroscopy* (abbreviated to PPS henceforth) is to address.

As is obvious from the nature of polarization, the polarization phenomenon is related with spatial (more accurately, *directional*) anisotropy of the plasma. As a typical example of anisotropy, which will be important in PPS as discussed in more detail later in this book, we consider anisotropic electron impact on atoms. The most extreme example would be excitation of atoms by a beam of monoenergetic electrons. We discuss this collision process in a classical picture here.

An electron traveling in the $z$-direction collides with an atom located at the origin. This classical atom consists of an ion core and an electron that is attracted to the core with a harmonic force. In the case that the incident electron has an energy just enough to excite the atom and the collision is head on, the electron would give up the whole of its momentum and energy to excite the atom, and it stops there. The atomic electron begins to oscillate in the $z$-direction. This excited atom is nothing but a classical electric dipole, and it emits dipole radiation. If observed in the $x$–$y$ plane, the radiation is polarized in the $z$-direction, or it is the $\pi$ light, the electric vector of which oscillates in the $z$-direction. See Appendix A. Figure 1.2 shows an example of experimental observations on real atoms; helium atoms in the ground state

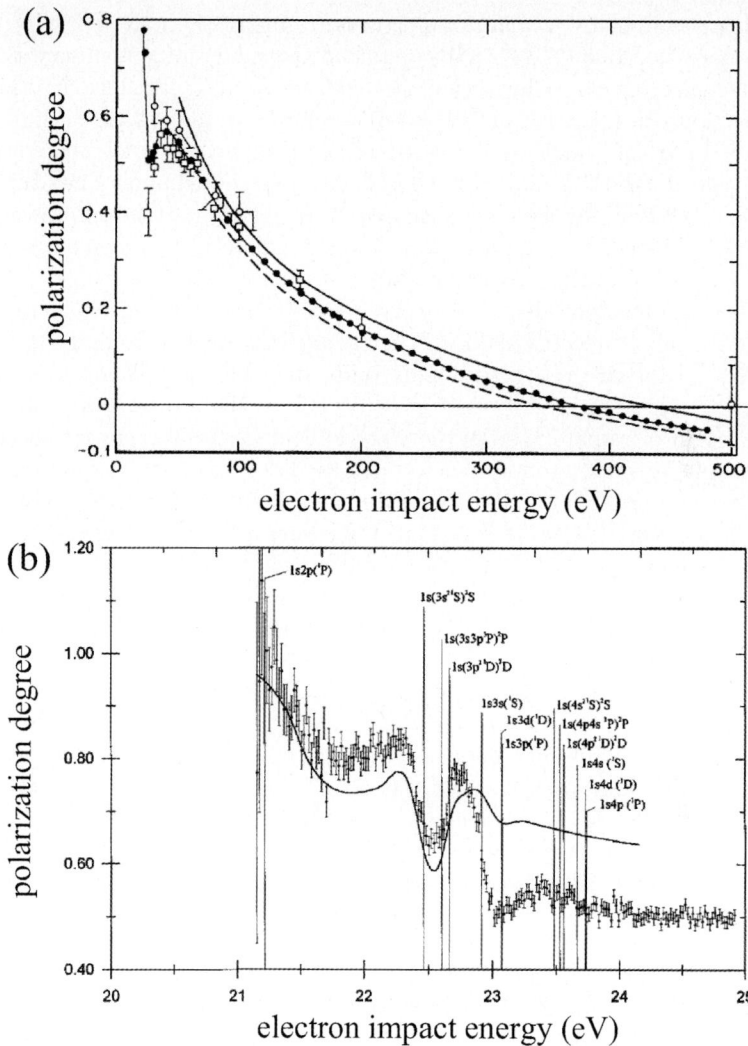

**Fig. 1.2.** Polarization degree of emitted radiation of neutral helium upon excitation from the ground state by an electron beam. (a) $1^1S \rightarrow n^1P$ with $n \geqslant 2$ for a broad energy region. (Quoted from [1], with permission from The American Physical Society). (b) $1^1S \rightarrow 2^1P$ close to the excitation threshold at 21.2 eV. The *full curve* represents the result of calculations convoluted with a 0.16 eV Gaussian function. In the figure, the positions of doubly excited levels are given near 22.5 eV and 23.5 eV; the former levels give rise to a structure because of the resonance effects. Singly excited level positions are also marked near 23 eV and 23.7 eV. The substantial deviation of the experimental polarization degree from the theoretical values in the higher energy region is obviously attributed to the cascading effects from these higher lying levels (Quoted from [2], with permission from The American Physical Society.)

$(1^1S_0)$ are excited by a beam of electrons to one of the $n^1P_1$ ($n \geqslant 2$) levels and a transition line $(1^1S_0 - n^1P_1)$ emitted by these atoms is observed [1,2]. The degree of linear polarization $P = (I_\pi - I_\sigma)/(I_\pi + I_\sigma)$ is determined, where $I_\pi$ is the intensity of the $\pi$ light, and $I_\sigma$ is that of the $\sigma$ light, the electric vector of which oscillates in the $x$–$y$ plane. Figure 1.2a shows the overall feature and Fig. 1.2b is the detailed structure just above the excitation threshold, $21.2\,\mathrm{eV}$, for the resonance line $(1^1S_0 - 2^1P_1)$ excitation. Toward the excitation threshold, the polarization degree tends to 1, in agreement with our above discussion in the classical picture.

When the incident electron is very fast and passes by our classical atom, it exerts a pulsed electric field on the atom. This field is, roughly speaking, directed within the $x$–$y$ plane. This pulse may be approximated as a half cycle of an electromagnetic wave propagating in the $z$-direction. It is noted that a beam of radiation lacks the electric field in its propagation direction. The "photo"-excited atomic electron will oscillate within this plane, and this atom again emits dipole radiation. This time, the radiation is the $\sigma$ light. As Fig. 1.2a suggests, within our picture, the polarization degree would go to $-1$ at very high energy.

Thus, an excited atom or ion keeps the memory of the direction of the collision by which it was produced and presents its memory in the form of polarization of the light it emits.

Since an atom (or an ion) in a plasma could be affected by various atomic interactions in its excitation and subsequent time development, the direction that the atom remembers may not be limited to that of the electron velocity. Atom and ion velocities, external fields, a radiation field, all these entities can enter into the memory of an atom and thus can be reflected in the polarization characteristics of the radiation it emits. Even recombination of electrons having an anisotropic velocity distribution could make the recombination continuum polarized and, in the case of recombination to an excited level, subsequent line emissions to still lower-lying levels are polarized, too. Only in the case when these atomic interactions are random in direction, or they are isotropic, we can expect the radiation to be unpolarized. In the conventional plasma spectroscopy, which we may call *intensity* spectroscopy, we implicitly assumed this situation. In the present context, the intensity spectroscopy provides information only of how *many* atoms were excited. The intensity distribution of the emission line in Fig. 1.1 gives us this information.

The above arguments constitute the starting point of PPS. If we utilize the polarization characteristics of radiation in interpreting the plasma, we should be able to deduce information of *how* these atoms or ions were excited in the plasma. Determination of anisotropic, therefore nonthermal, distribution function of electrons is an immediate example. As will be discussed in the subsequent chapters, atom collisions, electric and/or magnetic fields, a radiation field or even electromagnetic waves also affect the polarization characteristics of emission lines. All of these aspects are included in the framework of PPS.

## 1.2 History of PPS

The history of PPS may be traced back to 1924 when Hanle [3] reported a change of the polarization characteristics of the fluorescence light from a mercury vapor against applied magnetic field; the photo-excited atoms in a magnetic field perform Larmor precession, and the initial memory of excitation anisotropy is modified by the magnetic field during the lifetime of the atoms. See Appendix D for a more detailed explanation. In investigating the newly found Stark effect (see Chap. 2) by using a canal ray, Mark and Wierl [4] found that the intensity distribution among the polarized components of the Stark split Balmer $\alpha$ line depends on whether the ray passes through a low-pressure gas or a vacuum. This polarization may be interpreted as due to anisotropic collisional excitation of the canal ray atoms.

On the basis of the experimental and theoretical investigations of polarization of emission lines upon collisional excitation of atoms by electron impact [5], much progress was made in the 1950s in developing the theoretical framework, by which these excited atoms are treated in terms of the density matrix [6,7]. The density matrix is briefly discussed in Appendix C. The studies in the 1970s of interactions of photons with atoms, especially optical pumping [8,9], founded the theoretical basis of PPS.

Modern PPS research started in the middle 1960s. Spontaneous polarization of emission lines from plasma was discovered by three groups. The first was the observation of polarization of neutral helium lines from a high-frequency $rf$-discharge by Lombardi and Pebay-Peyroula in 1965 [10]. A little later, Kallas and Chaika [11], and Carrignton and Corney [12], almost simultaneously, reported their observations of the magnetic-field dependent polarization of neutral neon lines from DC discharge plasmas. Interestingly, they had little knowledge of other groups' work. This new phenomenon was named *the self alignment*. The polarization shown in Fig. 1.1 may be regarded as an example of self alignment. In these early observations, the origin of polarization of light, or of the alignment (this term will be explained in Chap. 4) in the upper-level "population", was attributed to directional collisional excitation by electrons, as mentioned later and discussed in Chaps. 5 and 6 in detail, or to radiation reabsorption in the anisotropic geometry, which will be discussed in Chap. 7.

In the 1970s–1980s, the self alignment phenomena of various origins were discovered and investigated vigorously on various discharge plasmas, mainly in the former Soviet Union. Gradually, it became recognized that PPS is a promising new technique, which would provide us with valuable information about the plasma, i.e., its anisotropy, to which no other ordinary techniques have an access. Thus, the target of PPS observations expanded to a variety of plasmas, and this trend continues now. These developments until a decade ago are summarized in Fujimoto and Kazantsev [13]. In astrophysical observations, polarization has been an important source of information about the magnetic field, the sprathermal electrons, and so forth in the solar atmosphere. Several monographs have been published recently [14–17].

An element always important in the PPS research is the instrumentation. For stationary discharge plasmas, an observation system based on the Hanle effect was developed, which was capable of determining polarization degrees as low as $10^{-4}$ [18]. For a variety of discharge conditions, self alignment produced by anisotropic electron impact, or by radiation reabsorption was observed, and even self alignment due to the ion drift motion was discovered [19]. By the use of the Hanle effect method, the lifetime of excited atoms and the alignment destruction rate coefficient (cross-section) by atom collisions were determined for many atomic species. Various possibilities of plasma diagnostics were demonstrated: obtaining the quadrupole moment of the electron velocity distribution [20], determining the energy input in a high-frequency discharge [21], determining the electric field [22]. The term *Plasma Polarization Spectroscopy* was first introduced by Kazantsev et al [23]. An interesting observation was on an atmospheric-pressure argon arc plasma; ionized argon lines showed polarization and this was quantitatively interpreted as due to the distorted Maxwell distribution of electron velocities [24].

An important target of PPS is the solar atmosphere; Atoms in the solar prominence is illuminated by the light from the solar disk, and the photoexcitation is anisotropic. The alignment thus produced is perturbed by the magnetic field present there. From the direction and the magnitude of the observed polarization of a helium emission line, for example, the direction and the strength of the magnetic field were deduced [25, 26]. Solar flares, in which anisotropic excitation of ions by electrons having a directional motion would produce alignment, were also a subject of PPS observation [27].

In laboratories, vacuum sparks and plasma focuses were also the target of PPS observations. Polarization was found on helium-like lines in the x-ray region [28]. However, the difficulty stemming from the observation geometry sometimes makes the interpretation complicated, and efforts to improve the instrumentation are being continued [29]. The $z$-pinch and the so-called X-pinch are being investigated vigorously [29–32].

The first PPS observation on a laser-produced plasma was made by Kieffer et al on helium-like aluminum lines [33, 34]. They interpreted the polarization as due to the anisotropic electron velocity distribution, which was caused by the nonlocal spatial transport of hot electrons from the underdense plasma to the overdense plasma. Another observation was performed by Yoneda et al. [35] on helium-like fluorine lines. The intensity distribution pattern of the resonance-series lines ($1^1S_0 - n^1P_1$) and the presence of the recombination continuum ($1^1S_0 - \varepsilon^1P_1$) clearly indicate that the observed plasma was in the recombining phase (see Chap. 3). Interesting findings were that the recombination continuum was polarized, and that the resonance-series lines were also polarized. The first fact indicates that the velocity distribution of the low-energy electrons that make radiative recombination is anisotropic: more directional to the direction of the target surface normal. This is against the general understanding that low-energy electrons are thermalized very rapidly. The second point indicates that owing, probably, to the anisotropic elastic collisions by electrons, $n^1P_1$ upper-level atoms are aligned: i.e., among the $M = 0, \pm 1$ magnetic sublevels, the $M = 0$ level is more populated. Here $M$ means the magnetic quantum number of the level having the total angular momentum quantum number $J$. $J$ is 1 in the present case. No interpretation of this experiment has been made so far, except for the discussion [36], which will be presented in Chap. 6 later in this book. A new experiment is performed [37]. Kawachi et al. [38] examined polarization of the neon-like germanium X-ray laser line of 19.6 nm. The transition was $2p^53s - 2p^53p$ ($J = 1 - 0$), so that the spontaneous emission of this line is never polarized. The observed polarization was ascribed to the alignment of the $2p^53s$ lower-level population, which was due to anisotropic radiation trapping, $2p^6 \leftrightarrow 2p^53s$. This experiment will be introduced in Chap. 10.

Magnetically confined plasmas including tokamak plasmas are also the target of PPS observations. MSE (motional Stark effect) is now a standard technique to determine the direction of the local magnetic field, and thus to determine the current distribution in the plasma [39, 40]. The Zeeman effect is also employed for plasma diagnostics [41]. The polarization resolved observation of the Zeeman profile of the Balmer $\alpha$ line was found quite useful [42]. Fujimoto et al. [43] first reported the polarization observation on carbon- and oxygen-ion emission lines from a tokamak plasma. They used a calcite plate incorporated into the spectrometer as the polarization resolving element. Anisotropic distributions of electron velocities were suggested as the origin of the observed polarizations. As shown in Fig. 1.1, magnetically confined plasmas are now a target of PPS observations. The full PPS formalism, which is described in [13] and also in Chap. 4 later, was implemented on the helium plasma in Fig. 1.1. An oblate-shaped distribution function was deduced from the intensity and polarization of several emission lines [44].

## 1.3 Classification of PPS Phenomena

As noted earlier, emission lines (and continua) can be polarized because of the anisotropy of the plasma. This anisotropy may be due to anisotropic collisional excitation as discussed in Sect. 1.1, or due to an external field, electric, or magnetic. Even electromagnetic waves could affect the polarization characteristics [45] as will be shown in Chap. 13. We classify the polarization phenomena into three classes:

Class 1: When an atom is placed in an electric field or a magnetic field it is subjected to the Stark effect or the Zeeman effect: an atomic level, and therefore a spectral line, is split into components and each of the components is polarized. When all the components are added together, the line is overall unpolarized. These phenomena are known for a long time and the formulation of these effects is well established. Still, new techniques are being developed for plasma diagnostics on the bases of these classical principles. When both the electric and magnetic fields are present at the same time with arbitrary strengths and relative directions, the problem is rather involved, and a prediction of the line profile as observed from an arbitrary direction is less straightforward. When a time-dependent electromagnetic field is applied, especially when the frequency is resonant with the energy separation of the Zeeman or Stark split sublevels, a new polarization phenomenon may emerge. This aspect is not well explored yet. If the applied field is static but extremely strong, the effects may not be a small perturbation, and the spectral line may show a new feature, including an appearance of overall polarization.

Class 2: An external field is absent. Atoms are subjected to anisotropic excitation: the directional electron collisions, photo-excitation by a laser beam, reabsorption of radiation (resonance scattering) in an anisotropic geometry, and so on. For the first anisotropy, the key is the velocity distribution of plasma electrons that excite the atoms. We simply call that EVDF (electron velocity distribution function) in the following. In this case, the immediate objective of PPS diagnostics is to deduce the "shape" of EVDF of the plasma in the velocity space. The presence of a weak magnetic field would make the produced atomic anisotropy rotate around the field direction, or it even defines the local axis of axial symmetry. The phenomena of this class are one of the main subjects to be developed in this book.

Class 3: This is the combination of Class 1 and Class 2. Anisotropic excitation under an electric field or a magnetic field, or even both of them. This Class is very difficult to treat, but, from the practical standpoint of plasma diagnostics of, say, $z$-pinch plasmas, this class should be explored and its formulation should be established. If the electric field is extremely strong, the problem of EVDF and that of the anisotropic excitation of atoms may not be separated, and they have to be treated self-consistently in a single framework.

## 1.4 Atomic Physics

Plasma spectroscopy is, from its nature, based on various elements in atomic physics; see Chap. 3 of Fujimoto [46]. This strong correlation with atomic physics is even more true with PPS. This is because polarization of radiation is due to intricate properties of an atom and its interaction with colliding perturbers, and further, due to the interaction of atom with the radiation field. Therefore, atomic physics constitutes an important element, or even a half, of PPS research.

Among the elements of atomic physics relevant to PPS, the area that is still under development is the field of atomic collisions involving polarization of atoms. Other elements, e.g., the density matrix formalism, which plays important roles in PPS, are well established. For readers who are unfamiliar with these concepts, the outline will be given in Appendices A–C. Among the collision processes, elastic and inelastic scattering of electrons on atoms or ions constitute the central problem. The classical picture introduced in Sect. 1.1 is too simplistic, and realistic theoretical treatments should be performed according to the particular problem that we face. For neutral atoms, studies of emission polarization upon electron or ion collisions have a long history. For ions as a target of collision experiment, experimental investigations have been quite limited for a long time because of the difficulty of producing enough number of ions. However, owing to the developments in the ion source technology, especially the invention of the device called EBIT (electron beam ion trap), the polarization study progressed substantially [47, 48]. On the theoretical side, thanks to the developments of computers, large-scale calculations have become possible, and calculations based on a new formalism are being made. Even so-called user-friendly codes, e.g., the FAC code, are becoming available [49]. Some workers put up their calculation results of cross sections on their home page, which is easily accessible. These circumstances are quite favorable for practicing PPS experiments on a variety of plasmas.

In the past PPS experiments, in many cases, polarization of virtually only one emission line was measured, and it was interpreted on the corona equilibrium assumption with a model anisotropic EVDF. However, intensity and polarization of several emission lines of atoms or ions in a plasma should give more comprehensive information about the plasma. A formulation for such an interpretation has been developed. This method is a generalization of the collisional-radiative (CR) model. The conventional collisional-radiative model has been the versatile tool in *intensity* plasma spectroscopy [46]. This new method is called the *population-alignment collisional-radiative* (PACR) model in [13]. This model will be introduced and discussed in Chap. 4.

Finally, the structure of the present book is outlined. Chapter 2 introduces the well-known effects of an electric or magnetic field on atoms, the Class 1 polarization. This chapter is intended for the reader to become familiar with these phenomena and, further, to be able to develop a new technique on the basis of the knowledge of these classical principles. Several recent examples of such developments are given. Chapter 3 is the summary of the collisional-radiative (CR) model. Neutral hydrogen is taken as an example of atoms and ions in a plasma. The objective of this chapter is twofold: the first is that the reader obtains the idea of what are the general properties of the excited-level populations in various situations of the plasma. The classification of plasmas into *the ionizing plasma* and *the recombining plasma* is introduced. The second objective is to establish the basis of the PACR model, which is to be developed in Chap. 4. As already noted, the PACR model is a generalization or an extension of the CR model. In Chap. 4, various cross sections relevant to the alignment are introduced, and the PACR formulation is established for the ionizing plasma and for the recombining plasma. In this chapter, the cross sections are treated semiclassically. Chapter 5 gives the quantum mechanical formulation of these cross sections. Chapter 6 discusses the physical meanings of various collision cross sections and rate coefficients introduced in

Chaps. 4 and 5. We also review briefly the present status of our knowledge of the cross section data. Chapter 7 deals with two polarization phenomena, which result from reabsorption of line radiation. They are creation and destruction of alignment. Both of the phenomena may be important in performing a PPS experiment on neutral atoms in a plasma in which radiation reabsorption is substantial. In Chap. 8, we review typical PPS experiments so far performed on plasmas that belong to the class of ionizing plasma, including discharge plasmas which have a long history of PPS research. Chapter 9 is devoted to the class of recombining plasma. In these chapters, we confine ourselves to the Class 2 polarization. Several other interesting facets of PPS experiments and formulation are introduced in Chap. 10. They are emission line polarization from a plasma confined by a gas, a polarized X-ray laser and an alternative approach to the PACR model. Chapter 11 is devoted to the problem of Class 3 polarization, i.e., anisotropic excitation in electric and magnetic fields. In Chap. 12, PPS observations of solar plasmas are introduced. Chapter 13 treats emission line polarization of hydrogen atoms under the influence of electromagnetic waves. In Chaps. 14 and 15, we look at several facets of instrumentation. In the visible–UV region, highly sophisticated devices have been developed. In the X-ray region, PPS experiments are extremely difficult, though information of anisotropy, e.g., the presence of beam electrons in a $z$-pinch plasma, is strongly needed. Several facets of instrumentation of X-ray PPS will be introduced. In Appendices, short summaries of the "tools" of PPS, e.g., the angular momentum, the density matrix, and the Hanle effect, are given for the purpose of convenience of the readers.

In the last decade, a series of international workshop has been held in every two-and-a-half years. These meetings are the forum among the researchers in plasma spectroscopy and in atomic physics, who are interested in PPS. The progress in PPS researches is reported and information exchanged. In the Reference section below, the Proceedings books of these workshops are given. An excellent review of PPS activities until the meeting of 2004 is given by Csanak [50]. The present book is, in a sense, an outcome from this series of workshop. A decade after the start of the workshops, it was felt that PPS has reached the stage of some maturity, and it was agreed among some of the participants that a monograph be published, which resulted in the present book.

# References

Proceedings of the series of international workshops provide a good perspective of the progress in this field:

Proceedings of the second workshop 1998 (Eds. T. Fujimoto and P. Beiersdorfer) NIFS-PROC-37 (National Institute for Fusion Science, Toki).
http://www.nifs.ac.jp/report/nifsproc.html

Proceedings of the third workshop 2001 (Eds. P. Beiersdorfer and T. Fujimoto) Report UCRL-ID-146907 (University of California Lawrence Livermore National Laboratory).
http://www-phys.llnl.gov/Conferences/Polarization/ http://www.osti.gov/energycitations/product.biblio.jsp?osti_id=15013530

Proceedings of the fourth workshop 2004 (Eds. T. Fujimoto and P. Beiersdorfer) NIFS-PROC-57 (National Institute for Fusion Science, Toki).
http://www.nifs.ac.jp/report/nifsproc.html

Papers in these Proceedings are referred to by the report number.

1. P. Hammond, W. Karras, A.G. McConkey, J.W. McConkey: Phys. Rev. A **40**, 1804 (1989)
2. C. Norén, J.W. McConkey, P. Hammond, K. Bartschat: Phys. Rev. A **53**, 1559 (1995)
3. W. Hanle: Z. Phys. **30**, 93 (1924)
4. H. Mark, R. Wierl: Z. Phys. **55**, 156 (1929)
5. For example: I.C. Percival, M.J. Seaton: Phil. Trans. R. Soc. A **251**, 113 (1958)
6. U. Fano: Rev. Mod. Phys. **29**, 74 (1957)
7. U. Fano, J.H. Macek: Rev. Mod. Phys. **45**, 553 (1973)
8. W. Happer: Rev. Mod. Phys. **44**, 169 (1972)
9. A. Omont: Prog. Quantum Electron. **5**, 69 (1977)
10. M. Lombardi, J.-C. Pebay-Peyroula: C. R. Acad. Sc. Paris **261**, 1485 (1965)
11. K. Kallas, M. Chaika: Opt. Spectrosc. **27**, 376 (1969)
12. C.G. Carrington, A. Corney: Opt. Comm. **1**, 115 (1969)
13. T. Fujimoto, S.A. Kazantsev: Plasma Phys. Control. Fusion **39**, 1267 (1997)
14. S.A. Kazantsev, J.-C. Hénoux: *Polarization Spectroscopy of Ionized Gases* (Kluwer Academic, Dordrecht, 1995)
15. S.A. Kazantsev, A.G. Petrashen, N.M. Firstova: *Impact Spectropolarimetric Sensing* (Kluwer Academic/Plenum, New York, 1999)
16. J.C. del Toro Iniesta: *Introduction to Spectropolarimetry* (Cambridge University Press, Cambridge, 2003)
17. E. Landi Degl'Innocenti, M. Landolfi: *Polarization in Spectral Lines* (Kluwer Academic, Dordrecht, 2004)
18. S.A. Kazantsev: Sov. Phys.-Usp. **26**, 328 (1983)
19. S.A. Kazantsev, A.G. Petrashen', N.T. Polezhaeva, V.N. Rebane, T.K. Rebane: JETP Lett. **45**, 17 (1987)
20. S.A. Kazantsev: JETP Lett. **37**, 158 (1983)
21. A.I. Drachev, S.A. Kazantsev, A.G. Rys, A.V. Subbotenko: Opt. Spectrosc. **71**, 527 (1991)
22. V.P. Demkin, S.A. Kazantsev: Opt. Spectrosc. **78**, 337 (1995)
23. S.A. Kazantsev, L.Ya. Margolin, N.Ya. Polynovskaya, L.N. Pyatnitskii, A.G. Rys, S.A. Edelman: Opt. Spectrosc. **55**, 326 (1983)
24. L.Ya. Margolin, N.Ya. Polynovskaya, L.N. Pyatnitskii, R.S. Timergaliev, S.A. Édel'man: High Temp. **22**, 149 (1983)
25. S. Sahal-Bréchot, V. Bommier, J.L. Leroy: Astron. Astrophys. **59**, 223 (1977)
26. V. Bommier, J.L. Leroy, S. Sahal-Bréchot: Astron. Astrophys. **100**, 231 (1981)
27. J.C. Hénoux, G. Chambe: J. Quant. Spectrosc. Radiative Transfer **44**, 193 (1990)

28. F. Walden, H.-J. Kunze, A. Petoyan, A. Urnov, J. Dubau: Phys. Rev. E **59**, 3562 (1999)
29. E.O. Baronova, M.M. Stepanenko, L. Jakubowski, H. Tsunemi: J. Plasma and Fusion Res. **78**, 731 (2002) (in Japanese)
30. A.S. Shlyaptseva, V.L. Kantsyrev, B.S. Bauer, P. Neill, C. Harris, P. Beiersdorfer, A.G. Petrashen, U.L. Safronova: UCRL-ID-146907 (2001) p.339
31. A.S. Shlyaptseva, V.L. Kantsyrev, N.D. Ouart, D.A. Fedin, P. Neill, C. Harris, S.M. Hamasha, S.B. Hansen, U.I. Safronova, P. Beiersdorfer, A.G. Petrashen: NIFS-PROC-57, (2004), p. 47
32. L. Jakubowski, M.J. Sadowski, E.O. Baronova: NIFS-PROC-57 (2004) p. 21
33. J.C. Kieffer, J.P. Matte, H. Pépin, M. Chaker, Y. Beaudoin, T.W. Johnston, C.Y. Chien, S. Coe, G. Mourou, J. Dubau: Phys. Rev. Lett. **68**, 480 (1992)
34. J.C. Kieffer, J.P. Matte, M. Chaker, Y. Beaudoin, C.Y. Chien, S. Coe, G. Mourou, J. Dubau, M.K. Inal: Phys. Rev. E **48**, 4648 (1993)
35. H. Yoneda, N. Hasegawa, S. Kawana, K. Ueda: Phys. Rev. E **56**, 988 (1997)
36. K. Kawakami, T. Fujimoto: UCRL-ID-146907 (2001) p. 187
37. J. Kim, D.-E. Kim: Phys. Rev. E **66**, 017401 (2002)
38. T. Kawachi, K. Murai, G. Yuan, S. Ninomiya, R. Kodama, H. Daido, Y. Kato, T. Fujimoto: Phys. Rev. Lett. **75**, 3826 (1995)
39. F.M. Levinton, R.J. Fonck, G.M. Gammel, R. Kaita, H.W. Kugel, E.T. Powell, D.W. Roberts: Phys. Rev. Lett. **63**, 2060 (1989)
40. D.J. Den Hartog et al.: UCRL-ID-146907 (2001) p. 205
41. M. Goto, S. Morita: Phys. Rev. E **65**, 026401 (2002)
42. A. Iwamae, M. Hayakawa, M. Atake, T. Fujimoto: Phys. Plasmas **12**, 042501 (2005)
43. T. Fujimoto, H. Sahara, T. Kawachi, T. Kallstenius, M. Goto, H. Kawase, T. Furukubo, T. Maekawa, Y. Terumichi: Phys. Rev. E **54**, R2240 (1996)
44. A. Iwamae, T. Sato, Y. Horimoto, K. Inoue, T. Fujimoto, M. Uchida, T. Maekawa: Plasma Phys. Control. Fusion **47**, L41 (2005)
45. R. More: UCRL-ID-146907 (2001), p. 201
46. T. Fujimoto: *Plasma Spectroscopy* (Oxford University Press, Oxford, 2004)
47. P. Beiersdorfer et al.: UCRL-ID-146907 (2001) pp. 299, 311, 329
48. P. Beiersdorfer et al.: NIFS-PORC-57 (2004) pp. 40, 87, 97
49. M.F. Gu: Astrophys. J. **582**, 1241 (2003)
50. G. Csanak: NIFS-PROC-57 (2004) p. 1

# Zeeman and Stark Effects

M. Goto

When an external magnetic field or electric field is present, emission lines from atoms or ions split into several components. The magnitude of the wavelength separation and the relative intensity among the split line components depend on the field strength. Such phenomena caused by the magnetic field and electric field are called the Zeeman effect and the Stark effect, respectively. The splitting of the emission lines is ascribed to the resolution of magnetic sublevels which are degenerate in the absence of an external field. The variation of relative intensity among the split line components is interpreted as a change of electric dipole moment between the magnetic sublevels of the transition; this change is caused by the wavefunction mixing. In this chapter, a quantitative treatment of these effects is introduced according to the perturbation theory.

## 2.1 General Theory

The determination of level energies of the resolved magnetic sublevels reduces to the eigenvalue problem of the Hamiltonian. Since perturbations due to external fields could be of a similar magnitude to the intrinsic perturbations, such as the spin–orbit interaction which is responsible for the fine structure splittings, they must be considered simultaneously. The Hamiltonian $H$ is then expressed as

$$H = H_0 + H_{\mathrm{FS}} + V, \tag{2.1}$$

where $H_0$ is the unperturbed Hamiltonian, $H_{\mathrm{FS}}$ is the perturbation resulting in the fine structures and $V$ is that due to the external fields. We employ the $|\alpha L S J M\rangle$ scheme as the base wavefunctions, where $L$, $S$, $J$, and $M$ are the orbital, spin, and total angular momentum quantum numbers, and the magnetic quantum number, respectively, and $\alpha$ represents all other parameters such as the principal quantum number by which the state is uniquely identified. This is a natural choice because $H_{\mathrm{FS}}$ is diagonal in this scheme and as a result $H'_0 = H_0 + H_{FS}$ is also diagonal: the elements of the matrix are the intrinsic energies of $J$ levels and therefore of the magnetic sublevels belonging to these $J$s.

Let us consider the Hamiltonian within the space of $n$ magnetic sublevels. The operator is then explicitly written as

$$H = H'_0 + V \tag{2.2}$$

$$= \begin{pmatrix} E_1^0 + V_{11} & V_{12} & \cdots & V_{1n} \\ V_{21} & E_2^0 + V_{22} & \cdots & V_{2n} \\ \vdots & \vdots & \ddots & \vdots \\ V_{n1} & V_{n2} & \cdots & E_n^0 + V_{nn} \end{pmatrix}, \tag{2.3}$$

where $E_i^0$ are the intrinsic energies of the $J$ levels, i.e., the magnetic sublevels under no external fields, and $V_{ij}$ are the perturbations due to the external field. The diagonalization of $H$ gives perturbed level energies as the eigenvalues, and wavelengths of possible line components are readily calculated from these level energies of the initial and final terms of the transition. The question whether individual lines are actually observed may be answered by the so-called selection rule.

When $V$ has nonvanishing nondiagonal elements, diagonalization gives rise to wavefunction mixing among the levels, and thus an eigenfunction is expressed as a linear combination of several base functions: the coefficients of the linear combination are obtained as the elements of the eigenvectors as a result of the diagonalization of $H$. In such circumstances the conventional selection rules are no longer valid. Instead, we calculate the spontaneous transition probabilities between such mixed levels and regard them as the relative intensities of the resolved line components. This is true when the population of the initial state is equally distributed over all the magnetic sublevels.

Each line component is polarized and thus the relative intensities depend on the observation direction. The polarization type of the emitted light is determined by the change of the magnetic quantum number $M$ in the transition. When the quantization axis ($z$-axis) is taken in the direction of the external field, transitions for $\Delta M = 0$ and $\Delta M = \pm 1$ give linearly polarized light in the $z$-direction ($\pi$-component) and circularly polarized light on the plane perpendicular to the $z$-axis ($\sigma$-component), respectively. Though the wavefunctions are generally mixed under an external field and some quantum numbers may no longer be good, $M$ remains always good and the statement concerning $\Delta M$ and the polarized components is valid irrespective of the existence of an external field. Here, the transition probability is calculated for each of the three polarized components. The observed intensities of the individual components are obtained as the product of the transition probability and the coefficients which depend on the angle $\theta$ between the field direction and the line of sight:

$$\sin^2\theta \qquad \text{for } \pi\text{-component,}$$
$$\frac{1+\cos^2\theta}{2} \qquad \text{for } \sigma\text{-component} \tag{2.4}$$

(see Appendix A). The spontaneous transition probability for a transition from magnetic sublevels $|i\rangle$ to $|f\rangle$ is denoted as $A_{if}^q$, where $q$ takes $-1$, $0$, or $1$ which corresponds to $\Delta M$. $A_{if}^q$ is explicitly written as

$$A_{if}^q = \frac{16\pi^3\nu^3 e^2}{3\varepsilon_0 hc^3}|\langle f|d_q|i\rangle|^2, \tag{2.5}$$

where $e$, $\varepsilon_0$, $h$, $c$, and $\nu$ are the elementary charge, dielectric constant of vacuum, Planck's constant, speed of light, and the frequency of the emitted light, respectively, and $d_q$ is the spherical component of the electric dipole moment $\boldsymbol{d}$.

We express the eigenfunctions under the perturbation of an external field as

$$|f\rangle = \sum_m C_m^f |f_m'\rangle \tag{2.6}$$

and

$$|i\rangle = \sum_n C_n^i |i_n'\rangle, \tag{2.7}$$

where $|f_m'\rangle$ and $|i_n'\rangle$ are the base functions of the final and initial states, respectively, and coefficients $C_m^f$ and $C_n^i$ are obtained as the elements of the eigenvectors. The factor $|\langle f|d_q|i\rangle|^2$ in (2.5) is then rewritten as

$$|\langle f|d_q|i\rangle|^2 = \left|\sum_{m,n} C_m^f C_n^i \langle f_m'|d_q|i_n'\rangle\right|^2, \tag{2.8}$$

and our problem is concisely expressed, an evaluation of the matrix elements $\langle f_m'|d_q|i_n'\rangle$.

The base functions can be explicitly written as

$$|f_m'\rangle = |\alpha_1 L_1 S_1 J_1 M_1\rangle \tag{2.9}$$

and

$$|i_n'\rangle = |\alpha_2 L_2 S_2 J_2 M_2\rangle, \tag{2.10}$$

and thus the term to be evaluated is rewritten as

$$\langle f_m'|d_q|i_n'\rangle = \langle\alpha_1 L_1 S_1 J_1 M_1|d_q|\alpha_2 L_2 S_2 J_2 M_2\rangle. \tag{2.11}$$

In the following discussion, the spin quantum number $S$ is assumed common to both the levels and is omitted in the expression of the wavefunctions because the only electric dipole transition for which the spin-change transition is forbidden is considered here and no interaction which gives rise to wavefunction mixing among different spin levels is taken into consideration.

With the help of the Wigner–Eckart theorem, the $M$-dependence of the matrix element is extracted as a coefficient [1] as

$$\langle \alpha_1 L_1 J_1 M_1 | d_q | \alpha_2 L_2 J_2 M_2 \rangle$$

$$= (-1)^{J_1 - M_1} \begin{pmatrix} J_1 & 1 & J_2 \\ -M_1 & q & M_2 \end{pmatrix} \langle \alpha_1 L_1 J_1 || d || \alpha_2 L_2 J_2 \rangle, \qquad (2.12)$$

where the coefficient

$$\begin{pmatrix} J_1 & 1 & J_2 \\ -M_1 & q & M_2 \end{pmatrix}$$

is the 3-$j$ symbol. The last term is the reduced matrix element which can be further reduced as

$$\langle \alpha_1 L_1 J_1 || d || \alpha_2 L_2 J_2 \rangle = (-1)^{S+1+L_1+J_2}$$

$$\times \sqrt{(2J_1 + 1)(2J_2 + 1)} \begin{Bmatrix} L_1 & J_1 & S \\ J_2 & L_2 & 1 \end{Bmatrix} \langle \alpha_1 L_1 || d || \alpha_2 L_2 \rangle, \qquad (2.13)$$

where

$$\begin{Bmatrix} L_1 & J_1 & S \\ J_2 & L_2 & 1 \end{Bmatrix}$$

is the 6-$j$ symbol. For the 3-$j$ and 6-$j$ symbols, see Appendix B. From these results (2.11) is rewritten as

$$\langle f'_m | d_q | i'_n \rangle = (-1)^{(J_1 - M_1) + (S+1+L_1+J_2)} \sqrt{(2J_1 + 1)(2J_2 + 1)}$$

$$\times \begin{pmatrix} J_1 & 1 & J_2 \\ -M_1 & q & M_2 \end{pmatrix} \begin{Bmatrix} L_1 & J_1 & S \\ J_2 & L_2 & 1 \end{Bmatrix} \langle \alpha_1 L_1 || d || \alpha_2 L_2 \rangle. \qquad (2.14)$$

The eventual reduced matrix element is related to the line strength $S'$ [2] as

$$\langle \alpha_1 L_1 || d || \alpha_2 L_2 \rangle = \pm a_0 \sqrt{\frac{S'}{2S+1}}, \qquad (2.15)$$

where $a_0$ is the Bohr radius and $S'$ is in atomic units. The positive and negative signs are valid for $L_1 = L_2 + 1$ and $L_1 = L_2 - 1$, respectively [2].

Substituting (2.8), (2.14), and (2.15) into (2.5), we obtain the transition probabilities or the relative intensities of the resolved line components.

## 2.2 Zeeman Effect

The perturbation $V$ due to a magnetic field $\boldsymbol{B}$ is

$$V = -\boldsymbol{\mu} \cdot \boldsymbol{B}, \tag{2.16}$$

where $\boldsymbol{\mu}$ is the magnetic moment of the atom. We take the quantization axis ($z$-axis) in the field direction. When the $L$–$S$ coupling scheme is valid, $V$ is rewritten as

$$
\begin{aligned}
V &= -\mu_{\mathrm{B}}(g_L \boldsymbol{L} + g_S \boldsymbol{S}) \cdot \boldsymbol{B} \\
&= -\mu_{\mathrm{B}} B (g_L L_z + g_S S_z),
\end{aligned}
\tag{2.17}
$$

where $\mu_{\mathrm{B}}$ is the Bohr magneton, $g_L (= 1)$ and $g_S (\simeq 2)$ are the orbital and spin $g$-factors, respectively, and $B = |\boldsymbol{B}|$.

For the Zeeman effect, the wavefunction mixing among the magnetic sublevels in the same $L$–$S$ term is important. We consider a Hamiltonian which consists of all the magnetic sublevels belonging to a single term $n\,^{2S+1}L$. We denote the wavefunction of the magnetic sublevels as $|JM\rangle$, and other common quantum numbers such as $n$, $S$, and $L$ are omitted in the expression.

The matrix elements of the perturbation term $V$ is calculated as

$$
\begin{aligned}
\langle JM|V|J'M'\rangle &= -\mu_{\mathrm{B}} B \langle JM|(g_L L_z + g_S S_z)|J'M'\rangle \\
&= -\mu_{\mathrm{B}} B \left( \sum_{M_L M_S} \langle JM|LSM_L M_S\rangle\langle LSM_L M_S| \right) (g_L L_z + g_S S_z) \\
&\quad \times \left( \sum_{M'_L M'_S} |LSM'_L M'_S\rangle\langle LSM'_L M'_S|J'M'\rangle \right),
\end{aligned}
\tag{2.18}
$$

where

$$\sum_{M_L M_S} |LSM_L M_S\rangle\langle LSM_L M_S| = 1 \tag{2.19}$$

is used. It is readily noticed that the component

$$\langle LSM_L M_S|(g_L L_z + g_S S_z)|LSM'_L M'_S\rangle$$

is zero unless $M_L = M'_L$ and $M_S = M'_S$, and consequently $M' = M$. When these conditions are satisfied, this component is calculated as

$$\langle LSM_L M_S|(g_L L_z + g_S S_z)|LSM_L M_S\rangle = g_L M_L + g_S M_S. \tag{2.20}$$

From these results, the nonzero elements are obtained as

$$
\begin{aligned}
\langle JM|V|J'M\rangle &= -\mu_{\mathrm{B}} B \sum_{M_L M_S} \langle LSM_L M_S|JM\rangle \\
&\quad \times \langle LSM_L M_S|J'M\rangle (g_L M_L + g_S M_S),
\end{aligned}
\tag{2.21}
$$

where the Clebsch–Gordan coefficient $\langle LSM_LM_S|JM\rangle$, for example, can be rewritten with the 3-$j$ symbols [1] as

$$\langle LSM_LM_S|JM\rangle = (-1)^{(L-S+M)}\sqrt{2J+1}\begin{pmatrix} L & S & J \\ M_L & M_S & -M \end{pmatrix} \qquad (2.22)$$

(see also Appendix B). The matrix elements of the perturbation term is thus obtained.

For the remaining part of the Hamiltonian, $H_0'$ in (2.2), the energies of the intrinsic fine structure levels are adopted, and the calculation of the total Hamiltonian $H$ is now completed.

For each of the initial and final terms all the base functions which constitute the Hamiltonian have the same quantum numbers $S$ and $L$, and therefore the reduced component of the electric dipole moment operator

$$\langle \alpha_1 L_1 ||d|| \alpha_2 L_2 \rangle$$

in (2.14) is common to all the transitions. This means if our interest is focused only on the relative intensities over the resolved line components, this quantity is common and not necessary to be evaluated.

We take CII ion as an example and consider the transitions between $[1s^2 2s^2]$ 3s $^2S_{1/2}$ and 3p $^2P_{1/2,\,3/2}$ terms. Figure 2.1a shows the level energy shifts against the magnetic-field strength for the initial and final terms and

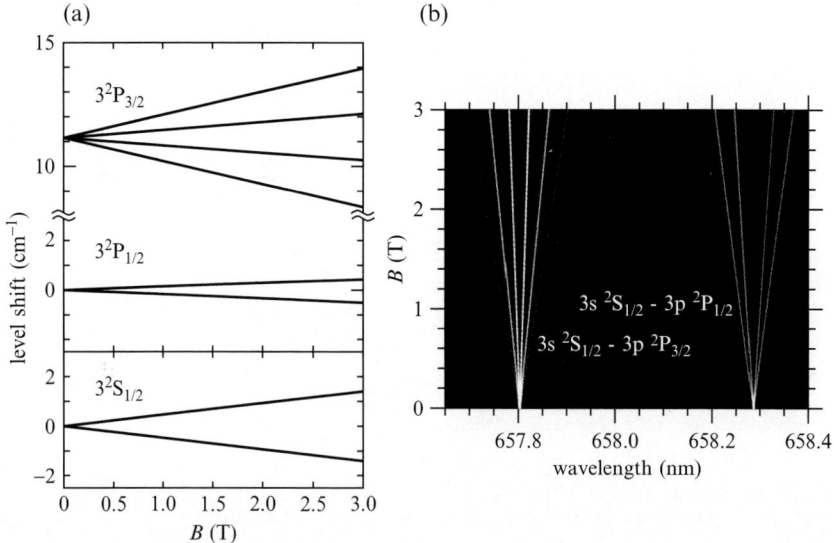

**Fig. 2.1.** (a) Energy level shifts against the magnetic field strength for 3p $^2P_{1/2,3/2}$ and 3s $^2S_{1/2}$ terms of CII and (b) line splitting for CII (3s $^2S_{1/2}$–3p $^2P_{1/2,3/2}$) lines where the tone of lines indicates the relative intensity

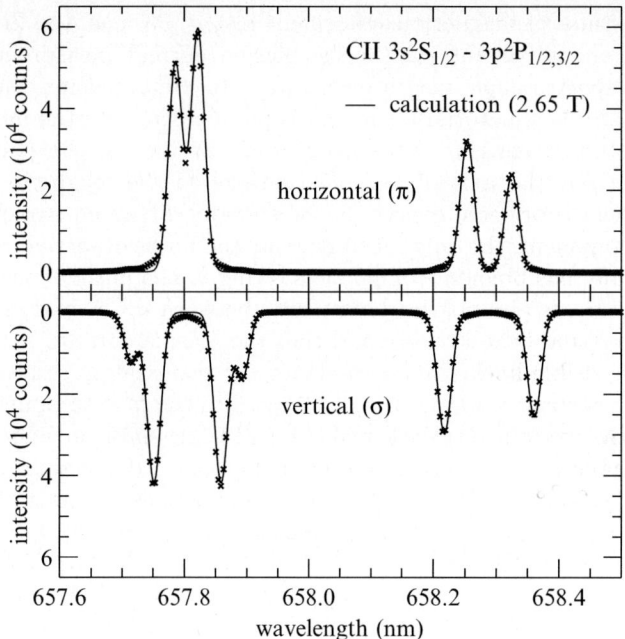

**Fig. 2.2.** Polarization-separated spectra of CII ($3s\ ^2S_{1/2}$–$3p\ ^2P_{1/2,3/2}$) transition observed at the Large Helical Device (LHD), National Institute for Fusion Science, Toki. *Solid lines* are the results of calculation with $B = 2.65$ T

Fig. 2.1b shows the consequent line splittings. The tone of the curves indicates the relative line intensity where the line of sight is assumed to be perpendicular to the field direction. In this range of the field strength the shifts are almost exactly proportional to the field strength; the anomalous Zeeman effect appears a good approximation. An actual observation result at the Large Helical Device (LHD) in National Institute for Fusion Science, Toki, is shown in Fig. 2.2 for which the polarization components are separated with a linear polarizer. From the splitting width of the line components the field strength of $B = 2.65$ T is deduced and the synthetic spectrum with this field strength is confirmed to agree with the measurement; in particular, the intensity distribution among the resolved components is well reproduced.

What should be noted here is that the line intensity distribution is asymmetric with respect to the original line positions and this result is against the expectation from the anomalous Zeeman effect scheme. The reason is the modification of the individual spontaneous transition probabilities due to the wavefunction mixing. It must therefore be kept in mind that even when the anomalous Zeeman effect approximation appears valid from the wavelength shifts, the detailed calculation in Sect. 2.1 could still be required to obtain the line intensity distribution.

In the study of magnetic confinement fusion plasmas, the Zeeman effect has played an important role in the plasma current measurement. Let us consider a spectroscopic measurement for a tokamak plasma where the line of sight is on the equatorial plane and is perpendicular to the magnetic axis. We then observe emission lines from atoms or ions in the plasma with a linear polarizer, the axis of which is parallel to the magnetic axis. If the magnetic field is oriented exactly in the toroidal direction, namely, it has no poloidal component, the only $\pi$-components should be observed. However, the field generally has poloidal components owing to the plasma current and the $\sigma$-components are also obtained. The ratio between the $\pi$- and $\sigma$-components gives the poloidal field strength and thus the plasma current.

Though visible lines are preferable for the polarization resolved measurement, their line emissions are usually located in the plasma boundary region. In the Texas experimental tokamak (TEXT), an emission line accompanying the magnetic dipole transition of highly charged titanium ion, TiXVII 383.4 nm $(2s^2 2p^2\,{}^3P_1 - {}^3P_2)$, has been used for the measurement of the plasma current in the central region [3–6]. Another method for the similar purpose is based on a visible measurement of neutral atoms which are injected in the central region with a monoenergetic lithium beam [7–9] or a pellet injection [10]. With these techniques the radial profile measurement has also been attempted.

Besides the plasma current measurement, the Zeeman effect is exploited for the study of neutral or low-ionized particle dynamics in the plasma boundary region. Some recent researches are the emission location determination from the field strength obtained from the line splittings due to the Zeeman effect [11–15]. In these studies the observation is conducted with a line of sight which passes through a poloidal cross section of the plasma, and the line emissions at the inboard- and outboard-side plasma boundaries are separated from their different splitting widths. The center wavelengths of the observed two Zeeman profiles are relatively shifted and the inward velocity of the atoms and ions are deduced from the shift width. In [15], a polarization-resolved measurement is applied to the Balmer $\alpha$ line of neutral hydrogen, and the difficulty that the Zeeman splittings are veiled by other broadenings such as the Doppler broadening is overcome.

## 2.3 Stark Effect

The perturbation term due to an electric field, $\boldsymbol{E}$, is

$$V = -\boldsymbol{E} \cdot \boldsymbol{d}, \qquad (2.23)$$

where $\boldsymbol{d}$ is the electric-dipole moment. When we take the quantization axis ($z$-axis) in the direction of $\boldsymbol{E}$, it is rewritten as

$$V = -Ed_z, \qquad (2.24)$$

where $E = |\boldsymbol{E}|$. In the Stark effect the wavefunction mixing among different $L$-levels plays an important role. Therefore we consider a Hamiltonian on the $|LJM\rangle$ basis assuming a common $S$. No interaction which gives rise to a mixing between different $S$ terms is taken into account. The matrix elements are calculated as

$$\langle LJM|V|L'J'M'\rangle = -E\langle LJM|d_z|L'J'M'\rangle$$
$$= -E(-1)^{(J-M)+(S+1+L+J')}\sqrt{(2J+1)(2J'+1)}$$
$$\times \begin{pmatrix} J & 1 & J' \\ -M & 0 & M' \end{pmatrix} \begin{Bmatrix} L & J & S \\ J' & L' & 1 \end{Bmatrix} \langle L||d||L'\rangle, \qquad (2.25)$$

where (2.12) and (2.13) are used. The reduced element $\langle L||d||L'\rangle$ can be calculated with (2.15). The remaining calculations to obtain the line splittings and the transition probabilities are the same as in Sect. 2.2.

We take the Balmer $\alpha$ line of neutral hydrogen as an example. Figure 2.3a shows the dependence of level energies of $n = 2$ and $n = 3$ on the electric-field strength. Though the effect on hydrogen and hydrogen-like ions is understood as a linear Stark effect, i.e., the level shifts are proportional to the field strength as shown in Fig. 2.3a, this is not the case when the field is so weak that the level shifts are smaller than or comparable with the intrinsic fine structure splittings as shown in Fig. 2.3b.

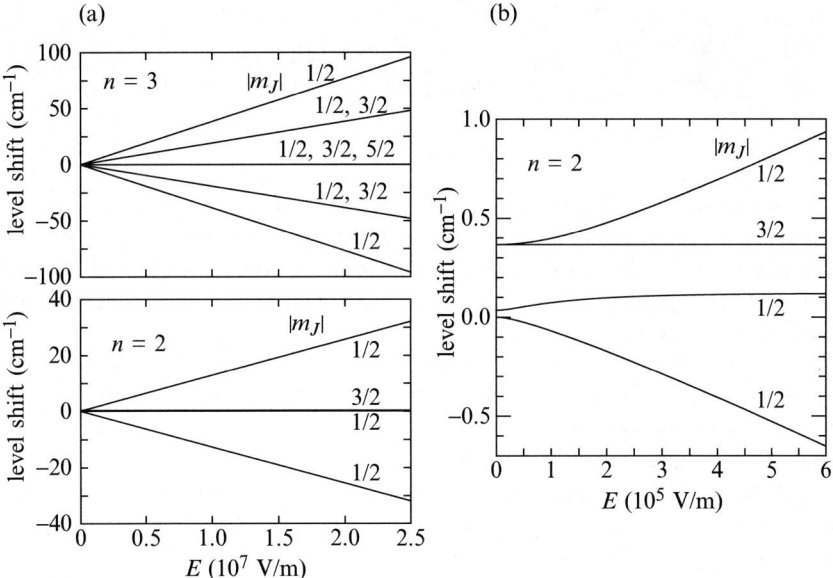

**Fig. 2.3.** Energy level shifts against electric field strength for $n = 3$ and $n = 2$ of neutral hydrogen (**a**) in a wide field strength range and (**b**) in a weak field strength range where the intrinsic fine structure and Stark splittings are comparable

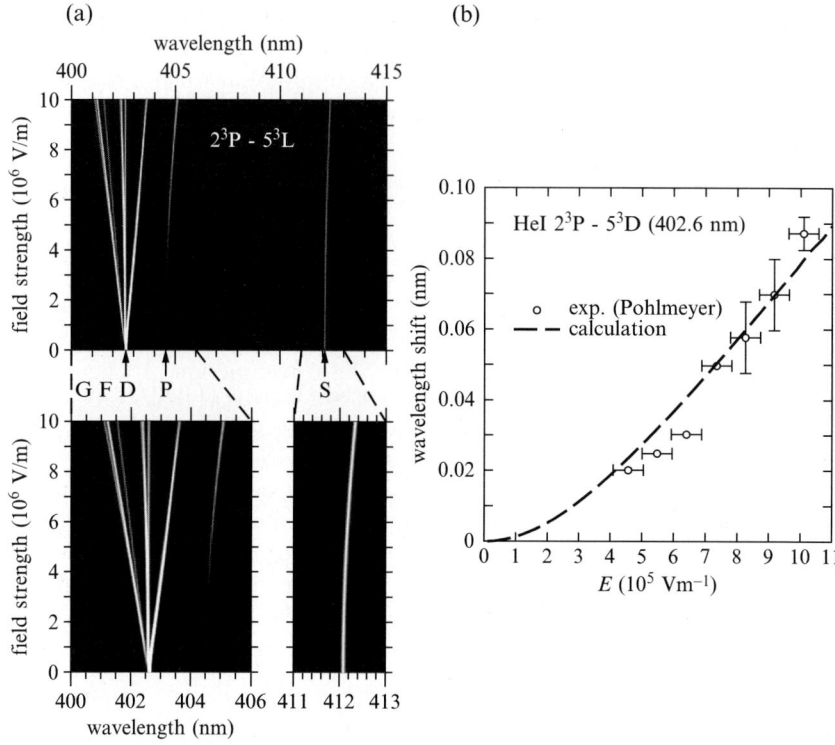

**Fig. 2.4.** (a) Stark effect for the $2\,^3\text{P}$–$5\,^3L$ ($L = 0$, 1, 2, 3, and 4) terms of neutral helium, where the tone of lines indicates the relative intensity and (b) comparison between measured and calculated line shift for the $2\,^3\text{P}$–$5\,^3\text{D}$ line, for which the experimental data are taken from [16]

Figure 2.4a is another example for transitions between $2\,^3\text{P}$ and $5\,^3L$ ($L = \text{S}$, P, D, F, and G) terms of neutral helium where the wavelength shifts and relative intensities are shown as a function of field strength. It is readily noticed that the effect is nonlinear with respect to the field strength. This remarkable difference between the helium and hydrogen cases originates from the apparent $L$-degeneracy of hydrogen. The detailed discussion for this is found in literature of atomic physics like [1,2] and omitted here. In Fig. 2.4b the calculation result for the $2\,^3\text{P}$–$5\,^3\text{D}$ transition[1] is confirmed to show a good agreement with an experiment [16].

As shown in Fig. 2.4a, a new line emerges at around 405 nm and becomes stronger with the increasing electric-field strength. This line obviously corresponds to the forbidden transition, $2\,^3\text{P}$–$5\,^3\text{P}$, and is caused owing to the

---

[1] Strictly speaking, this notation is not correct because the $L$–$S$ coupling scheme breaks down under an electric field. It is actually understood that "the line which is continuously connected to the original line of the $2\,^3\text{P}$–$5\,^3\text{D}$ transition."

wavefunction mixing of the term $5\,^3$D into $5\,^3$P. Therefore, though it is called forbidden line, the mechanism itself is the electric dipole transition.

Naturally, the Stark effect can be used for the electric field measurement. An example is introduced which is the measurement of the field strength in the plasma sheath region close to the metal surface against a plasma flow. The method is based on the laser-induced fluorescence (LIF) technique for neutral helium [17]. The excitation $2\,^1$S $\rightarrow$ $3\,^1$D ($\lambda = 504.2$ nm), which is optically forbidden, is induced by a linearly polarized laser light and the subsequent line emission accompanying the spontaneous transition down to $2\,^1$P ($\lambda = 667.8$ nm) is observed.

The excitation takes place through two mechanisms. One is the electric dipole scheme which is made possible by the wavefunction mixing between the $3\,^1$P and $3\,^1$D terms due to the Stark effect. Since the mixing degree changes depending on the electric-field strength, the excitation rate also depends on the field strength. The other is the electric quadrupole scheme which has an almost constant transition probability irrespective of the presence of the external field. Figure 2.5 is a schematic diagram for these processes and the subsequent line emissions where the quantization axis ($z$-axis) is taken in the direction of the laser light beam. As indicated in the figure, the electric quadrupole excitation yields only the $\sigma$-light, and this suggests the polarization degree, or the longitudinal alignment $\alpha$ (see (4.7) later), of the emission line depends on the strength of the static electric field in the plasma.

Figure 2.6 shows an example of the polarization-resolved line intensities and the longitudinal alignment obtained for a line of sight perpendicular to the

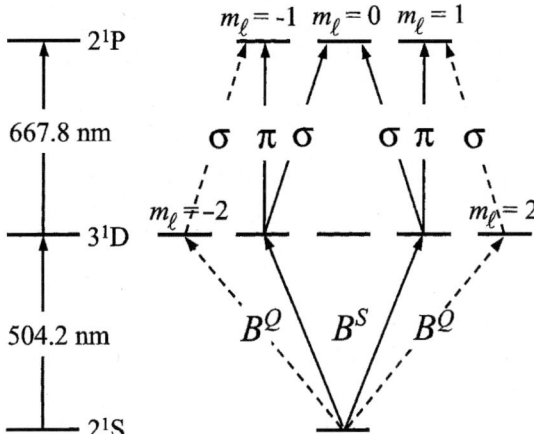

**Fig. 2.5.** Energy levels related to the LIF observation. The direction of the electric field oscillation of the pump laser is perpendicular to the quantization axis. The absorption coefficients $B^{\rm S}$ and $B^{\rm Q}$ correspond to the transition caused from the wavefunction mixing due to the Stark effect and the electric quadrupole transition, respectively (Quoted from [17], with permission from Elsevier.)

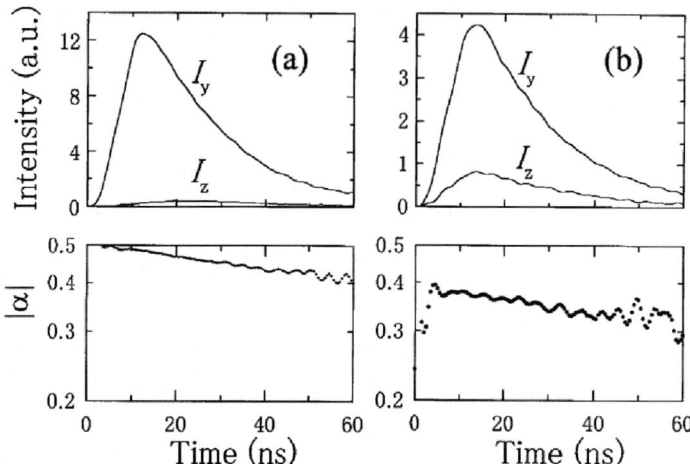

**Fig. 2.6.** Temporal variation of polarization-resolved line intensities and the longitudinal alignment of the 667.8 nm line in the plasma region (**a**) and in the sheath region (**b**) (Quoted from [17], with permission from Elsevier.)

quantization axis (a) in the plasma region and (b) in the sheath region. Here, $I_y$ and $I_z$ correspond to the $\sigma$- and $\pi$-light, respectively. The result (a) shows the radiation is almost completely polarized just after the laser irradiation, and this indicates that the excitation is dominated by the electric quadrupole transition; namely, the electric field is weak. The decay of the longitudinal alignment corresponds to the relaxation of population imbalance among the magnetic sublevels (disalignment) probably due to electron collisions. Meanwhile, in the result (b), both the polarized components are observed from the beginning, and a static electric field of $8.1\,\mathrm{kV\,m^{-1}}$ is deduced from the obtained longitudinal alignment value at the beginning.

Another example is the plasma current measurement for the fusion plasma of magnetic confinement. When neutral hydrogen beam is injected into a plasma, atoms in the beam feel an electric field of

$$\boldsymbol{E} = \boldsymbol{v} \times \boldsymbol{B}, \tag{2.26}$$

where $\boldsymbol{v}$ is the atom velocity, and exhibit the Stark effect. This is called the motional Stark effect [18–22]. Figure 2.7 shows an example of the measured Stark splitting for the Balmer $\alpha$ line in LHD. The principle of the measurement is the same as the Zeeman effect case: the polarization separated measurement gives the direction of the induced electric field, and from this result and the beam direction the magnetic field direction is deduced. This is used as a standard measurement method in various fusion devices.

The Stark effect also plays an important role in the line broadening, the so-called Stark broadening. Atoms and ions in a high-density plasma feel various strengths of electric field from surrounding charged particles, and the energy

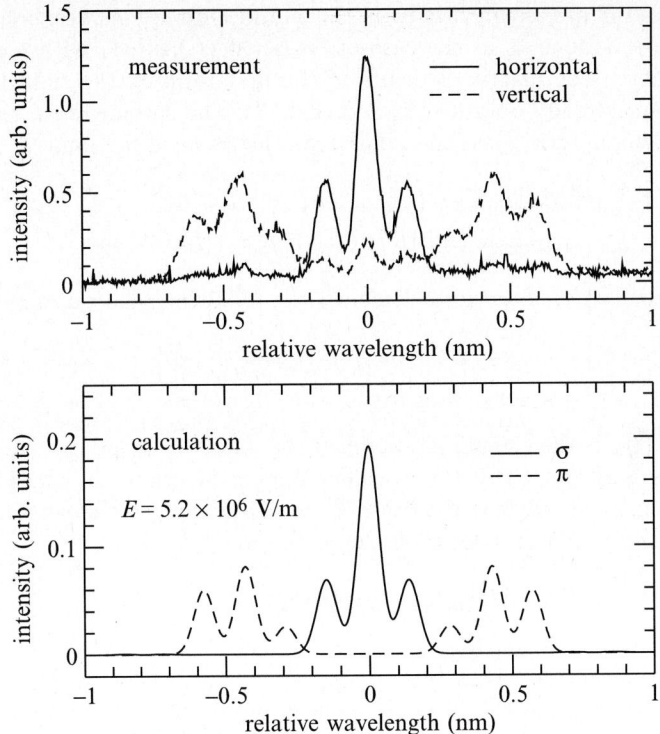

**Fig. 2.7.** Motional Stark effect of the Balmer $\alpha$ line of neutral hydrogen measured on LHD and a calculation result with $E = 5.2 \times 10^6 \, \mathrm{Vm}^{-1}$

levels are broadened rather than shifted or split. Consequently, the emission lines related to the perturbed energy levels are broadened. Practically, the profile or the width of the Stark broadening has been utilized as a measure of the electron density. The details concerning the Stark broadening are found in [23, 24], for example.

## 2.4 Combination of Electric and Magnetic Fields

We consider the case where both the magnetic and electric fields are present at the same time and they are perpendicular to each other. This is the situation where, in the motional Stark effect, for example, the influence from the magnetic field cannot be neglected.

The perturbation term is expressed as

$$V = -\boldsymbol{E} \cdot \boldsymbol{d} - \boldsymbol{\mu} \cdot \boldsymbol{B}$$
$$= -E d_z - \mu_{\mathrm{B}} B (g_L L_x + g_S S_x), \tag{2.27}$$

where $\boldsymbol{E}$ and $\boldsymbol{B}$ are assumed to be in the $z$- and $x$-directions, respectively. The same base functions as in the case of Stark effect, $|LJM\rangle$ with common $S$, are used here. The matrix elements of the first term on the right-hand side of (2.27) are already obtained in Sect. 2.3. For the second term, the matrix elements remain only when the related two levels have the same $L$, i.e.,

$$\langle LJM|\left[-\mu_B B\left(g_L L_x + g_S S_x\right)\right]|L'J'M'\rangle$$
$$= -\mu_B B\langle LJM|\left(g_L L_x + g_S S_x\right)|L'J'M'\rangle\delta_{LL'}. \qquad (2.28)$$

For the calculation of nonzero elements, the following relation is used:

$$|LJM\rangle = \sum_m |LJm\rangle_x r^J_{mM}\left(-\frac{\pi}{2}\right), \qquad (2.29)$$

where $|LJm\rangle_x$ is the eigenfunction when the quantization axis is taken in the $x$-direction, and $r^J_{mM}(\alpha)$ is the so-called Wigner function [25] which indicates the rotation by an angle $\alpha$ with respect to the $y$-axis (see Appendix B). The nonzero matrix elements are then calculated as

$$-\mu_B B\langle LJM|(g_L L_x + g_S S_x)|LJ'M'\rangle$$
$$= -\mu_B B \left(\sum_m r^J_{mM}\left(-\frac{\pi}{2}\right)\langle LJm|_x\right)(g_L L_x + g_S S_x)$$
$$\times \left(\sum_{m'} r^{J'}_{m'M'}\left(-\frac{\pi}{2}\right)|LJ'm'\rangle_x\right)$$
$$= -\mu_B B \sum_{mm'} r^J_{mM}\left(-\frac{\pi}{2}\right)r^J_{m'M'}\left(-\frac{\pi}{2}\right)$$
$$\times \langle LJm|_x(g_L L_x + g_S S_x)|LJ'm'\rangle_x.$$

The elements

$$-\mu_B B\langle LJm|_x(g_L L_x + g_S S_x)|LJ'm'\rangle_x$$

are identical with (2.18) and derived already.

The following procedure, i.e., the diagonalization of the Hamiltonian to determine the level energies and coefficients for the wavefunction mixing, is the same as in Sect. 2.3. For the line intensities, however, some attentions should be paid. So far, the electric dipole moment operator, $\boldsymbol{d}$, has been decomposed into three spherical components, $d_q$, which correspond to one linearly polarized component in the external field direction ($q = 0$) and two circularly polarized components in the plane perpendicular to the external field direction ($q = \pm 1$), respectively. Then the line intensity of the three polarized components have been calculated under the assumption of axisymmetry with respect to the external field direction. In the present situation, however, the axisymmetry breaks down and we instead consider the cartesian components, $d_x$, $d_y$, and $d_z$ of the electric dipole moment, which correspond

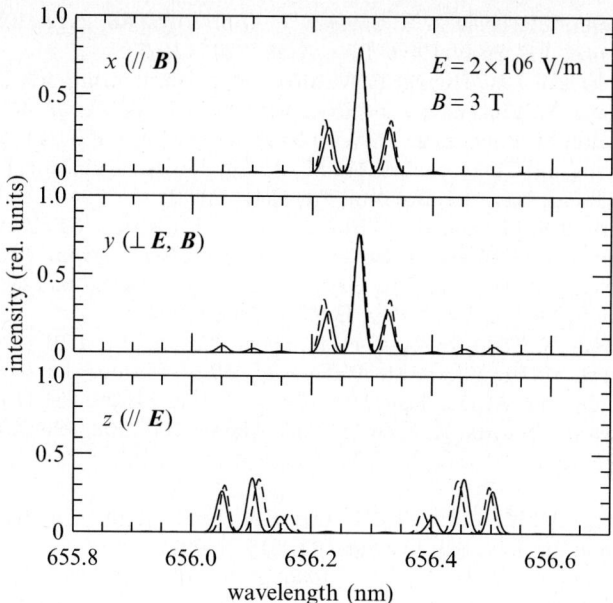

**Fig. 2.8.** Calculation result of line profile for the Balmer $\alpha$ line of neutral hydrogen under crossed electric ($=2.6 \times 10^6\,\mathrm{Vm^{-1}}$) and magnetic ($=3\,\mathrm{T}$) fields. *Dashed lines are the results without the magnetic field*

to the linearly polarized light in the respective axis directions. The cartesian and spherical components have the relationship

$$d_x = -\frac{1}{\sqrt{2}}(d_1 - d_{-1}), \tag{2.30}$$

$$d_y = -\frac{i}{\sqrt{2}}(d_1 + d_{-1}), \tag{2.31}$$

$$d_z = d_0. \tag{2.32}$$

When the cartesian components are substituted in (2.5) instead of the spherical components, the transition probability of linearly polarized light emission in each of the cartesian axes is obtained. Figure 2.8 shows an example for $E = 2 \times 10^6\,\mathrm{Vm^{-1}}$ and $B = 3\,\mathrm{T}$. The results for the case where the second term in (2.27) is neglected are also shown with the *dashed lines* in the same figure to clarify the effect of the magnetic field.

# References

1. I.I. Sobel'man, *Introduction to the Theory of Atomic Spectra* (Pergamon Press, Oxford, 1972)
2. P.H. Heckmann, E. Träbert, *Introduction to the Spectroscopy of Atoms* (North-Holland, Amsterdam, 1989)

3. U. Feldman, J.F. Seely, N.R. Sheeley, J. Appl. Phys. **56**, 2512 (1984)

4. V.L. Jacobs, J.F. Seely, Phys. Rev. A **36**, 3267 (1987)

5. D. Wróblewski, L.K. Huang, H.W. Moos, Rev. Sci. Instrum. **59**, 2341 (1988)

6. L.K. Huang, M. Finkenthal, D. Wróblewski et al., Phys. Rev. A **45**, 1089 (1992)

7. L.K. Huang, M. Finkenthal, D. Wróblewski et al., Phys. Fluids B **2**, 809 (1990)

8. L.K. Huang, M. Finkenthal, H.W. Moos, Rev. Sci. Instrum. **62**, 1142 (1991)

9. D.M. Thomas, Rev. Sci. Instrum. **74**, 1541 (2003)

10. J.L. Terry, E.S. Marmar, R.B. Howell, Rev. Sci. Instrum. **61**, 2908 (1990)

11. J.L. Weaver, B.L. Welch, H.R. Griem et al., Rev. Sci. Instrum. **71**, 1664 (2000)

12. B.L. Welch, J.L. Wiaver, H.R. Griem et al., Phys. Plasma **8**, 1253 (2001)

13. M. Goto, S. Morita, Phys. Rev. E **65**, 026401 (2002)

14. T. Shikama, S. Kado, H. Zushi et al., Phys. Plasmas **11**, 4701 (2004)

15. A. Iwamae, M. Hayakawa, M. Atake et al., Phys. Plasmas **12**, 042501 (2005)

16. B.A. Pohlmeyer, A. Dinklage, H.J. Kunze, J. Phys. B **29**, 221 (1996)

17. K. Takiyama, T. Oda, K. Sato, J. Nucl. Mater. **290–293**, 976 (2001)

18. F.M. Levinton, R.J. Fonck, G.M. Gammel et al., Phys. Rev. Lett. **63**, 2060 (1989)

19. D. Wróblewski, K.H. Burrell, L.L. Lao et al., Rev. Sci. Instrum. **61**, 3552 (1990)

20. F.M. Levinton, Rev. Sci. Instrum. **63**, 5157 (1992)

21. F.M. Levinton, S.H. Batha, M. Yamada et al., Phys. Fluids B **5**, 2554 (1993)

22. W. Mandl, R.C. Wolf, M.G. von Hellermann et al., Plasma Phys. Control. Fusion **35**, 1373 (1993)

23. H.R. Griem, *Spectral Line Broadening by Plasma* (Academic, New York, 1974)

24. I.I. Sobel'man, L.A. Vainshtein, E.A. Yukov, *Excitation of Atoms and Broadening of Spectral Lines* (Springer, Berlin Heidelberg New York, 1995)

25. J.J. Sakurai, *Modern Quantum Mechanics* (Addison-Wesley, Redwood City, 1985)

# 3

# Plasma Spectroscopy

T. Fujimoto

The population of excited levels (and the ground state) of atoms and ions has been the subject of conventional intensity spectroscopy, because the upper level population of the emission line gives its observed intensity. In this chapter, we review the collisional-radiative (CR) model which is a versatile tool in dealing with the problem of the population. In this framework, we define the ionizing plasma and the recombining plasma. Many laboratory plasmas fall into either class in this category. We examine various features of plasmas in each class. We introduce the concept of the ionization balance and examine the population characteristics of the ionization-balance plasma. This chapter forms the basis of the population-alignment collisional-radiative model (PACR model), by which we are able to deal with the Class 2 polarization phenomena, to be discussed in the succeeding chapter.

## 3.1 Collisonal-Radiative Model: Rate Equations for Population

Since detailed discussions on this subject are given in another monograph [1], only the *main results* are given in this chapter. For more details, the reader is referred to Chaps. 4 and 5 in [1].

Figure 3.1 shows a schematic energy-level diagram of an atom or an ion: we will be concerned with the population of level $p$ or $r$ in the ionization stage $(z - 1)$. $z$ is the charge state of the next ionization-stage ions, or $ze$ is the effective core charge felt by the optical electron when this electron is far from the ion core, where $e$ is the elementary charge. Here the optical electron is defined as the atomic electron that plays the dominant role in making a transition and emitting a photon. Let $p = 1$ mean the ground state. The energy difference between $p$ and $r$ is denoted by $E_{z-1}(p, r)$, and the ionization potential of $p$ by $\chi_{z-1}(p)$ (positive quantity). The population of level $p$ is denoted as $n_{z-1}(p)$ and the statistical weight as $g_{z-1}(p)$. Whenever confusion is unlikely to occur, the subscript $z - 1$ will be omitted.

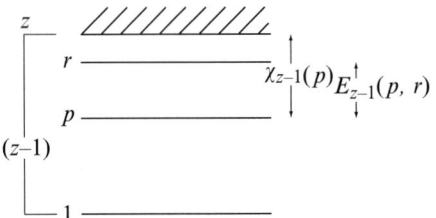

**Fig. 3.1.** Energy-level diagram of atoms or ions

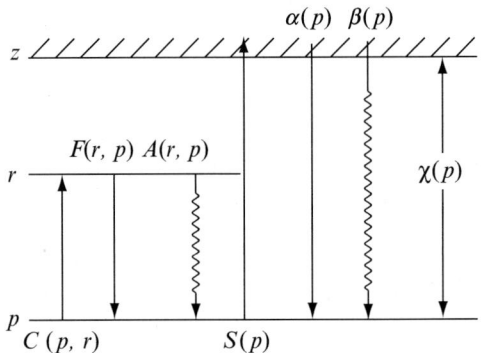

**Fig. 3.2.** Atomic processes

The transition processes considered in this chapter are shown in Fig. 3.2. The spontaneous transition probability is given as $A(r, p)$, which is in units of $s^{-1}$. The rate coefficient for excitation by electron impact is denoted by $C(p, r)$ $[m^3\,s^{-1}]$, that for deexcitation, which is the inverse process to excitation, by $F(r, p)$ $[m^3\,s^{-1}]$, and that for ionization by $S(p)$ $[m^3\,s^{-1}]$. The rate coefficient for radiative recombination into level $p$ is denoted by $\beta(p)$ $[m^3\,s^{-1}]$, and that for three-body recombination, which is the inverse process to ionization, is by $\alpha(p)$ $[m^6\,s^{-1}]$. Since three-body recombination involves two incident electrons, its rate coefficient has different units from those for other reactions.

In this chapter, we assume that our plasma has a thermal electron velocity distribution function (EVDF), i.e., the Maxwell distribution, so that we can define an electron temperature $T_e$. The plasma radiation, line and continuum, is unpolarized. The electron density is denoted by $n_e$. Suppose, by using a spectrometer we observe an emission line from the plasma for transition $r - p$. (Remember that $r$ is the lower level and $p$ is the upper level here. As noted in Chap. 1, we follow the convention that the lower level comes first.) The observed intensity is expressed in terms of the upper-level population $n(p)$:

$$\Phi(p, r) = n(p)A(p, r)h\nu dV\,(d\Omega/4\pi), \qquad (3.1)$$

where $dV$ is the volume of the plasma which we observe and $d\Omega$ is the solid angle subtended by our optics, e.g., when we use a condenser lens it determines d$\Omega$. The quantity of (3.1) is called *the radiant flux* or *the radiant power* in radiometry and has units of W. Suppose the transition probability is known. Then, the problem of the intensity of spectral lines reduces to the problem of the populations. Here we assume that the plasma is optically thin and the emitted photons do not suffer absorption in the plasma. We now construct the rate equation which describes the temporal development of the population in the plasma

$$\frac{d}{dt}n(p) = \sum_{r<p} C(r,p)n_e n(r)$$

$$- \left[ \sum_{r<p} A(p,r) + \left\{ \sum_{r<p} F(p,r) + \sum_{r>p} C(p,r) + S(p) \right\} n_e \right] n(p)$$

$$+ \sum_{r>p} [A(r,p) + F(r,p)n_e] n(r)$$

$$+ [\beta(p) + \alpha(p)n_e] n_e n_z. \tag{3.2}$$

We use the convention that $r < p$ means that level $r$ lies energetically below level $p$, and a summation sign with $r < p$ means the summation over level $r$ lying below level $p$, which is under consideration. We call quantities like $C(r,p)n_e$ or $A(p,r)$, having units of $s^{-1}$ *the rate* or *the probability*, and those like $C(r,p)n_e n(r)$ or $A(p,r)n(p)$, having units of $m^{-3} s^{-1}$, *the flux*. The first line represents the excitation flux by electron impact from lower-lying levels including the ground state, the third line represents the populating flux, collisional and radiative, from higher-lying levels, and the fourth line is the populating flux by direct recombination. The second line is the depopulating flux from this level. Equation (3.2) is coupled with similar equations for other excited levels, for the ground-state atom density $n_{z-1}(1)$ and for the ion density $n_z$. (The words *atom* and *ion* are used here to indicate ions (atoms) in two successive ionization stages.)

Although a problem remains of how many levels should be included in the set of equations, these equations could be solved numerically. However, we take an alternative approach, which is more general though approximate. It turns out that in the majority of practical problems, our approximate method gives quite accurate results.

For the purpose of illustrating our method, we take neutral hydrogen or hydrogen-like ions as an example. In this case, $p$ or $r$ is understood to represent the principal quantum number of the level. We define the relaxation time $t_{rl}(p)$ for $n(p)$ from (3.2) as

$$t_{rl}(p) = \left[ \sum_{r<p} A(p,r) + \left\{ \sum_{r<p} F(p,r) + \sum_{r>p} C(p,r) + S(p) \right\} n_e \right]^{-1}. \tag{3.3}$$

This is a measure of the time constant in which this population reaches its stationary-state value provided that the populating flux into this level is constant. It is obvious that, at low plasma densities, the relaxation time is determined by the radiative decay rate or the natural lifetime, while at high densities, it is given by the collisional depopulation rate. In the latter rate, the dominant contribution comes from excitation to the adjacent higher-lying level except at very low temperatures, i.e., the collisional depopulating rate is well approximated by $C(p, p+1)n_e$.

We consider the relaxation time of the ground-state *atom* population. It is given by an equation similar to (3.2) with the first line absent and, in the second line, the $\sum A$ term and the $\sum F$ term absent. Unless the density of the plasma is very high, a significant fraction, or even almost all, of the excitation flux out of the ground state returns to this state by radiative decay. Thus the $\sum C$ term in the second line could be eliminated in effect. As a rough approximation the relaxation time of the ground-state population may be given as

$$t_{\rm rl}(1) \simeq [S(1)n_e]^{-1}. \tag{3.4}$$

The *ion* density $n_z$ has a similar relaxation time under normal conditions. It can be concluded that in many practical cases

$$t_{\rm rl}(p) \ll t_{\rm rl}(1) \quad \text{for} \quad p \geqslant 2 \tag{3.5}$$

is valid. It is also expected that the total number of populations in excited levels is much smaller than the sum of the ground-state population and the ion density.

$$\sum_{p \geqslant 2} n_{z-1}(p) \ll [n_{z-1}(1) + n_z]. \tag{3.6}$$

It turns out that, except for very extreme cases, (3.5) and (3.6) are well satisfied.

We now consider what (3.5) and (3.6) mean. Unless $n(1)$ or $n_z$, and/or our plasma parameters ($T_e$ and $n_e$) undergo a very rapid change so that the excited-level populations do not have enough time to finish their relaxation, we can expect that, at a certain time, the excited-level populations have already reached their stationary-state values that are given by $n_{z-1}(1)$ and $n_z$, as well as by $T_e$ and $n_e$, *at that instance*. We call this situation the quasi-steady state (QSS).

The above considerations suggest that, in the coupled rate equations (3.2), we may approximate the time derivative of the excited-level populations to zero:

$$\frac{\mathrm{d}}{\mathrm{d}t}n(p) = 0 \quad \text{for} \quad p = 2, 3, \ldots, \tag{3.7}$$

while the time derivative should be retained for the ground-state population $n_{z-1}(1)$ and the ion density $n_z$. This is equivalent to formulating our problem as follows. Our system is divided into two subsystems: The populations of the whole excited levels constitute the first subsystem which is "sandwiched" by the second subsystem which consists of the ground-state population and the ion density.

The set of the coupled linear equations, (3.2) with (3.7), for the first subsystem can be expressed in the matrix form

$$
\begin{pmatrix} \cdots \\ \cdots \\ \cdots \\ \cdots \end{pmatrix}
\begin{pmatrix} n(2) \\ \cdot \\ n(p) \\ \cdot \end{pmatrix}
=
\begin{pmatrix} \cdot \\ \cdot \\ \cdot \\ \cdot \end{pmatrix} n(1)
+
\begin{pmatrix} \cdot \\ \cdot \\ \cdot \\ \cdot \end{pmatrix} n_z . \tag{3.8}
$$

The dimension of the matrix, or the number of levels, should be limited at some appropriate values, say 10, 20, or even 100. The elements of the square matrix on the left-hand side and the column matrices on the right-hand side are terms on the right-hand side of (3.2) and are functions of $T_e$ through the collisional rate coefficients and of $n_e$. It is obvious from the structure of (3.8) that this equation is readily solved for $p \geqslant 2$ as a sum of two terms, each of which is proportional to $n(1)$ and $n_z$, respectively. We assume the solution in the form

$$
n(p) = R_0(p) n_z n_e + R_1(p) n(1) n_e \tag{3.9}
$$
$$
\equiv n_0(p) + n_1(p). \tag{3.9a}
$$

Equation (3.9) indicates that an excited-level population is the sum of the two components; the first component being proportional to the *ion* density $n_z$ and the second to the ground-state *atom* density $n(1)$. We name these components of populations *the recombining plasma component* and *the ionizing plasma component*, respectively. Figure 3.3 schematically shows this situation.

We substitute (3.9) into the coupled linear equations (3.8). We then obtain two sets of coupled equations, one for $R_0(p)$ and another for $R_1(p)$. It is straightforward to obtain solutions to these equations by matrix inversion. The solutions $R_0(p)$ and $R_1(p)$ are called *the population coefficients* and they

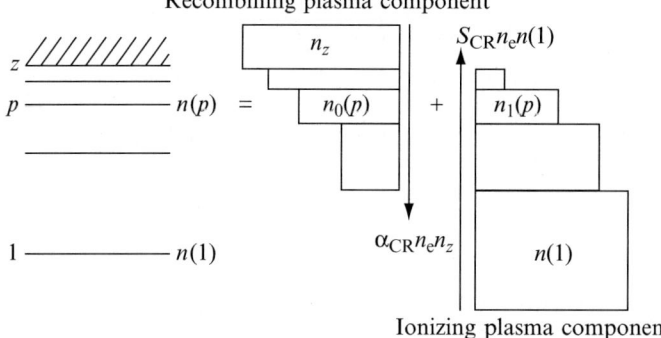

**Fig. 3.3.** The structure of the excited-level populations in the collisional-radiative model. The population $n(p)$ is the sum of the ionizing plasma component $n_1(p)$ which is proportional to the ground-state population $n(1)$ and the recombining plasma component $n_0(p)$ which is proportional to the "ion" density $n_z$

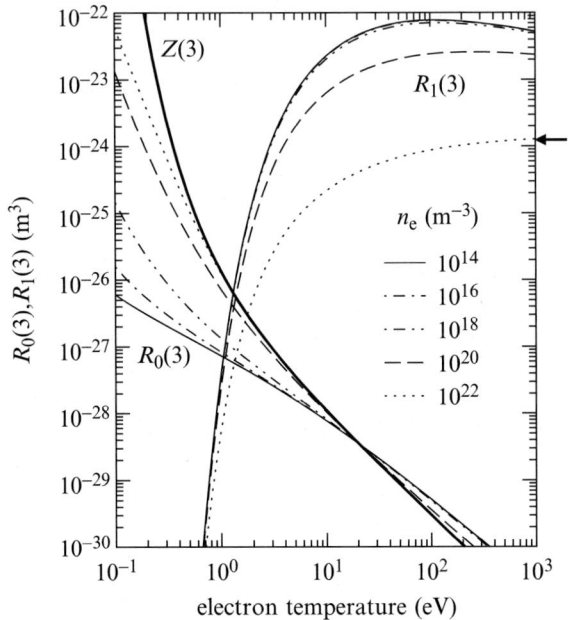

**Fig. 3.4.** The population coefficients $R_0(3)$ and $R_1(3)$ for the upper level of the Balmer $\alpha$ line. Also shown are the Saha–Boltzmann coefficient $Z(3)$ (the *thick solid line*) and $3^{-4}/n_e$ for $n_e = 10^{22}\,\mathrm{m}^{-3}$ (*the arrow*)

are functions of $T_e$ and $n_e$. Figure 3.4 shows an example of the population coefficients for neutral hydrogen for level $p = 3$; this is the upper level of the Balmer $\alpha$ line. The theoretical framework described above is called the collisional-radiative (CR) model.

## 3.2 Ionizing Plasma and Recombining Plasma

As (3.9) or Fig. 3.3 suggests, it would be natural to investigate the ionizing plasma component and the recombining plasma component separately, first. We briefly look at the features of these components. We assume high temperatures, but in treating the recombining plasma component a low temperature case is also emphasized.

### 3.2.1 Ionizing Plasma Component

Figure 3.4 shows that $R_1(3)$ has a strong temperature dependence and, at densities higher than $10^{20}\,\mathrm{m}^{-3}$, also a density dependence. Figure 3.5 shows the density dependence of $n_1(p)/g(p)$ at a particular temperature, $T_e = 11.0\,\mathrm{eV}$ $(1.28 \times 10^5\,\mathrm{K})$ for several excited levels on the assumption of $n(1) = 1\,\mathrm{m}^{-3}$.

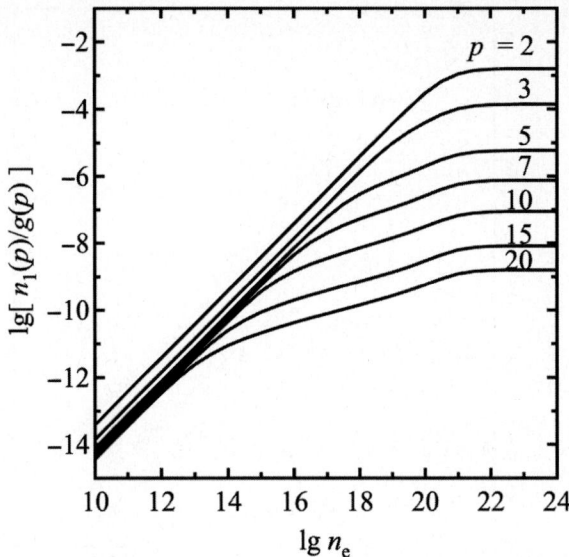

**Fig. 3.5.** The populations of the ionizing plasma component for $n(1) = 1\,\mathrm{m}^{-3}$. $T_e = 1.28 \times 10^5\,\mathrm{K}$ (Quoted from [2], with permission from The Physical Society of Japan.)

By remembering that $g(3) = 2 \times 3^2$, the reader can recognize the consistency between these figures. The density-independent $R_1(3)$ at low densities in Fig. 3.4 is expressed in Fig. 3.5 as $n_1(3)/g(3)$ that is proportional to $n_e$. At higher densities, $R_1(3)$ decreases, or $n_1(3)/g(3)$ deviates from the linear increase, starting at about $n_e = 10^{18}\,\mathrm{m}^{-3}$. The latter finally saturates, or reaches the high-density-limit value, at still higher densities.

This feature is common to all the excited levels. A major difference among the different levels is the boundary $n_e$ for the deviation. Figure 3.6 shows the population distribution among the excited levels at several densities. At a low density ($n_e = 10^{12}\,\mathrm{m}^{-3}$) at which virtually all the excited-level populations are proportional to $n_e$ in Fig. 3.5, the population distribution is rather flat. Figure 3.7a shows the sketch of the population fluxes in the energy level diagram at this density. The blank arrows represent the collisional transitions, i.e., excitation, deexcitation, and ionization by electron impact, and the hatched arrows represent the radiative transitions. The width of an arrow is proportional to the magnitude of the flux. The population fluxes concerning level $p = 5$ are given in some more detail than for other levels. It is seen that, at this density, all the levels are populated directly by excitation from the ground state with a small contribution from the cascading transitions from still higher levels, and depopulated through the radiative decay. Thus, the levels are in corona equilibrium and the population is given approximately by

$$n_1(p) \simeq \frac{C(1,p)n_e n(1)}{\sum_{r<p} A(p,r)} \tag{3.10}$$

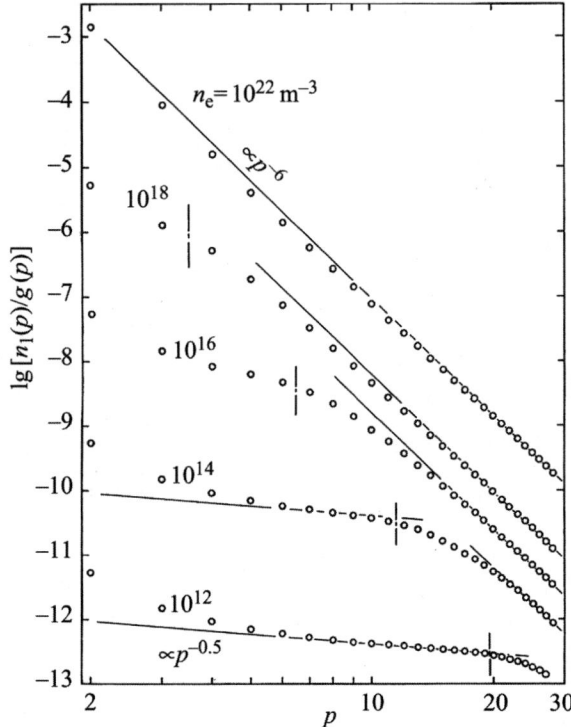

**Fig. 3.6.** Population distribution over excited levels for several $n_e$s. $T_e = 1.28 \times 10^5$ K. Griem's boundary, (3.15), is shown with the *dash-dotted lines*. The approximations, (3.11) and (3.14), are shown (Quoted from [2], with permission from The Physical Society of Japan.)

(remember (3.2)). We name the $n_e$-region in which this level is in corona equilibrium *the corona phase*. It can be shown that $C(1,p)$ is approximately proportional to $p^{-3}$ and $\sum A(p,r)$ is proportional to $p^{-4.5}$, both for large $p$, so that the population distribution is proportional to $p^{1.5}$ or

$$n_1(p)/g(p) \propto p^{-0.5}. \tag{3.11}$$

Figure 3.6 shows this approximation, which is good for high-lying levels.

With an increase in $n_e$, in Fig. 3.5, populations begin to deviate from the linear increase, starting from very high-lying levels: these levels enter into *the saturation phase*. In Fig. 3.6, the deviation is expressed as the steeper slope of the population distribution for levels higher than the boundary given by the dash-dotted line. This suggests that the population kinetics for these latter levels is different from (3.10). Figure 3.7b shows the sketch of the population fluxes at $n_e = 10^{18}$ m$^{-3}$, where the boundary lies between $p = 3$ and 4 (see also Fig. 3.6). It is seen that the higher-lying level is populated mainly by excitation from the adjacent lower-lying level, and it is depopulated by excitation to the

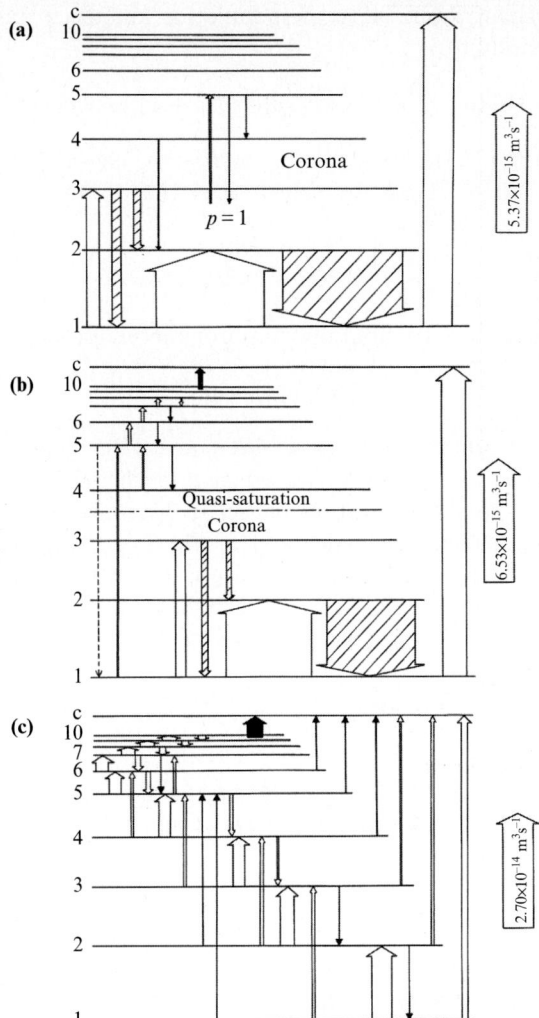

**Fig. 3.7.** The dominant flows of electrons in the energy-level diagram for the ionizing plasma component. The *open arrows* are for collisional transitions and the *hatched arrows* for radiative transitions. The widths of the arrows, or the magnitudes of the fluxes, are scaled according to the number given on the right-hand side, which is the collisional-radiative ionization rate coefficient. The excitation flux through level $p = 10$, resulting in ionization, is shown with the *filled arrow*. $T_e = 1.28 \times 10^5$ K. The energy level positions are not to scale. **(a)** $n_e = 10^{12}$ m$^{-3}$. All the levels are in the corona phase. **(b)** $n_e = 10^{18}$ m$^{-3}$. Griem's boundary, given by the *dash-dotted line*, divides the levels into the corona phase and the saturation phase. **(c)** $n_e = 10^{22}$ m$^{-3}$. All the levels are in the saturation phase, and the ladder-like excitation-ionization scheme is established (Quoted from [2], with permission from The Physical Society of Japan.)

adjacent higher-lying level. Thus, the population kinetics is, instead of (3.10), given approximately by

$$n_1(p-1)C(p-1,p)n_e \simeq n_1(p)C(p,p+1)n_e(\simeq \text{const.}) \qquad (3.12)$$

(remember again (3.2)). We call this multistep excitation mechanism *the ladder-like excitation*. It can be shown that the excitation rate coefficient is approximately given by

$$C(p,p+1) \propto p^4, \qquad (3.13)$$

so that, the population distribution is approximately proportional to $p^{-4}$ or

$$n_1(p)/g(p) \propto p^{-6}. \qquad (3.14)$$

We call this distribution *the minus 6th power distribution* (see Fig. 3.6). The above observation suggests that when a level makes a transition from the corona phase to the saturation phase, in Fig. 3.5, $n_1(p)/g(p)$ begins to deviate from the linear dependence and in Fig. 3.4, $R_1(3)$ begins to depend on $n_e$. The boundary between the corona phase and the saturation phase is given from the depopulating terms on the right-hand side of (3.2) by

$$\sum_{r<p} A(p,r) \simeq \left[\sum_{r<p} F(p,r) + \sum_{r>p} C(p,r) + S(p)\right]n_e, \qquad (3.15)$$

or even approximately by

$$\sum_{r<p} A(p,r) \simeq C(p,p+1)n_e. \qquad (3.15a)$$

For more detailed discussions, the reader is referred to Chap. 4 of [1]. This boundary is called Griem's boundary: this is the boundary between low density and high density, or, at the same time, between the low-lying levels and the high-lying levels, as expressed by $p_G$. Numerically this boundary is approximately given by

$$n_e/z^7 \simeq 6.7 \times 10^{22}/p_G^{8.5} \text{ m}^{-3} \qquad (3.16)$$

or

$$p_G \simeq 480(n_e/z^7)^{-2/17}. \qquad (3.16a)$$

The dash-dotted lines in Figs. 3.6 and 3.7b give Griem's boundary as given by (3.15).

Figure 3.7c shows the sketch of the population fluxes at $n_e = 10^{22} \text{ m}^{-3}$, virtually the high-density limit. At this density, Griem's boundary level is below $p = 2$, the first excited level, and the ladder-like excitation is established for the all excited levels, or it starts from the ground state all through the excited levels. Then, all the populations become independent of $n_e$, as shown

in Fig. 3.5, and the population distribution is given by (3.14) for the all excited levels. Figure 3.6 shows that this very crude approximation, (3.14), is
surprisingly good for the actual population distribution. At sufficiently high
temperatures, this relationship can even be extended down to the ground
state $p = 1$. In Fig. 3.4, this approximation, $R_1(3) = 3^{-4}/n_e$, is shown with
the horizontal arrow for $n_e = 10^{22}\,\mathrm{m}^{-3}$.

Figure 3.8a shows the "phase diagram" of the populations. The abscissa is
the reduced electron density $n_e/z^7$ and the ordinate is the principal quantum
number of excited levels. It can be shown that neutral hydrogen and hydrogen-
like ions follow the scaling law against the nuclear charge $z$. According to this
law, the reduced electron density $n_e/z^7$ and temperature $T_e/z^2$ express general
characteristics, though approximately, of the atoms and ions in the plasma. In
the case of neutral hydrogen $z$ is 1. The boundary "GRIEM" is given by (3.15)
or (3.16a), and the name of the phase and the population distribution is given
together with the dominant population kinetics of level $p$ under consideration.
The line "BYRON" and the lower part than this boundary are absent in
practical situations.

### 3.2.2 Recombining Plasma Component

Figure 3.4 shows the population coefficient $R_0(3)$ as a function of $T_e$ for several
$n_e$s. The figure also includes the Saha–Boltzmann coefficient

$$Z(p) = \frac{g(p)}{2g_z} \left( \frac{h^2}{2\pi m k T_e} \right)^{3/2} \exp\left[ \frac{\chi(p)}{kT_e} \right] \tag{3.17}$$

for $p = 3$, where $g_z(=1)$ is the statistical weight of the ground-state ion $z$, $m$
the electron mass, and other symbols have usual meanings. If the population
is given by

$$n(p) = Z(p)n_z n_e, \tag{3.18}$$

this population is in thermodynamic equilibrium with respect to the ion density $n_z$ and the electron density $n_e$. We say in this situation that level $p$ is in
local thermodynamic equilibrium (LTE). It is sometimes believed that a level
enters into LTE when, and only when, electron density is higher than a certain
value. Figure 3.4 shows, however, that, for temperatures higher than several
eV, the recombining plasma component $n_0(p)$ is close to the LTE values even
at low densities. More exactly, at extremely high temperatures, $n_0(p)$ at low
densities is slightly higher than the LTE values, and for temperatures below
20 eV it is lower than the LTE values. At high densities, it tends exactly to
the LTE values at any temperatures if the temperature is higher than 0.5 eV.
Provided that the contribution from the ionizing plasma component $n_1(p)$ is
negligibly small, level $p$ is in LTE in the above range of plasma parameters.
This feature, except for the boundary temperature at high density, is common
to all the excited levels.

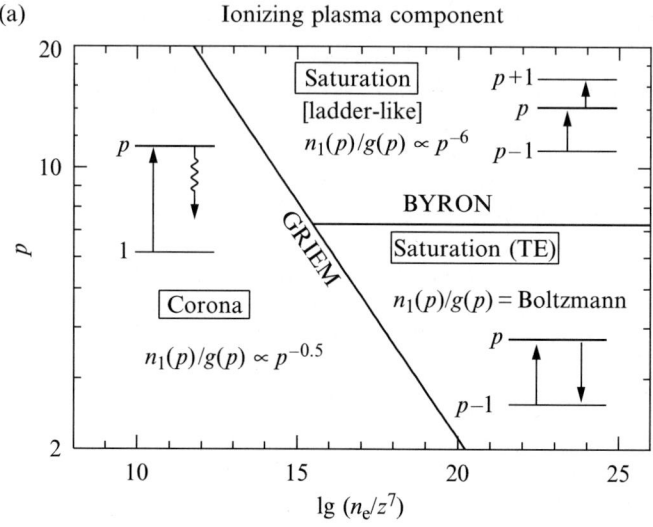

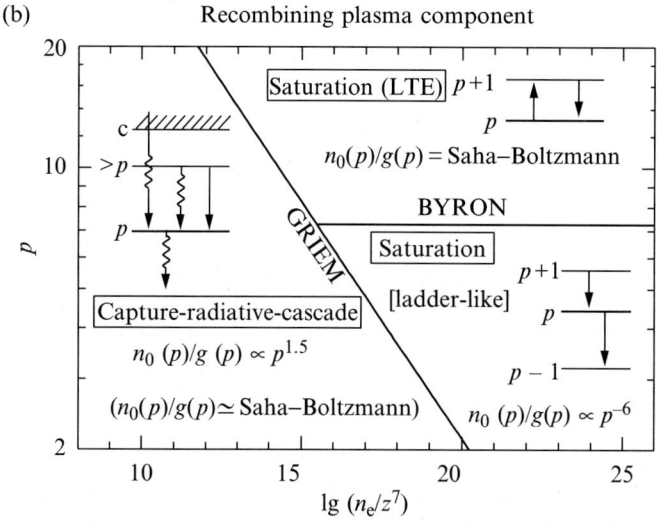

**Fig. 3.8.** The "map" of the excited-level populations of neutral hydrogen and hydrogen-like ions in plasma. The abscissa is the (reduced) electron density and the ordinate is the principal quantum number of excited levels. (**a**) The ionizing plasma component; (**b**) the recombining plasma component. Griem's boundary divides the whole area into a low-density region and a high-density region. Byron's boundary divides the high-density region into low-lying levels and high-lying levels. In each area, the name of the phase, the population distribution, and the dominant population kinetics are shown for level $p$ with which we are concerned (Quoted from [3], with permission from The Physical Society of Japan.)

Figure 3.9 shows the population fluxes for the recombining plasma component. The meanings of the arrows are the same as in Fig. 3.7. The hatched arrows starting from the ionization limit represent radiative recombination. Figure 3.9a is for the low-density limit. Roughly speaking, an excited-level population is determined by the balance between the radiative recombination into this level and the radiative decay from this level. It is noted that Fig. 3.9 is for $T_e = 10^3\,\mathrm{K}$ (0.09 eV), quite low temperature. However, the basic feature of the population kinetics is much the same as that at high temperatures which we are now considering. The major difference is the $p$-dependence of the radiative recombination flux: at low temperatures $\beta(p) \propto p^{-1}$, while at high temperatures $\beta(p) \propto p^{-2.5}$.

At low densities, the population kinetics is given by

$$n_0(p) \simeq \frac{\beta(p)n_z n_e + \sum_{r>p} n_0(r)A(r,p)}{\sum_{r<p} A(p,r)} \qquad (3.19)$$

(remember (3.2)). It can be shown that, at high temperatures,

$$\frac{\beta(p)}{\sum_{r<p} A(p,r)} \simeq \frac{2}{3} Z(p), \qquad (3.20)$$

holds and, by noting that the higher-lying levels have near LTE populations,

$$n_0(r) \simeq Z(r)n_z n_e,$$

we have

$$\frac{\sum_{r>p} Z(r)A(r,p)}{\sum_{r<p} A(p,r)} \simeq \frac{1}{3} Z(p). \qquad (3.21)$$

Equation (3.19) with (3.20) and (3.21) leads to $n_0(p)$ which is close to (3.18). The above intricate interrelationships between the rate coefficients and the transition probabilities are the origin of the near LTE populations at low densities as noted above.

At high densities, as shown in Fig. 3.4, the population $n_0(p)$ is given exactly by (3.18). This is the result of the strong population coupling between excited levels and, further, with the continuum-state electrons. This coupling, however, is due mainly to the stepwise excitation–deexcitation chain between the adjacent levels continuing to the low-energy continuum-state electrons, not due to the balance between the direct ionization and three-body recombination as is sometimes assumed.

We now turn to the low temperature case, which is more important in practical situations. Figure 3.10 shows the population for several levels as functions of density. This is for $T_e = 10^3\,\mathrm{K}$, which was the temperature of Fig. 3.9. In this figure the population per unit statistical weight is further divided by ion and electron densities. The reader can recognize the consistency between $n_0(3)/g(3)n_z n_e$ in this figure and $R_0(3)$ in Fig. 3.4. In the low-density limit, the population starts with a much lower value than the LTE population, as

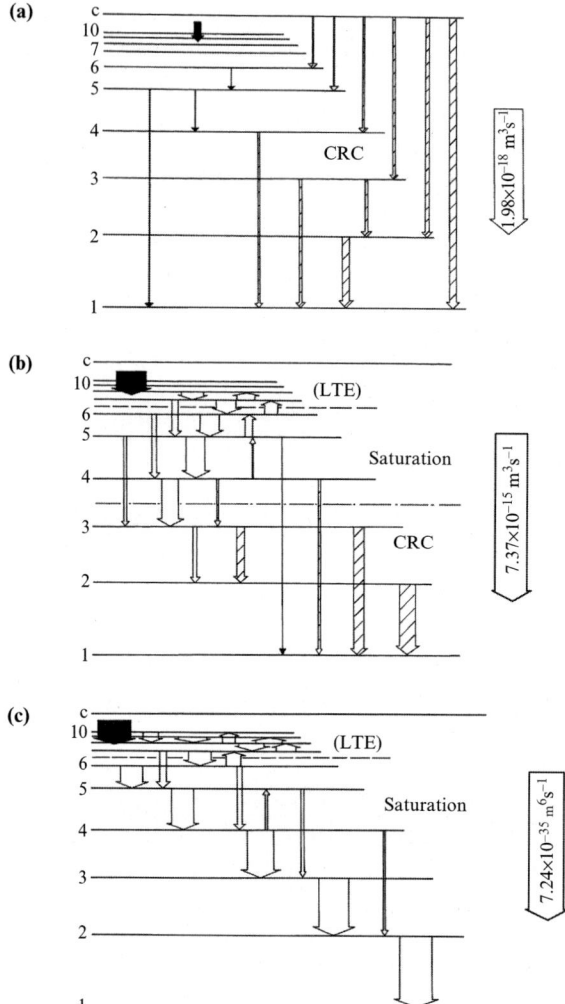

**Fig. 3.9.** The sketch of the dominant flows of electrons among the energy-level diagram for the recombining plasma component. The meanings of the arrows are the same as in Fig. 3.7, and the recombination flux through a high-lying level $p = 10$ is given with the filled arrow. The collisional-radiative recombination rate coefficient is given on the right-hand side. $T_e = 10^3$ K. **(a)** The low-density limit. All the levels are in the CRC phase. **(b)** $n_e = 10^{20}$ m$^{-3}$. Levels below Griem's boundary, (3.15) and given by the dash-dotted line, are in the CRC phase. The higher-lying levels are in the saturation phase and they are further divided by Byron's boundary, (3.22) and given by the dashed line, into the levels in LTE and the lower-lying levels where the ladder-like deexcitation scheme is established. **(c)** The high-density limit. All the levels are divided by Byron's boundary into high-lying levels in LTE and the low-lying levels in the flow of the ladder-like deexcitation. The collisional-radiative recombination rate coefficient has been further divided by $n_e$ (Quoted from [4], with permission from The Physical Society of Japan)

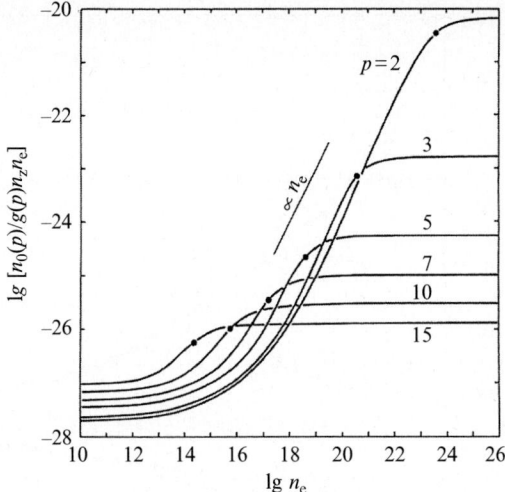

**Fig. 3.10.** The populations for the recombining plasma component. In the ordinate, the population per unit statistical weight is further divided by ion and electron densities. Griem's boundary, (3.15), is shown with the *closed circles* (Quoted from [4], with permission from The Physical Society of Japan.)

shown in Fig. 3.4. Figure 3.9a was for this situation. Figure 3.11 shows the population distribution: (a) is a plot in a similar format to Fig. 3.6 and (b) is the conventional Boltzmann plot. It is seen that, in the low-density limit, (3.19) with $\beta(p) \propto p^{-1}$ and the neglect of the second term in the numerator leads to an approximate distribution $n_0(p)/g(p)n_z n_e \propto p^{1.5}$, which is seen to be reasonably good for the actual distribution in Fig. 3.11a.

With an increase in $n_e$, in Fig. 3.10 the "population" begins to increase gradually. Then its slope becomes steep until it saturates. These low-density regions altogether are named the capture-radiative-cascade (CRC) phase. The region of higher density is named *the saturation phase*. The electron density as given by (3.15) is plotted with the closed circle on each curve. In Fig. 3.11, this boundary is shown with the dash-dotted line as the boundary between levels. This figure shows that, with the increase in $n_e$, this boundary comes down. It is obvious that this boundary gives the boundary between these phases. This boundary is nothing but Griem's boundary which was defined for the ionizing plasma component. Figure 3.8b shows schematically these phases and Griem's boundary.

Figure 3.11 shows that, at sufficiently high densities where all the levels are in the saturation phase, populations of high-lying levels tend to their LTE values, while low-lying levels can never enter into LTE. Therefore another boundary divides the levels into two groups: the higher-lying levels are in LTE and the lower-lying levels have lower populations than the LTE values. We call this boundary level Byron's boundary $p_B$, which is given by the dotted

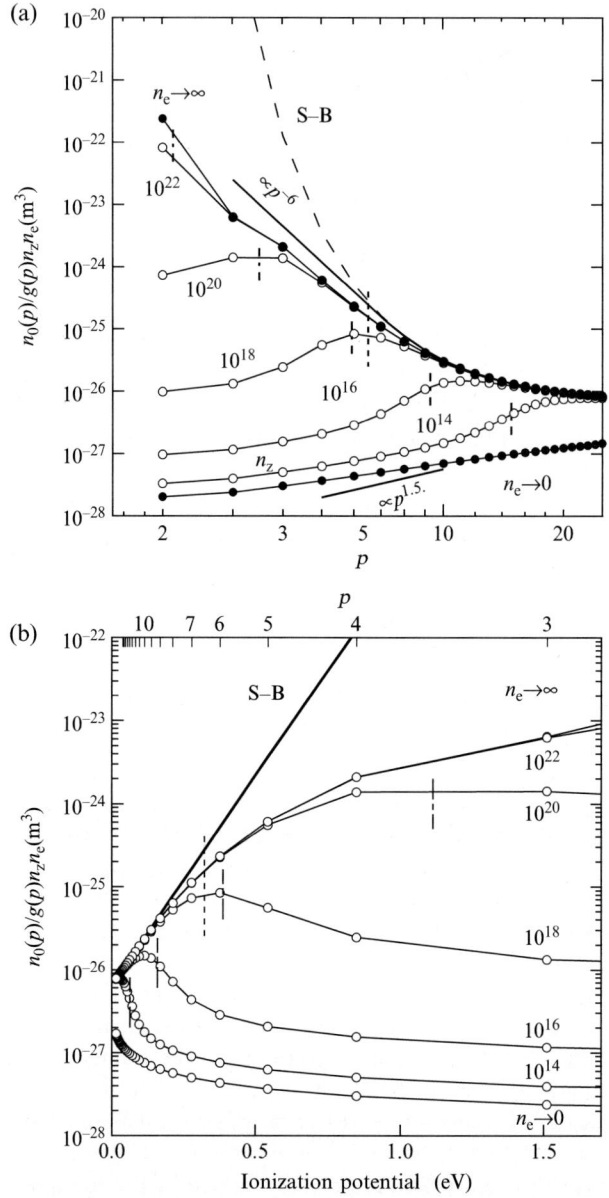

**Fig. 3.11.** Population distributions over excicted levels of the recombining plasma component. $T_e = 10^3$ K. Griem's boundary is given with the *dash-dotted lines* and Byron's boundary is given by the *dotted line*. "S–B" denotes the populations in LTE. (**a**) is in the same format as in Fig. 3.6. (**b**) is in the conventional Boltzmann plot (Quoted from [4], with permission from The Physical Society of Japan.)

line in Fig. 3.11 and labelled "BYRON" in Fig. 3.8b. For the higher-lying levels than Byron's boundary level relation $C(p, p+1) > F(p, p-1)$ holds, while for the lower-lying levels $C(p, p+1) < F(p, p-1)$ holds. Therefore, Byron's boundary is given by

$$C(p, p+1) = F(p, p-1) \tag{3.22}$$

and approximately by

$$p_{\mathrm{B}} \simeq \left( \frac{z^2 R}{3kT_{\mathrm{e}}} \right)^{1/2}, \tag{3.23}$$

where $R$ is one Rydberg (13.6 eV) (remember (3.2), this is the matter of the second line on the right-hand side. Since, the higher-lying levels than Byron's boundary enter into LTE at higher densities than Griem's boundary, the saturation phase for these levels is also called the LTE phase (see Fig. 3.8b).

Figure 3.9b shows the sketch of the population kinetics at $n_{\mathrm{e}} = 10^{20}$ m$^{-3}$, where Griem's boundary is between $p = 3$ and 4,[1] and Byron's boundary is between 6 and 7. It is seen that for the levels lying lower than Byron's boundary but higher than Griem's boundary the ladder-like deexcitation flow is established. In this case, the population distribution is given again approximately by the minus 6th power distribution.

$$n_0(p)/g(p) \propto p^{-6}. \tag{3.24}$$

Figure 3.11a shows that this feature is approximately valid. With a further increase in $n_{\mathrm{e}}$, Griem's boundary comes down further, and reaches $p = 2$ at a high-enough density (see Fig. 3.8b). Figure 3.9c shows the population kinetics at the high-density limit. Byron's boundary level divides all the levels into two groups; the higher-lying levels are strongly coupled with each other and thus to the continuum states, and the lower-lying levels are in the ladder-like deexcitation flow which eventually reaches the ground state. It can be shown that, in Fig. 3.11a, the slope of the LTE population, (3.18), at $p = p_{\mathrm{B}}$ as given by (3.23) is equal to –6, so that the minus 6th power distribution starts at this boundary as the smooth extrapolation from the LTE populations. This approximate distribution is given with the solid line. This crude approximation is seen to be reasonably good for the actual population distribution. For level 3, for example, Byron's boundary temperature is 0.5 eV from (3.23). In Fig. 3.4, at higher temperatures than this boundary temperature, $R_0(3)$ for high density should come close to $Z(p)$. This is actually seen to hold for densities higher than Griem's boundary, $n_{\mathrm{e}} \simeq 2 \times 10^{20}$ m$^{-3}$, at this temperature.

### 3.2.3 Ionizing Plasma and Recombining Plasma

In treating the excited-level populations, the rate equation for the ground-state *atom* density, $n(1)$, and that for the *ion* density, $n_z$, have been left as they

---

[1] In Figs. 3.5 and 3.6, Griem's boundary level at this density is around $p = 2$. This difference comes from the difference in temperatures.

were. Since we have obtained the solutions for the excited-level populations we substitute them into these rate equations. Equation (3.2) for $p = 1$ reduces to

$$
\begin{aligned}
\frac{d}{dt}n(1) = -& \left[ \sum_{r\geqslant 2} C(1,r) + S(1) \right] n(1)n_e \\
+& \sum_{r\geqslant 2} [A(r,1) + F(r,1)n_e] \cdot [R_0(r)n_z + R_1(r)n(1)] \, n_e \\
+& [\beta(1) + \alpha(1)n_e] \, n_z n_e.
\end{aligned} \tag{3.25}
$$

After rearrangments of the terms this equation is rewritten as

$$
\frac{d}{dt}n(1) = -S_{CR}n(1)n_e + \alpha_{CR}n_z n_e \tag{3.26}
$$

with

$$
S_{CR} = \sum_{r\geqslant 2} C(1,r) + S(1) - \sum_{r\geqslant 2} R_1(r)\left[F(r,1)n_e + A(r,1)\right], \tag{3.27}
$$

$$
\alpha_{CR} = \alpha(1)n_e + \beta(1) + \sum_{r\geqslant 2} R_0(r)\left[F(r,1)n_e + A(r,1)\right]. \tag{3.28}
$$

Since we assumed (3.6) the sum of the *atom* density and the *ion* density is conserved:

$$
\frac{d}{dt}n(1) + \frac{d}{dt}n_z = 0. \tag{3.29}
$$

Equation (3.26) with (3.29) describes ionization–recombination of the ensemble of atoms (ions) which are immersed in a plasma. The rate coefficients in this equation express the effective rate of ionization and that of recombination. They include the ionization and recombination processes via excited levels. $S_{CR}$ is called the *collisional-radiative (CR) ionization rate coefficent* and $\alpha_{CR}$ is the *CR recombination rate coefficient*. Like the population coefficients $R_0(p)$ and $R_1(p)$, these coefficients are functions of $n_e$ and $T_e$. It is to be noted that, in the expression of (3.27) for $S_{CR}$, only $R_1(p)$ appears and $R_0(p)$ is absent. For $\alpha_{CR}$ of (3.28), vice versa. This is an important feature of our formulation as schematically depicted in Fig. 3.3; The ionizing plasma component of the populations gives the effective ionization flux of the plasma $S_{CR}n(1)n_e$, and the recombining plasma component gives the effective recombination flux $\alpha_{CR}n_z n_e$. This feature can be understood from Figs. 3.7 and 3.9; in the high-density limit, for example, the majority of the ionization flux passes through excited levels and whole the recombination flux passes through the excited levels below Byron's boundary. Figure 3.12 shows several examples of $S_{CR}$ and $\alpha_{CR}$ for neutral hydrogen. The above statements can also be understood from the close similarity between $R_1(3)$ in Fig. 3.4 and $S_{CR}$ in Fig. 3.12 and the even closer similarity between $R_0(3)$ and $\alpha_{CR}$.

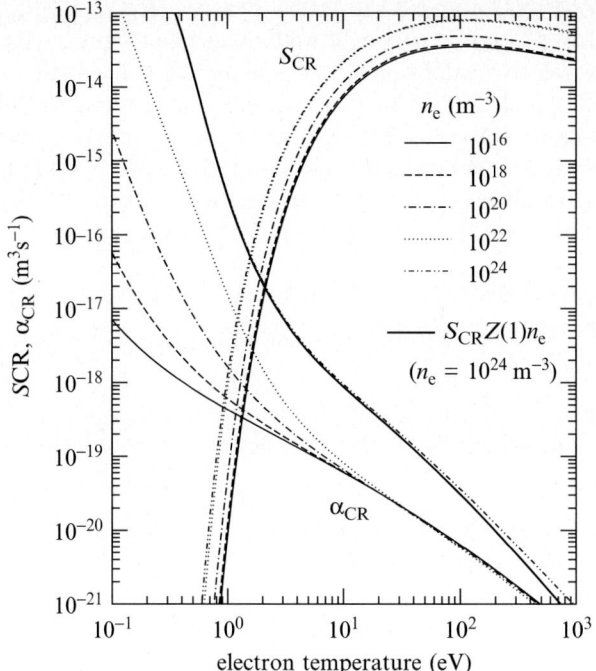

**Fig. 3.12.** Collisional-radiative ionization rate coefficient and recombination rate coefficient for several $n_e$s. For the purpose of comparison with $\alpha_{CR}$ at high density, $S_{CR}Z(1)n_e$ is also plotted for $n_e = 10^{24}\,\mathrm{m}^{-3}$

If the plasma is stationary, so that the time derivatives in (3.26) vanishes, and spatially homogeneous, so that particle transport does not affect the ionization–recombination relationship of the plasma, the plasma is in *ionization balance*, and the ionization ratio is given from (3.26) with $\mathrm{d}/\mathrm{d}t = 0$:

$$\left[\frac{n_z}{n(1)} = \frac{S_{CR}}{\alpha_{CR}}\right]_{IB} , \tag{3.30}$$

where the subscript IB means that this relationship is valid in ionization balance. We have several important features for the ionization-balance plasma as given below.

We start with a low density plasma. As understood from Fig. 3.12, the ionization ratio has a strong temperature dependence. We now keep Fig. 3.3 in mind. It can be shown that a plasma in ionization balance has a remarkable feature in population characteristics: with a change in electron temperature $[n_z/n(1)]$ varies by many orders of magnitude, i.e. more than ten orders for $0.8\,\mathrm{eV} \leqslant T_e \leqslant 10^3\,\mathrm{eV}$ (see Fig. 3.12); still the ratio of the ionizing plasma component and the recombining plasma component for all the excited levels is of the order of 10 at low densities. This feature is understood from the comparison of Figs. 3.4 and 3.12: At low density, $[R_1(3)/R_0(3)]$ behaves very similarly to $[S_{CR}/\alpha_{CR}]$. However, if we look at these figures closely, we recognize that the relative values of $S_{CR}$ is smaller than $R_1(3)$ by about a factor of 10. See (3.30) and (3.9). This fact suggests that, so long as the plasma is in ionization balance, the conventional practice of assuming corona equilibrium, (3.10), for excited-level populations is barely justified. We further remember that, even in low densities, $R_0(p)$ is rather close to $Z(p)$ (see Fig. 3.4), so that the total population, (3.9), is higher than the LTE population by about an order.

In the high-density limit, as Fig. 3.12 shows, $S_{CR}$ and $\alpha_{CR}$ are related by the thermodynamic equilibrium relationship:

$$\alpha_{CR} = S_{CR} Z(1) n_e. \qquad (3.31)$$

So that, the ionization ratio is in Saha–Boltzmann equilibrium or

$$n(1) = Z(1) n_z n_e. \qquad (3.32)$$

It can be shown that, in this limit,

$$R_0(p) + R_1(p) Z(1) n_e = Z(p) \qquad (3.33)$$

holds for all the excited levels, irrespective of temperature, leading to the LTE populations for these levels. In Fig. 3.4, the deviation of $R_0(3)$ at $n_e = 10^{22} \, \text{m}^{-3}$, virtually the high-density limit, from $Z(3)$ at low temperatures is compensated by the ionizing plasma component on the left-hand side of (3.33) so that the total population, (3.9), reaches $Z(3) n_z n_e$. Equations (3.31)–(3.33) indicate that (a) all the populations are given by the Saha–Boltzmann relationship

$$n(p) = Z(p) n_z n_e,$$

and that (b) the relative contribution from the ionizing plasma component and the recombining plasma component depends on $p$ or $T_e$ (see Fig. 3.4): At higher temperatures, the recombining plasma component alone gives the Saha–Boltzmann population, while at lower temperatures, the ionizing plasma component is substantial or even predominant in the LTE populations.

In actual plasmas, it is rather difficult for the ensemble of atoms or ions to attain ionization balance; a plasma may be time dependent or it is significantly spatially inhomogeneous. In such cases the real ionization ratio can deviate from (3.30), sometimes by many orders of magnitude. In such cases, the above arguments suggest that, if the plasma is far from ionization balance, one of the two plasma components predominates over another. If the actual ionization ratio is much smaller than the ionization balance value: $n_z/n(1) \ll [n_z/n(1)]_{IB}$, then it is expected that only the ionizing plasma component would give the actual populations, and the recombining plasma component is negligibly small. In the opposite case: $n_z/n(1) \gg [n_z/n(1)]_{IB}$, only the recombining plasma component is dominant. We call the former class of plasmas *the ionizing plasma* and the latter *the recombining plasma*.

Examples of the ionizing plasma are low-pressure discharges like a positive column of glow discharge and a pulsed discharge during the formation of the discharge plasma. Examples of the recombining plasma are afterglow plasmas including flowing afterglows and laser-produced plasma in the decaying phase.

# References

1. T. Fujimoto, *Plasma Spectroscopy* (Oxford University Press, Oxford, 2004)
2. T. Fujimoto, J. Phys. Soc. Japan **47**, 273 (1979)
3. T. Fujimoto, J. Phys. Soc. Japan **49**, 1569 (1980)
4. T. Fujimoto, J. Phys. Soc. Japan **49**, 1561 (1980)

# 4

# Population-Alignment Collisional-Radiative Model

T. Fujimoto

In this chapter, we construct the theoretical framework to deal with the Class 2 polarization. We assume a situation in which our plasma has an anisotropic EVDF (electron velocity distribution function). We assign a quantity *the alignment* to each level (more exactly, the ensemble of atoms (and ions) in this level) besides its number density or population. The presence of an alignment in a level makes emission lines originating from this level linearly polarized. The alignment is created by collisional excitation by electrons having an anisotropic EVDF. We construct rate equations for the alignment and for the population and solve them in the quasi-steady-state (QSS) approximation. We call this method the population-alignment collisional-radiative (PACR) model. We are thus able to interpret the observed polarization and intensity of emission lines in terms of the anisotropic EVDF.

The theoretical framework developed in the photon–atom interaction studies constitutes the theoretical basis of the PACR model. In particular, the formulations described below are based on Omont [1] and applied to the present problem in Fujimoto et al. [2]. The part of the ionizing plasma component has been presented in Fujimoto and Kazantsev [3].

In the PACR model, various kinds of cross sections are involved concerning the alignment. In Sect. 4.2, we adopt a heuristic approach, i.e., the semiclassical impact parameter method, to express the cross sections. In the succeeding Chap. 5 a quantum mechanical formulation is developed, which is more appropriate for electron impact.

## 4.1 Population and Alignment

We consider an ensemble of atoms (and ions) having no hyperfine structure and restrict ourselves to situations in which the following three assumptions hold:

1. Axial symmetry is present, and we define the $z$-axis, the quantization axis, as the symmetry axis of our plasma.

2. There is no coherence among different Zeeman multiplets. We specify an atomic state by its total angular momentum quantum number, $J$, and its projection onto the $z$-axis, $M$, besides the other indices, $\alpha$, necessary to specify the state. In the following, an atomic level (Zeeman multiplet) $\alpha J$ is sometimes denoted by $p$ or $r$. It should be noted, however, that, in Chap. 3, $p$ or $r$ was used to denote the principal quantum number of a hydrogen-like level: a different usage from this chapter. We further assume the absence of coherence among the different magnetic sublevels in a Zeeman multiplet, i.e., no Zeeman coherence.

3. Electric and magnetic fields are absent. When a weak magnetic field is present, this field affects little the formulation developed in the following in this chapter. Rather, this field defines the quantization axis $z$ and makes the system axially symmetric around it.

The above assumptions imply that the atomic system is described as an incoherent superposition of "level states," and that its density matrix (See Appendix C and, for example, Blum [4].) reduces to a sum of density matrices for each level $\alpha J$ or $p$,

$$\rho(p) = \sum_{MN} \rho_{M,N}(p)|\alpha JM\rangle\langle\alpha JN|, \tag{4.1}$$

where $\rho_{M,N}(p)$ with $N \neq M$ is the coherence and with $N = M$ is the "population" of the magnetic sublevel $M$. In place of $|\alpha JM\rangle\langle\alpha JN|$ we introduce the irreducible tensorial set

$$T_q^{(k)}(p) = \sum_{MN} (-)^{J-N}\langle JJM-N|kq\rangle|\alpha JM\rangle\langle\alpha JN|, \tag{4.2}$$

where $(-)$ means $(-1)$ and $\langle JJM-N|kq\rangle$ is the Clebsch–Gordan coefficient. As is stated in assumption 2, we will restrict our consideration to the situations in which $\rho_{M,N}(p) = 0$ for $N \neq M$, so that we will only have the cases with $q = 0$. We then expand (4.1) in terms of (4.2).

$$\rho(p) = \rho_0^0(p)T_0^{(0)}(p) + \rho_0^2(p)T_0^{(2)}(p) + \cdots, \tag{4.3}$$

where the expansion coefficients are given by

$$\rho_q^k(p) = \sum_{MN} (-)^{J-N}\langle JJM-N|kq\rangle\rho_{M,N}(p). \tag{4.4}$$

It is noted that the odd rank terms are dropped in (4.3) because, owing to the symmetry of our situation, they do not appear in the formulation below. This is connected with the fact that, for an atom, a collision of a perturber incident on it from the $+z$-direction cannot be distinguished from a collision from the $-z$-direction, except for the recoil motion, which we ignore in our discussion. Therefore, an orientation $\rho_q^1(p)$ does not appear. In the following we retain only the first two terms in this equation and neglect those of higher rank. We thus assign two quantities to each level $p$: the "population" $\rho_0^0(p)$

and the alignment $\rho_0^2(p)$. We note here that the conventional population is given by $n(p) = \sqrt{2J_p + 1}\rho_0^0(p)$.

The population was the central problem of the conventional intensity spectroscopy and it was treated in Chap. 3 on the assumption of a Maxwell distribution for EVDF (electron velocity distribution function). The alignment is a measure of the "population" imbalance among the magnetic sublevels in a level, and it gives rise to linear polarization of emission lines. For the purpose of simplicity, we use $a(p)$ in place of $\rho_0^2(p)$. Table 4.1 gives examples of $a(p)$ for several values of $J$. Figure 4.1 shows, as an example of $J = 2$, the "population" distribution having (a) a positive alignment and (b) a negative

**Table 4.1.** Examples of alignments for several values of $J$

| $J$ | $\rho_0^2$ or $a$ |
|---|---|
| 1 | $(1/\sqrt{6})(\rho_{11} - 2\rho_{00} + \rho_{-1-1})$ |
| $\frac{3}{2}$ | $(1/2)\left(\rho_{\frac{3}{2}\frac{3}{2}} - \rho_{\frac{1}{2}\frac{1}{2}} - \rho_{-\frac{1}{2}-\frac{1}{2}} + \rho_{-\frac{3}{2}-\frac{3}{2}}\right)$ |
| 2 | $(1/\sqrt{14})(2\rho_{22} - \rho_{11} - 2\rho_{00} - \rho_{-1-1} + 2\rho_{-2-2})$ |
| $\frac{5}{2}$ | $\left(1/2\sqrt{21}\right)\left(5\rho_{\frac{5}{2}\frac{5}{2}} - \rho_{\frac{3}{2}\frac{3}{2}} - 4\rho_{\frac{1}{2}\frac{1}{2}} - 4\rho_{-\frac{1}{2}-\frac{1}{2}} - \rho_{-\frac{3}{2}-\frac{3}{2}} + 5\rho_{-\frac{5}{2}-\frac{5}{2}}\right)$ |
| 3 | $\left(1/2\sqrt{21}\right)(5\rho_{33} - 3\rho_{11} - 4\rho_{00} - 3\rho_{-1-1} + 5\rho_{-3-3})$ |

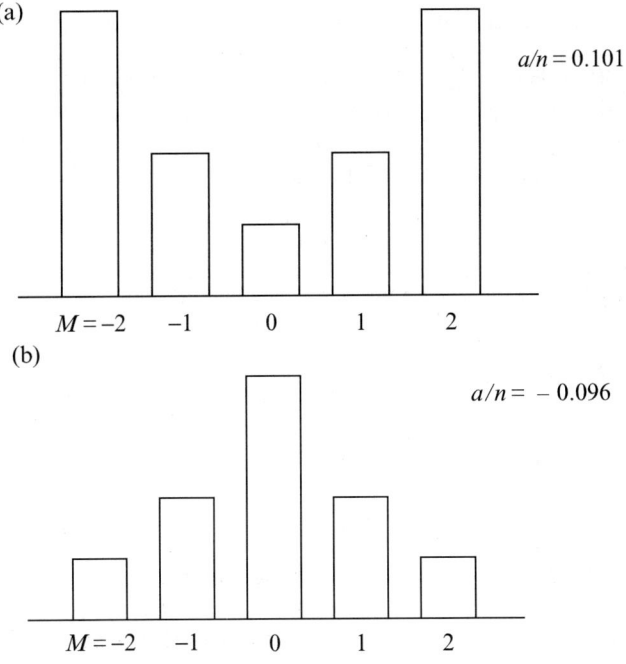

**Fig. 4.1.** Examples of (**a**) a positive alignment and (**b**) a negative alignment. The relative alignment $a(p)/n(p)$ is given in the figure

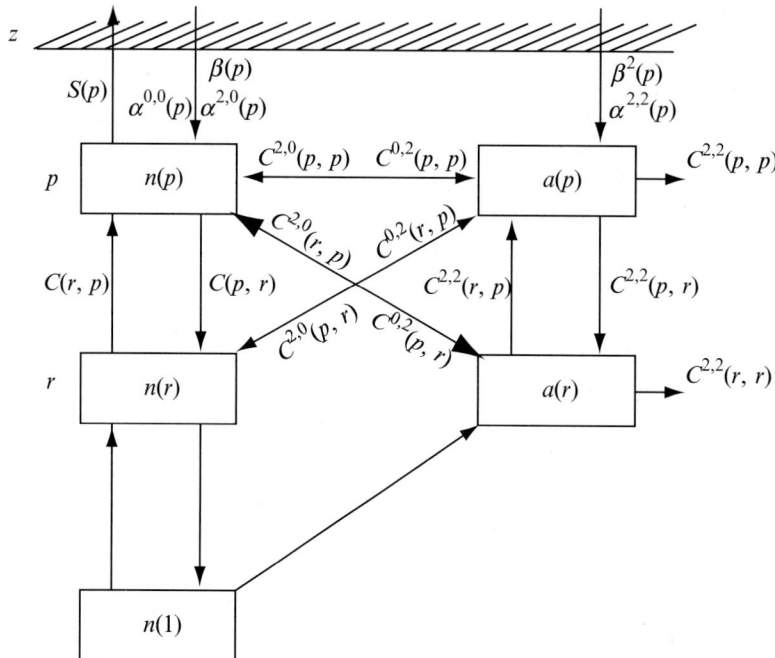

**Fig. 4.2.** Populations and alignments of levels and collision processes between them

alignment. We assign to each level these two quantities, $n(p)$ and $a(p)$, both in units of $[\text{m}^{-3}]$. Figure 4.2 shows the structure of our system. In our new model we deal with the system of $a(p)$ for excited levels besides the system of $n(p)$. We assume that the ground state atoms (ions), denoted as "1" in Fig. 4.2, are not aligned.

Suppose we observe an emission line $p \to s$ from a particular direction with a linear polarizer, the direction of its transmission axis is expressed with a unit vector $\boldsymbol{e}$. The observed intensity is given as

$$I(\boldsymbol{e}) = \frac{1}{8\pi} n(p) A(p, s) h\nu$$
$$\times \left[ 1 - (-)^{J_p + J_s} \frac{\sqrt{6}}{2} (1 - 3\cos^2 \eta)(2J_p + 1) \begin{Bmatrix} J_p & J_p & 2 \\ 1 & 1 & J_s \end{Bmatrix} \frac{a(p)}{n(p)} \right] \mathrm{d}V \mathrm{d}\Omega,$$

$$(4.5)$$

where $\eta$ is the angle of $\boldsymbol{e}$ with respect to the quantization axis $z$, $\{:::\}$ is the 6-$j$ symbol, $\mathrm{d}V$ and $\mathrm{d}\Omega$ are the same as in (3.1). See Appendix B. Here we assume that the solid angle subtended by our optics is sufficiently small, so that $\eta$ can be given a definite value. This equation corresponds to (3.1), and in the absence of the alignment, the second term, this equation reduces to

(3.1) apart from the factor $1/2$, which comes from one of the two polarized components. We name $a(p)/n(p)$ the *relative alignment*.

In the electron impact experiment mentioned in the beginning of Chap. 1, the $\pi$ light is observed with the direction of the polarizer $\eta = 0$, so that $(1 - 3\cos^2\eta) = -2$, and for the $\sigma$ light $\eta = \pi/2$, so that $(1 - 3\cos^2\eta) = 1$. In the presence of an alignment the intensity that is proportional only to the population is

$$
I_0 \equiv \frac{1}{4\pi} n(p) A(p, s) h\nu \, dV \, d\Omega
$$
$$
= \frac{2}{3} (I_\pi + 2I_\sigma) . \tag{4.6}
$$

See Appendix E concerning this problem. Instead of the polarization degree $P$ we quantify the degree of linear polarization in terms of *the longitudinal alignment*

$$
A_{\mathrm{L}} = \frac{I_\pi - I_\sigma}{I_\pi + 2I_\sigma} = \frac{2P}{3 - P}, \tag{4.7}
$$

which is directly proportional to the relative alignment,

$$
A_{\mathrm{L}}(p, s) = (-)^{J_p + J_s} \sqrt{\frac{3}{2}} (2J_p + 1) \left\{ \begin{array}{ccc} J_p & J_p & 2 \\ 1 & 1 & J_s \end{array} \right\} \frac{a(p)}{n(p)} . \tag{4.8}
$$

We generalize our method of the collisional-radiative (CR) model which was introduced in Chap. 3 so as to incorporate the system of $a(p)$, and call this new method the *population-alignment collisional-radiative (PACR) model*. In this model, we include such transition processes as creation of alignment in a level from a population in another level, transfer of alignment from a level to another level, and so on. Figure 4.2 shows these collision processes.

## 4.2 Excitation, Deexcitation and Elastic Collisions: Semiclassical Approach

The time development of the system due to a relaxation mechanism is described in terms of the relaxation matrix $G$,

$$
(d\rho/dt)_{\mathrm{relax}} = -G\rho. \tag{4.9}
$$

In the dyadic representation this equation can be written as

$$
(d\rho_{ij}/dt)_{\mathrm{relax}} = -\sum_{rs} G_{ij,rs}\rho_{rs} \tag{4.10}
$$

with

$$
G_{ij,rs} = \langle\langle ij^+|G|rs^+\rangle\rangle, \tag{4.11}
$$

where $|\cdots\rangle\rangle$ denotes a vector in the Liouville space.

We now consider the effects of collisions in the semiclassical approximation. We assume a rectilinear path. Since this approximation is inappropriate for electrons, we call the colliding particles "perturbers." A collision is defined by the relative velocity $\boldsymbol{v}$ and the impact parameter $\boldsymbol{b}$ $(\boldsymbol{b} \cdot \boldsymbol{v} = 0)$. The time development of the density matrix is described in terms of the $S$ matrix, which is equal to the time evolution operator $U(\infty, -\infty)$, and is given by

$$(\mathrm{d}\rho/\mathrm{d}t)_{\mathrm{coll}} = -G_{\mathrm{c}}\rho = 2\pi n_{\mathrm{p}} \langle v \int b\,\mathrm{d}b[S(\boldsymbol{b}, \boldsymbol{v})\rho S(\boldsymbol{b}, \boldsymbol{v})^{\dagger} - \rho]\rangle_{\mathrm{Av}}\,, \qquad (4.12)$$

where $n_{\mathrm{p}}$ is the density of perturbers and the average is over the velocities and the directions of $\boldsymbol{b}$.

### 4.2.1 Monoenergetic Beam Perturbers and Cross Sections

We first consider monoenergetic beam perturbers with velocity $\boldsymbol{v}$ in the $z'$-direction, which is different from the $z$-direction defined as the quantization axis of the system. We define an intermediate "superoperator" matrix $\widetilde{\Pi}(b, v)$ by averaging over the directions of $\boldsymbol{b}$ (perpendicular to $\boldsymbol{v}$).

$$\widetilde{\Pi}(b, v)\rho = \rho - \left\{ S_{\mathrm{c}}\rho S_{\mathrm{c}}^{\dagger} \right\}_{b\mathrm{Av}}\,, \qquad (4.13)$$

where the index c means that the matrix $S$ is referred to these collision axes $z'$ and therefore the atomic system is also referred to $z'$. The matrix elements of $\widetilde{\Pi}$ are

$$\langle\langle \alpha'J'M', (\alpha'J'N')^{+}|\widetilde{\Pi}|\alpha JM, (\alpha JN)^{+}\rangle\rangle$$
$$= \delta - \langle \alpha'J'M'|S_{\mathrm{c}}|\alpha JM\rangle\langle \alpha'J'N'|S_{\mathrm{c}}|\alpha JN\rangle^{*}\delta_{M'-N', M-N}\,, \quad (4.14)$$

where $\delta = \delta_{\alpha\alpha'}\delta_{JJ'}\delta_{MM'}\delta_{NN'}$. The matrix element in the irreducible basis is defined and given by

$$\widetilde{\Pi}_{q}(k'k) \equiv \langle\langle \alpha'J'(\alpha'J')^{+}; k'q|\Pi|\alpha J(\alpha J)^{+}; kq\rangle\rangle$$
$$= \sum_{MM'} (-)^{J+J'+M+M'} \langle J'J'M'(q-M')|k'q\rangle\langle JJM(q-M)|kq\rangle$$
$$\times \langle\langle \alpha' J' M', (\alpha' J' M'-q)^{+}|\widetilde{\Pi}|\alpha J M, (\alpha J M-q)^{+}\rangle\rangle\,, \qquad (4.15)$$

where, because of the symmetry, $q$ does not change by the collisions. It is noted that in this $z'$-system there can be Zeeman coherence, or nonzero density matrices, $\widetilde{\rho}_{\alpha JM, \alpha JM-q}$ with $q \neq 0$, where $\rho$ is defined in the $z'$-system. Then, (4.15) with (4.14) reduces to

$$\widetilde{\Pi}_{q}(k'k) = \sum_{MM'} (-)^{J+J'+M+M'} \langle J' J' M' (q-M')|k'q\rangle\langle\langle (J J M (q-M)|kq\rangle$$
$$\times [\delta_{\alpha\alpha'}\delta_{JJ'}\delta_{MM'} - \langle \alpha'J'M'|S_{\mathrm{c}}|\alpha JM\rangle$$
$$\times \langle \alpha'J'(M'-q)|S_{\mathrm{c}}|\alpha J(M-q)\rangle^{*}]\,. \qquad (4.16)$$

Equation (4.12) in the $z'$-system is rewritten as

$$\mathrm{d}\widetilde{\rho}_q^{k'}(\alpha' J')/\mathrm{d}t = 2\pi n_\mathrm{p} v \int [-\widetilde{\Pi}_q(k'k)]b\,\mathrm{d}b\,\widetilde{\rho}_q^{k}(\alpha J), \qquad (4.17)$$

where $\widetilde{\rho}_q^{k}(\alpha J)$ has been defined in the $z'$-system. The initial state $\alpha J$ is denoted by $r$ and the final state by $p$. We define the cross section by

$$\sigma_q^{k,k'}(r,p) = 2\pi \int [\pm\widetilde{\Pi}_q(k'k)]b\,\mathrm{d}b\,, \qquad (4.18)$$

where the plus sign applies to the case of $\alpha' J' = \alpha J$ $(p = r)$ or "elastic" collisions, and the minus sign to $\alpha' J' \neq \alpha J$ $(p \neq r)$ or "inelastic" collisions. This convention is chosen for the convenience of expressing the rate equations, which will be given later.

We now write down the cross section, (4.18), from (4.16). Let the integral cross section for transition from magnetic sublevel $\alpha J M$ to magnetic sublevel $\alpha' J' M'$ be $Q_{\alpha J M, \alpha' J' M'}$, which is defined by

$$Q_{\alpha J M, \alpha' J' M'} = 2\pi \int |\langle \alpha' J' M' | S_\mathrm{c} | \alpha J M \rangle|^2 b\,\mathrm{d}b. \qquad (4.19)$$

For excitation or deexcitation of $\alpha J \to \alpha' J'$ $(r \to p)$ we have

$$\sigma_0^{0,0}(r,p) = (2J+1)^{-1/2}(2J'+1)^{-1/2} \sum_{MM'} Q_{\alpha J M, \alpha' J' M'} \qquad (4.20\mathrm{a})$$

$$\sigma_0^{0,2}(r,p) = (2J+1)^{-1/2} \sum_{MM'} (-)^{J'-M'} \langle J' J' M' - M' | 20 \rangle Q_{\alpha J M, \alpha' J' M'}$$
$$(4.20\mathrm{b})$$

$$\sigma_0^{2,0}(r,p) = (2J'+1)^{-1/2} \sum_{M} (-)^{J-M} \langle J J M - M | 20 \rangle \sum_{M'} Q_{\alpha J M, \alpha' J' M'}$$
$$(4.20\mathrm{c})$$

$$\sigma_0^{2,2}(r,p) = \sum_{MM'} (-)^{J+J'+M+M'} \langle J' J' M' - M' | 20 \rangle \langle J J M - M | 20 \rangle Q_{\alpha J M, \alpha' J' M'}$$
$$(4.20\mathrm{d})$$

and for $q = 1$ or $2$ we have

$$\sigma_q^{2,2}(r,p) = \sum_{MM'} (-)^{J+J'+M+M'} \langle J' J' M' (q-M') | 2q \rangle \langle J J M (q-M) | 2q \rangle$$
$$\times 2\pi \int \langle \alpha' J' M' | S_\mathrm{c} | \alpha J M \rangle \langle \alpha' J' (M'-q) | S_\mathrm{c} | \alpha J (M-q) \rangle^* b\,\mathrm{d}b.$$
$$(4.20\mathrm{e})$$

It is to be noted that the coherence transfer cross section, (4.20e), cannot be expressed in terms of the magnetic-sublevel-to-magnetic-sublevel cross sections.

We define the total cross section for depopulation from magnetic sublevel $\alpha JM$ by inelastic collisions by

$$D_{\alpha JM} = 2\pi \int \sum_{M'} \left[ \delta_{MM'} - |\langle \alpha JM'|S_{\rm c}|\alpha JM\rangle|^2 \right] b\,{\rm d}b. \qquad (4.21)$$

For "elastic" collisions of $\alpha J \to \alpha J$ $(r \to r)$, the cross sections are given by

$$\sigma_0^{0,0}(r,r) = (2J+1)^{-1} \sum_M D_{\alpha JM} \qquad (4.22a)$$

$$\sigma_0^{0,2}(r,r) = (2J+1)^{-1/2} \sum_M (-)^{J-M} \langle JJM-M|20\rangle D_{\alpha JM}$$

$$+ (2J+1)^{-1/2} \sum_{M'\neq M} \left[ (-)^{J-M} \langle JJM-M|20\rangle \right.$$

$$\left. - (-)^{J-M'} \langle JJM'-M'|20\rangle \right] Q_{\alpha JM,\alpha JM'} \qquad (4.22b)$$

$$\sigma_0^{2,0}(r,r) = (2J+1)^{-1/2} \sum_M (-)^{J-M} \langle JJM-M|20\rangle D_{\alpha JM} \qquad (4.22c)$$

$$\sigma_0^{2,2}(r,r) = \sum_M \langle JJM-M|20\rangle^2 D_{\alpha JM} + \sum_{M'\neq M} \langle JJM-M|20\rangle$$

$$\times \left[ \langle JJM-M|20\rangle - (-)^{M'-M} \langle JJM'-M'|20\rangle \right] Q_{\alpha JM,\alpha JM'} \qquad (4.22d)$$

and for $q = 1$ or $2$ we have

$$\sigma_q^{2,2}(r,r) = \sum_{MM'} (-)^{M'-M} \langle JJM'(q-M')|2q\rangle \langle JJM(q-M)|2q\rangle$$

$$\times 2\pi \int \left[ \delta_{MM'} - \langle \alpha JM'|S_{\rm c}|\alpha JM\rangle \langle \alpha J(M'-q)|S_{\rm c}|\alpha J(M-q)\rangle^* \right] b\,{\rm d}b. \qquad (4.22e)$$

### 4.2.2 Axially Symmetric Distribution

For the sake of simplicity of expressions we abbreviate (4.17) to

$$\mathrm{d}\widetilde{\rho}_q^{k'}(p)/\mathrm{d}t = -\widetilde{g}_q^{k,k'}(r,p)\widetilde{\rho}_q^k(r), \qquad (4.23)$$

where $\alpha J$ has been denoted by $r$ and $\alpha'J'$ by $p$.

Until this point we have considered collisions by the mono-energetic beam perturbers. Now we consider the situation in which the angular distribution of $\boldsymbol{v}$ is axially symmetric around the $z$-axis, the quantization axis which is referred to the plasma as defined in the beginning of this chapter. The density matrix in the $z$-system is expressed as

$$\rho_q^k(p) = \sum_{q'} R_{qq'}^{(k)} \widetilde{\rho}_{q'}^k(p) \ , \tag{4.24}$$

and the inverse transformation is

$$\widetilde{\rho}_{q'}^k(p) = \sum_{q''} (R^{-1})_{q'q''}^{(k)} \rho_{q''}^k(p) \ ,$$

$$= \sum_{q''} (-)^{q''-q'} R_{-q'',-q'}^{(k)} \rho_{q''}^k(p) \ , \tag{4.25}$$

where $R_{qq'}^{(k)}$ is the rotation matrix from the "$z$-system" to the "$z'$-system" with the Euler angle $(\phi, \theta, 0)$. See Appendix B. The time development of the irreducible component in the $z$-system is given in terms of that in the $z'$-system as

$$\frac{\mathrm{d}}{\mathrm{d}t} \rho_q^{k'}(p) = -\sum_{q'} R_{qq'}^{(k')} \widetilde{g}_{q'}^{k,k'}(r,p) \, \widetilde{\rho}_{q'}^k(r)$$

$$= -\sum_{q'} \sum_{q''} (-)^{q''-q'} R_{qq'}^{(k')} R_{-q'',-q'}^{(k)} \widetilde{g}_{q'}^{k,k'}(r,p) \, \rho_{q''}^k(r)$$

$$= -\sum_{q'} \sum_{q''} \sum_K (-)^{q''-q'} R_{q-q'',0}^{(K)} \langle k' \, k \, q \, -q'' | K(q \, -q'') \rangle$$

$$\times \langle k' \, k \, q' \, -q' | K 0 \rangle \widetilde{g}_{q'}^{k,k'}(r,p) \, \rho_{q''}^k(r). \tag{4.26}$$

Since we assume that the velocity distribution of the perturbers is symmetric around the $z$-axis, we integrate the effects of collisions over $\phi$. By using the relationship

$$R_{MM'}^{(K)} = \mathrm{e}^{-\mathrm{i}\phi M} r_{MM'}^{(K)}(\theta) \mathrm{e}^{-\mathrm{i}\gamma M'} \tag{4.27}$$

where $\gamma(=0)$ is the third Euler angle, and the fact that our velocity distribution is axially symmetric, we drop, on the right-hand side of (4.26), the terms other than those for which $q'' = q$ to obtain

$$\frac{\mathrm{d}}{\mathrm{d}t} \rho_q^{k'}(p) = -2\pi \sum_{q'} (-)^{q-q'} \widetilde{g}_{q'}^{k,k'}(r,p) \sum_K \langle k' \, k \, q \, -q | K 0 \rangle \langle k' \, k \, q' \, -q' | K 0 \rangle$$

$$\times R_{00}^{(K)}(0, \theta, 0) \rho_q^k(r) \ . \tag{4.28}$$

Let the velocity distribution over the polar angle $\theta$ be expressed as $f_v(\theta)$. The integration over the angle leads to the following expression for the time development of the system,

$$\frac{\mathrm{d}}{\mathrm{d}t} \rho_q^{k'}(p) = -g_q^{k,k'}(r,p) \rho_q^k(r) \ . \tag{4.29}$$

Here the relaxation matrix is given as

$$g_q^{k,k'}(r,p) = \sum_{q'} (-)^{q-q'} \widetilde{g}_{q'}^{k,k'}(r,p) \sum_K \langle k'\,k\,q-q|K0\rangle \langle k'\,k\,q'-q'|K0\rangle$$

$$\times \int f_v(\theta) P_K(\cos\theta) \sin\theta\, d\theta, \tag{4.30}$$

where $P_K(\cos\theta)$ are the Legendre polynomials, and we have used the equality of the rotation matrix and the Legendre polynomial. This equation is nothing but (4.42) of Omont [1]. We expand the velocity distribution in terms of Legendre polynomials,

$$f_v(\theta) = \sum_K f_K P_K(\cos\theta) \tag{4.31}$$

with

$$f_K = \frac{2K+1}{2} \int f_v(\theta) P_K(\cos\theta) \sin\theta\, d\theta. \tag{4.32}$$

Equation (4.29) is now rewritten as

$$\frac{d}{dt}\rho_q^{k'}(p) = -\sum_{q'} (-)^{q-q'} \widetilde{g}_{q'}^{k,k'}(r,p) \sum_K \langle k'\,k\,q-q|K0\rangle \langle k'\,k\,q'-q'|K0\rangle$$

$$\times \frac{2}{2K+1} f_K \rho_q^k(r) \tag{4.33}$$

According to our earlier assumption we restrict ourselves to the irreducible spherical components with rank $k$ and $k'$ lower than 2, i.e., the population and the alignment. As we have assumed, there is no coherence among the magnetic sublevels in a level, i.e., $q = 0$, so that relaxation matrix elements $g_q^{k,k'}(r,p)$ with $q \neq 0$ are unnecessary. The matrix elements given by (4.30) are written as

$$\begin{aligned}
g_0^{0,0} &= \widetilde{g}_0^{0,0} 2f_0 \\
g_0^{0,2} &= \widetilde{g}_0^{0,2} 2f_2/5 \\
g_0^{2,0} &= \widetilde{g}_0^{2,0} 2f_2/5 \\
g_0^{2,2} &= \widetilde{g}_0^{2,2}[2f_0/5 + 4f_2/35 + 12f_4/105] \\
&\quad + \widetilde{g}_1^{2,2}[2f_0/5 + 2f_2/35 - 8f_4/105] \\
&\quad + \widetilde{g}_2^{2,2}[2f_0/5 - 4f_2/35 + 2f_4/105].
\end{aligned} \tag{4.34}$$

We note that $\widetilde{g}_q^{k,k'}$ is expresed in terms of the cross section $\sigma_q^{k,k'}$. See (4.17), (4.18), and (4.29).

We now consider the situation in which the perturbers are not monoenergetic, but they have a distribution over $v$; the distribution is expressed by $f(v,\theta)$. Here we note that the normalization condition is

$$\iiint f(v,\theta)v^2\,\mathrm{d}v\,\sin\theta\,\mathrm{d}\theta\mathrm{d}\phi = 1,$$

or

$$\iint f(v,\theta)v^2\,\mathrm{d}v\,\sin\theta\,\mathrm{d}\theta = 1/2\pi.$$

The distribution function $f_v(\theta)$ in (4.30)–(4.32) is replaced by $f(v,\theta)$, and the moments of the distribution function $f_K$ may be written as $f_K(v)$.

### 4.2.3 Rate Equation in the Irreducible-Component Representation

We define the rate coefficients for transition $\alpha J \to \alpha' J'$ or $r \to p$ as

$$c^{0,0}(r,p) = \int \sigma_0^{0,0}(r,p)4\pi f_0(v)v^3\,\mathrm{d}v$$

$$c^{0,2}(r,p) = \int \sigma_0^{0,2}(r,p)[4\pi f_2(v)/5]v^3\,\mathrm{d}v$$

$$c^{2,0}(r,p) = \int \sigma_0^{2,0}(r,p)[4\pi f_2(v)/5]v^3\,\mathrm{d}v \qquad (4.35)$$

$$c^{2,2}(r,p) = \int [\sigma_0^{2,2}(r,p) + \sigma_1^{2,2}(r,p) + \sigma_2^{2,2}(r,p)][4\pi f_0(v)/5]v^3\,\mathrm{d}v$$

$$+ \int [2\sigma_0^{2,2}(r,p) + \sigma_1^{2,2}(r,p) - 2\sigma_2^{2,2}(r,p)][4\pi f_2(v)/35]v^3\,\mathrm{d}v$$

$$+ \int [6\sigma_0^{2,2}(r,p) - 4\sigma_1^{2,2}(r,p) + \sigma_2^{2,2}(r,p)][4\pi f_4(v)/105]v^3\,\mathrm{d}v\ .$$

Spontaneous radiative transition processes are isotropic, and we have only the two corresponding rates: for $r \to p$ $(r \neq p)$

$$A^{0,0}(r,p) = (2J_r+1)^{1/2}(2J_p+1)^{-1/2}A(r,p) \qquad (4.36\mathrm{a})$$

$$A^{2,2}(r,p) = (-)^{J_r+J_p+1}(2J_r+1)\begin{Bmatrix} J_r & J_r & 2 \\ J_p & J_p & 1 \end{Bmatrix} A(r,p), \qquad (4.36\mathrm{b})$$

where $A(r,p)$ is the usual Einstein $A$ coefficient. For $p \to p$, (4.22a) and (4.22d) suggest

$$A^{0,0}(p,p) = A^{2,2}(p,p) = \sum_r A(p,r)\ , \qquad (4.37)$$

where we have utilized the facts that all the magnetic sublevels have an equal radiative decay probability and that $\sum_{(M)}\langle JJM-M|20\rangle^2 = 1$.

We now construct a set of rate equations for the ensemble of atoms according to (4.29). We assume that the collisional and radiative relaxation processes are additive. For "population" we have

$$\mathrm{d}\rho_0^0(p)/\mathrm{d}t = \sum_{r\neq p}\left[c^{0,0}(r,p)n_\mathrm{p}+A^{0,0}(r,p)\right]\rho_0^0(r) - \left[c^{0,0}(p,p)n_\mathrm{p}+A^{0,0}(p,p)\right]\rho_0^0(p)$$

$$+ \sum_{r\neq p}c^{2,0}(r,p)n_\mathrm{p}\rho_0^2(r) - c^{2,0}(p,p)n_\mathrm{p}\rho_0^2(p), \qquad (4.38)$$

and for alignment

$$
\begin{aligned}
\mathrm{d}\rho_0^2(p)/\mathrm{d}t = &\sum_{r \neq p} c^{0,2}(r,p)n_\mathrm{p}\rho_0^0(r) - c^{0,2}(p,p)n_\mathrm{p}\rho_0^0(p) \\
&+ \sum_{r \neq p} \left[ c^{2,2}(r,p)n_\mathrm{p} + A^{2,2}(r,p) \right] \rho_0^2(r) \\
&- \left[ c^{2,2}(p,p)n_\mathrm{p} + A^{2,2}(p,p) \right] \rho_0^2(p).
\end{aligned}
\tag{4.39}
$$

### 4.2.4 Rate Equation in the Conventional Representation

Up to this point, our formulation has been in terms of $\rho_0^0(p)$ and $\rho_0^2(p)(\equiv a(p))$. We now adopt the conventional definition of population $n(p)=\sqrt{2J_p + 1}\rho_0^0(p)$, so that, instead of the cross section $\sigma_q^{k,k'}(r,p)$ as defined by (4.18) we employ cross section $Q_q^{k,k'}(r,p)$ in the present convention. We now redefine the cross sections in place of (4.20a–e) and (4.22a–e). The difference in the definitions is only with the multiplication factor and trivial, but the present convention is more natural in practical applications.

For excitation or de-excitation of $\alpha J \rightarrow \alpha' J' (\alpha' J' \neq \alpha J)$ or $r \rightarrow p$ $(p \neq r)$ we have

$$
Q_0^{0,0}(r,p) = (2J+1)^{-1} \sum_{MM'} Q_{\alpha JM, \alpha' J' M'}
\tag{4.40a}
$$

$$
Q_0^{0,2}(r,p) = (2J+1)^{-1} \sum_{MM'} (-)^{J'-M'} \langle J'J'M' -M'|20\rangle Q_{\alpha JM, \alpha' J' M'}
\tag{4.40b}
$$

$$
Q_0^{2,0}(r,p) = \sum_{M} (-)^{J-M} \langle JJM -M|20\rangle \sum_{M'} Q_{\alpha JM, \alpha' J' M'}
\tag{4.40c}
$$

and for $q = 0, 1$ and $2$

$$
Q_q^{2,2}(r,p) = \sigma_q^{2,2}(r,p) .
\tag{4.40d,e}
$$

It is noted that the present notation of the cross sections is different from those in earlier literatures, e.g., Fujimoto and Kazantsev [2]. The former $Q_{\alpha' J' M', \alpha J M}$ is now written as $Q_{\alpha JM, \alpha' J' M'}$, and the former $Q^{20}(r,p)$ is now $Q^{0,2}(r,p)$.

For "elastic" collisions of $\alpha J \rightarrow \alpha J$, or $p \rightarrow p$, the cross sections are given by

$$
Q_0^{0,0}(p,p) = (2J+1)^{-1} \sum_{M} D_{\alpha JM}
\tag{4.41a}
$$

$$
\begin{aligned}
Q_0^{0,2}(p,p) = &(2J+1)^{-1} \sum_{M} (-)^{J-M} \langle JJM -M|20\rangle D_{\alpha JM} + (2J+1)^{-1} \\
&\times \sum_{M' \neq M} \left[ (-)^{J-M} \langle JJM -M|20\rangle - (-)^{J-M'} \langle JJM' -M'|20\rangle \right] \\
&\times Q_{\alpha JM, \alpha JM'}
\end{aligned}
\tag{4.41b}
$$

$$Q_0^{2,0}(p,p) = \sum_M (-)^{J-M} \langle JJM-M|20 \rangle D_{\alpha JM} \qquad (4.41c)$$

and for $q = 0$, 1 and 2

$$Q_q^{2,2}(p,p) = \sigma_q^{2,2}(p,p) . \qquad (4.41d,e)$$

Instead of (4.29) we redefine rate coefficients for transition $\alpha'J' \to \alpha J$ or $r \to p$ (for $p \neq r$ or $p = r$) as

$$C^{0,0}(r,p) = \int Q_0^{0,0}(r,p) 4\pi f_0(v) v^3 \, dv, \qquad (4.42a)$$

$$C^{0,2}(r,p) = \int Q_0^{0,2}(r,p)[4\pi f_2(v)/5]v^3 \, dv, \qquad (4.42b)$$

$$C^{2,0}(r,p) = \int Q_0^{2,0}(r,p)[4\pi f_2(v)/5]v^3 \, dv, \qquad (4.42c)$$

$$C^{2,2}(r,p) = \int [Q_0^{2,2}(r,p) + Q_1^{2,2}(r,p) + Q_2^{2,2}(r,p)][4\pi f_0(v)/5]v^3 \, dv$$

$$+ \int [2Q_0^{2,2}(r,p) + Q_1^{2,2}(r,p) - 2Q_2^{2,2}(r,p)][4\pi f_2(v)/35]v^3 \, dv$$

$$+ \int [6Q_0^{2,2}(r,p) - 4Q_1^{2,2}(r,p) + Q_2^{2,2}(r,p)][4\pi f_4(v)/105]v^3 \, dv. \qquad (4.42d)$$

It is noted that the above expressions of the rate coefficients may look different from the conventional definition, e.g., $C(r,p) = \int f(v)v\sigma(r,p) \, dv$. This is due to the difference in the expressions for the distribution functions; in the conventional expression the EVDF is assumed isotropic and the normalization condition is $\int f(v) \, dv = 1$, while in our present formulation, the corresponding normalization is $\int 4\pi f_0(v)v^2 \, dv = 1$.

Equation (4.40a) with (4.19) shows that $Q^{0,0}(r,p)$, so that $C^{0,0}(r,p)$, is the excitation (de-excitation) cross section, or the rate coefficient, averaged over the magnetic sublevels which are assumed statistically populated. This is nothing but the excitation (de-excitation) cross section, or the rate coefficient, in the conventional sense. We rewrite $C^{0,0}(r,p)$ as $C(r,p)$ in the following. Figure 4.2 adopts this nomenclature. In a similar sense, $C^{0,0}(p,p)$ of (4.41a) with (4.21) is the depopulation rate coefficient by inelastic collisions averaged over the magnetic sublevels. This is the total depopulation rate coefficient from this level, consisting of excitation, deexcitation and ionization. We rewrite this rate coefficient as

$$C^{0,0}(p,p) = \sum_{r \neq p} C(p,r) + S(p), \qquad (4.43)$$

where $S(p)$ denotes the ionization rate coefficient from level $p$ (see Fig. 4.2). Note that, in this and subsequent chapters, no distinction of the nomenclature is made between excitation and de-excitation: $C(p,r)$ is used for both of them.

$A^{0,0}(r,p)$ in the rate equation, (4.38) and (4.39), is replaced by $A(r,p)$ in the present convention, and (4.37) is unchanged.

In the present section, the colliding particles were "perturbers." We now assume that the formulas derived in this section are also valid for electron impact. Electrons are now colliding particles, so that the velocity distribution is now EVDF and $n_p$ is understood as $n_e$.

## 4.3 Ionization and Recombination

We now turn to other collision processes, which have not been considered above but are important in the actual situations.

Ionization has already been included in depopulation, (4.43). When different sublevels have different ionization rates, alignment can be created in this level. This effect, together with a similar effects by excitation and de-excitation, has been taken into account through $C^{0,2}(p,p)$ in (4.41b) (see Fig. 4.2). The alignment destruction by ionization has already been included in (4.22d) and by $C^{2,2}(p,p)$ or $C^{2,2}(r,r)$ so that it is not explicitly shown in Fig. 4.2.

We now turn to recombination. Suppose, the cross section for radiative recombination into individual $M$ levels, $Q_{\alpha JM}$, are known. The conventional radiative recombination cross section, $Q_0^0(p)$, and the alignment creation cross section, $Q_0^2(p)$, are constructed in a similar way to (4.40a) and (4.40b), respectively;

$$Q_0^0(p) = \sum_M Q_{\alpha JM}, \tag{4.44a}$$

$$Q_0^2(p) = \sum_M (-)^{J-M} \langle J\,J\,M-M|20\rangle Q_{\alpha JM}. \tag{4.44b}$$

The rate coefficient $\beta(p)$ and the alignment creation rate coefficient $\beta^2(p)$ are defined exactly in the same way as (4.42a) and (4.42b), respectively.

$$\beta(p) = \int Q_0^0(p) 4\pi f_0(v) v^3 \, dv, \tag{4.45a}$$

$$\beta^2(p) = \int Q_0^2(p) [4\pi f_2(v)/5] v^3 \, dv. \tag{4.45b}$$

In calculating these rate coefficients, we may assume that the ion motion is negligible in comparison with the electron motion. Since the cross section has no threshold and the cross section value diverges toward the zero energy, this assumption may not be justified under certain conditions, e.g., under a strong ion-plasma-oscillation condition.

Three-body recombination may be expressed as

$$A^{z+}(1) + e(\boldsymbol{v}_1) + e(\boldsymbol{v}_2) \rightarrow A^{(z-1)+}(p) + e(\boldsymbol{v}'), \tag{4.46}$$

where "1" for $A^{z+}$ stands for the ground state of ion $z$, which is assumed un-aligned, and the two electrons with the initial velocities $\boldsymbol{v}_1$ and $\boldsymbol{v}_2$ participate in this process. One of them is captured and the other carries away the excess energy. Therefore, an expression in terms of a cross section, which is appropriate for two-body reaction processes, cannot be applied. However, we may define an expression similar to a cross section and a rate coefficient for this process. The rate coefficient for creation of an ion in the magnetic sublevel $\alpha JM$ may be expressed in terms of a cross section and EVDF,

$$\alpha_{\alpha JM} = \int d\boldsymbol{v}_1 \int d\boldsymbol{v}_2 v_2 Q_{\alpha JM}(\boldsymbol{v}_1, \boldsymbol{v}_2) f(\boldsymbol{v}_1) f(\boldsymbol{v}_2), \qquad (4.47)$$

where the cross section is defined for the two particular velocities, $\boldsymbol{v}_1$ and $\boldsymbol{v}_2$. In this expression we regard the pair of the ion core and the first electron in (4.46) as constituting "the target" and the second electron as "the incident" electron.

The conventional rate coefficient for three-body recombination $\alpha(p)$ is for a thermal EVDF. The formulation is derived in [5] (p. 79). See Fig. 4.3, which is the reproduction of Fig. 3B.1 of [5]. In the present PPS formalism we have

$$\alpha^{0,0}(p) = 2\pi \left(\frac{h}{m}\right)^3 \frac{g_{z-1}(p)}{g_z(1)} \int dv_1 v_1^2 f_0(v_1) \int dv_2 v_2 f_0(v_2) \sigma_{p,v_1}(v'), \quad (4.48)$$

in place of (3B.6) of [5], where $\sigma_{p,v_1}(v')$ is the cross section for "ionization" from level $p$ by an incident electron having speed $v'$ producing a continuum electron having speed $v_1$ in a unit speed range $dv_1$; in this process the "continuum" electron has energy $E' = mv_1^2/2$ and the "scattered" electron has $\varepsilon = mv_2^2/2$. Of course, the energy conservation is fulfilled: $mv'^2/2 = \chi(p) + mv_1^2/2 + mv_2^2/2$, where $\chi(p)$ is the ionization potential of level $p$.

Formulation of other rate coefficients $\alpha^{2,0}(p)$ and $\alpha^{2,2}(p)$, necessary for constructing the PPS rate equations, is not known yet. The former is related with the angular correlation of the two incoming electrons in (4.46). As will

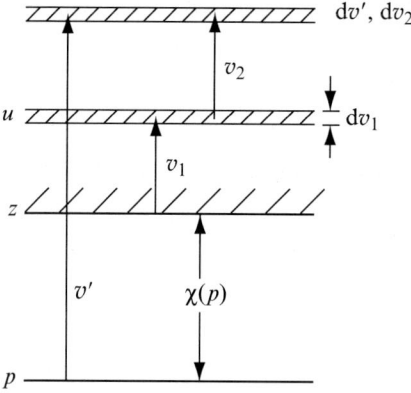

**Fig. 4.3.** Energy relationship of ionization and three-body recombination. Two electrons having speed $v_1$ and $v_2$ recombine with the ion $z$ to form an atom (ion) $p$

be demonstrated in Sect. 6.1, in order for the two electrons to recombine with the ion core, they should have a large relative angle. This fact suggests that the probability of three-body recombination tends to be small for an anisotropic EVDF. This effect is expressed by $\alpha^{2,0}(p)$. However, the above difficulty poses little inconvenience in interpreting practical PPS experiments on recombining plasma. The reason will be discussed later in Sect. 6.1.

Dielectronic recombination could create an alignment in excited levels along with a population. As shown in Sect. 6.2 the formalism is different from that for radiative recombination, but the rate coefficients can be expressed in a similar way to radiative recombination. Therefore, we understand that we include the contribution from dielectronic recombination in $\beta(p)$ and $\beta^2(p)$. The above rate coefficients are summarized in Fig. 4.2.

## 4.4 Rate Equations

We now construct rate equations for population and alignment. We follow the spirit of the collisional-radiative model developed in Chap. 3. The population was divided into the two components, i.e., the ionizing plasma component and the recombining plasma component,

$$n(p) = n_0(p) + n_1(p) . \tag{3.9a}$$

See Fig. 3.3. Likewise, we divide the alignment into the two components,

$$a(p) = a_0(p) + a_1(p). \tag{4.49}$$

Here $a_1(p)$ is the ionizing plasma component, which is proportional to the ground-state population, $n(1)$ (see (3.9)). We have assumed that the ground-state ions are unaligned, i.e., $a(1) = 0$ (see Fig. 4.2). This assumption may be violated under certain extreme conditions, but should be valid in ordinary situations. The first term, $a_0(p)$, is the recombining plasma component which is proportional to $n_z$, which is also assumed unaligned.

In Chap. 3, it was straightforward to deduce from (3.8) the system of coupled equations for each of $n_1(p)$ and $n_0(p)$. In the present case, our system is rather complicated as illustrated in Fig. 4.2. We write down the rate equations for each component and introduce the quasi-steady-state (QSS) approximation.

### 4.4.1 Ionizing Plasma Component

In place of (4.38) the rate equation for the population is given as

$$\frac{d}{dt}n_1(p) = \sum_{r \neq p}[A(r,p) + C(r,p)n_e]n_1(r)$$

$$- \left[\sum_r A(p,r) + \left(\sum_{r \neq p}C(p,r) + S(p)\right)n_e\right]n_1(p)$$

$$+ \sum_{r \neq p}C^{2,0}(r,p)n_e a_1(r)$$

$$- C^{2,0}(p,p)n_e a_1(p). \tag{4.50}$$

The first two lines on the right-hand side are exactly the same as the rate equation for the conventional collisional-radiative model, (3.2), where the Maxwell distribution was assumed for electrons. In the present formulation we use $C(r,p)$ for excitation and also for de-excitation, for which $F(r,p)$ was used in Chap. 3. We understand $n_1(1)$ to be $n(1)$. The third line represents, as can be seen from (4.42c) with (4.40c), the creation of population in this level from population imbalance among the magnetic sublevels, or the alignment, in other levels. The last line is a correction term to the second line due to the presence of alignment and unequal depopulation rates (4.41c) among the magnetic sublevels in this level.

In place of (4.39) the rate equation for the alignment is given by

$$
\frac{d}{dt}a_1(p) = \sum_{r\neq p} C^{0,2}(r,p)n_e n_1(r)
$$
$$
-C^{0,2}(p,p)n_e n_1(p)
$$
$$
+\sum_{r\neq p}\left[A^{2,2}(r,p)+C^{2,2}(r,p)n_e\right]a_1(r)
$$
$$
-\left[\sum_r A(p,r)+C^{2,2}(p,p)n_e\right]a_1(p). \tag{4.51}
$$

The first line on the right-hand side represents the creation of alignment in this level from population in other levels. The second line consists of two contributions as shown in (4.41b); the first part corresponds to the creation of alignment by unequal depopulation rates of the magnetic sublevels in this level, and the second part is the alignment creation by elastic collisions. The third line corresponds to the transfer of alignment from other levels to this level. The last line represents the decay of alignment; the interpretation of the first term, the total radiative decay rate, is straightforward. The second (collisional) term involves the three cross sections $Q_0^{2,2}(p,p)$, $Q_1^{2,2}(p,p)$ and $Q_2^{2,2}(p,p)$ as shown in (4.42d). This process will be examined later on.

## 4.4.2 Recombining Plasma Component

The rate equation for the population may be written as

$$
\frac{d}{dt}n_0(p) = \left[\beta(p)+\alpha^{0,0}(p)n_e\right]n_z n_e
$$
$$
+\sum_{r\neq p}\left[A(r,p)+C(r,p)n_e\right]n_0(r)
$$
$$
-\left[\sum_r A(p,r)+\left\{\sum_{r\neq p}C(p,r)+S(p)\right\}n_e\right]n_0(p)
$$
$$
+\alpha^{2,0}(p)n_z n_e^2
$$
$$
+\sum_{r\neq p}C^{2,0}(r,p)n_e a_0(r)-C^{2,0}(p,p)n_e a_0(p) \tag{4.52}
$$

The major difference of this equation from (4.50) is that, in place of the direct excitation, which was included in the first line in (4.50), contributions from radiative, dielectronic, and three-body recombination appears in the first line on the right-hand side. The fourth line is the correction term to the first line, which is due to the anisotropy of EVDF. In this equation we understand that $n_0(1) = 0$. Other terms are essentially the same as in (4.50), except that the populations and alignments appearing in this equation are the recombining plasma components.

The rate equation for the alignment is

$$
\begin{aligned}
\frac{\mathrm{d}}{\mathrm{d}t} a_0(p) = {} & [\beta^2(p) + \alpha^{2,2}(p)n_\mathrm{e}]n_z n_\mathrm{e} \\
& + \sum_{r \neq p} C^{0,2}(r,p)n_\mathrm{e}n(r) - C^{0,2}(p,p)n_\mathrm{e}n_0(p) \\
& + \sum_{r \neq p} \left[A^{2,2}(r,p) + C^{2,2}(r,p)n_\mathrm{e}\right]a_0(r) \\
& - \left[\sum_r A(p,r) + C^{2,2}(p,p)n_\mathrm{e}\right]a_0(p).
\end{aligned}
\tag{4.53}
$$

The difference from (4.51) is essentially the same as for (4.52).

# References

1. A. Omont: Prog. Quantum Electron. **5**, 69 (1977)
2. T. Fujimoto, H. Sahara, G. Csanak, S. Grabbe: NIFS-DATA-38 (National Institute for Fusion Science, Nagoya, 1996)
3. T. Fujimoto and S.A. Kazantsev: Plasma Phys. Control. Fusion **39**, 1267 (1997)
4. K. Blum: *Density Matrix Theory and Applications* (Plenum, New York, 1981)
5. T. Fujimoto: *Plasma Spectroscopy* (Oxford University Press, Oxford, 2004)

# Definition of Cross Sections for the Creation, Destruction, and Transfer of Atomic Multipole Moments by Electron Scattering: Quantum Mechanical Treatment

G. Csanak, D.P. Kilcrease, D.V. Fursa, and I. Bray

The *rates* for the creation, destruction, and transfer of atomic multipole moments by heavy-particle scattering have been studied for many years by Omont [1], by D'yakonov and Perel [2], and by Petrashen, Rebane, and Rebane [3–12] using semiclassical scattering theory along with the straight-line trajectory assumption for the scattering particle. This method has been adopted for electron scattering by Fujimoto et al. [13] and Fujimoto and Kazantsev [14]. These latter authors have given definitions of alignment creation, destruction, and transfer *cross sections* for both elastic and inelastic electron scattering within the semiclassical straigth-line trajectory approximation. The same practice was followed in the preceding Chap. 4. In the case of inelastic scattering, Kazantsev et al. [15, 16] gave a quantum-mechanical definition of the alignment-creation cross section, which was extended recently for elastic scattering by Csanak et al. [17]. The purpose of this chapter is to use the methods of Csanak et al. [17] to give general definitions for both elastic and inelastic scattering for the creation, destruction, and transfer cross sections of atomic multipole moments via the use of pure quantum-mechanical methods.

## 5.1 General Theory

Let us assume first (as in Csanak et al. [17]) that the incident electron is a *distinguishable* particle and can be described nonrelativistically, thus both exchange and spin-orbit coupling effects are neglected for the scattering electron. (Both of those effects can be considered for the target electrons.) We will assume (following Csanak et al. [17]) that an incident electron is described by the wave-packet,[1]

$$|\Phi_{\boldsymbol{k}_{\mathrm{in}},m_{\mathrm{s}}}\rangle = (2\pi)^{-3} \int \mathrm{d}\boldsymbol{k} A(\boldsymbol{k})|\boldsymbol{k},m_{\mathrm{s}}\rangle \exp\left(-\frac{\mathrm{i}}{\hbar}E_k t\right) , \qquad (5.1)$$

---

[1] Our approach here as there is based on the wave packet formalism of Rodberg and Thaler [18]. The reader is advised to have this book available for ready reference while reading this chapter and have studied its Chaps. 1, 2, and 7.

where $\boldsymbol{k}$ refers to the wave vector (of magnitude $k$) and $m_{\mathrm{s}}$ to the spin-projection of the incident electron, $E_k = \hbar^2 k^2 / 2m$, and $|\boldsymbol{k}, m_{\mathrm{s}}\rangle$ refers to a plane-wave state with wave vector $\boldsymbol{k}$ and spin-projection $m_{\mathrm{s}}$ normalized as

$$\langle \boldsymbol{k}, m_{\mathrm{s}} | \boldsymbol{k}', m_{\mathrm{s}}' \rangle = (2\pi)^3 \delta(\boldsymbol{k} - \boldsymbol{k}') \delta_{m_{\mathrm{s}}, m_{\mathrm{s}}'} .$$

We shall assume that the $A(\boldsymbol{k})$ function is strongly peaked around the value of $\boldsymbol{k}_{\mathrm{in}}$ wave vector with width $\delta k \ll k_{\mathrm{in}}$ and is normalized according to the formula

$$(2\pi)^{-3} \int \mathrm{d}\boldsymbol{k} |A(\boldsymbol{k})|^2 = 1 . \tag{5.2}$$

The spatio-temporal representation of the wave packet can be given in the form

$$\Phi(\boldsymbol{r}, \sigma, t) = (2\pi)^{-3} \int \mathrm{d}\boldsymbol{k} A(\boldsymbol{k}) \exp(\mathrm{i}\boldsymbol{k} \cdot \boldsymbol{r}) \chi_{m_{\mathrm{s}}}(\sigma) \exp\left(-\frac{\mathrm{i}}{\hbar} E_k t\right) ,$$

where $\boldsymbol{r}$ is the spatial variable, $\sigma$ refers to the spin-variable, and $\chi_{m_{\mathrm{s}}}(\sigma)$ refers to the spin-function of the electron with spin-projection $m_{\mathrm{s}}$.

If the atomic state of energy $E_{\alpha J}$ is described by the state-vector

$$|\alpha J M\rangle , \tag{5.3}$$

where $J$ refers to the total angular momentum of the atom, $M$ to its projection along the $z$-axis (the direction of the incidence of the electron), and $\alpha$ refers to all other quantum numbers, then the initial (noninteracting) electron plus atom system is described by the state-vector

$$|\Phi_{\alpha J M, \boldsymbol{k}_{\mathrm{in}}, m_{\mathrm{s}}}\rangle = (2\pi)^{-3} \int \mathrm{d}\boldsymbol{k} A(\boldsymbol{k}) |\boldsymbol{k}, m_{\mathrm{s}}\rangle |\alpha J M\rangle \exp\left(-\frac{\mathrm{i}}{\hbar} E_k^{\alpha J} t\right) , \tag{5.4}$$

where

$$E_k^{\alpha J} = E_k + E_{\alpha J} \tag{5.5}$$

is the total energy of the electron plus atom system.

We shall envision that the electron scattering process occurs like in an experiment, we are conducting a *Gedanken-experiment*, and that the conditions are the same as stipulated in Chap. 2 of Rodberg and Thaler [18], namely, the individual particles in the beam will each be described by a wave packet, and we shall assume that each particle is described by the same wave packet. We shall also assume that the wave packets will be small in their spatial extension compared to the dimensions of the laboratory, that the energy is well defined (i.e., as already mentioned we shall assume that $\delta k \ll k_{\mathrm{in}}$) and we shall assume that the packet will not spread appreciably during the course of the "experiment." Under these conditions the incident wave packet can be given to good approximation in the form, (see Rodberg and Thaler [18], p. 15)

$$\Phi(\boldsymbol{r}, \sigma, t) = (2\pi)^{-3} \exp\left(-\mathrm{i}\frac{\hbar^2 k_{\mathrm{in}}^2}{2m} t\right) \int \mathrm{d}\boldsymbol{k} A(\boldsymbol{k}) \exp\left[\mathrm{i}\boldsymbol{k} \cdot (\boldsymbol{r} - \boldsymbol{v}_{\mathrm{in}} t)\right] \chi_{m_{\mathrm{s}}}(\sigma) ,$$

$$(5.6)$$

where $\boldsymbol{v}_{\mathrm{in}} = \hbar\boldsymbol{k}_{\mathrm{in}}/m$ is the initial speed of the wave packet.

We shall now let the incident electron interact with the target (the atom), and we look at the system in the asymptotic future. Here we can use the completeness of the $|\alpha'' J'' M''; \boldsymbol{k}_1'' m_{\mathrm{s}}''\rangle$ basis set in the form

$$\mathcal{S}|\alpha J M; \boldsymbol{k}_1 m_{\mathrm{s}}\rangle = (2\pi)^{-3} \sum_{\alpha'' J'' M''} \sum_{m_{\mathrm{s}}''} \int \mathrm{d}\boldsymbol{k}_1'' |\alpha'' J'' M''; \boldsymbol{k}_1'' m_{\mathrm{s}}''\rangle$$

$$\times \langle \alpha'' J'' M''; \boldsymbol{k}_1'' m_{\mathrm{s}}'' |\mathcal{S}|\alpha J M; \boldsymbol{k}_1 m_{\mathrm{s}}\rangle ,$$

where $\mathcal{S}$ is the scattering operator. Since we will be interested only in the final states belonging to the $\alpha', J'$ manifold, we shall write the above completeness relation in the form

$$\mathcal{S}|\alpha J M; \boldsymbol{k}_1 m_{\mathrm{s}}\rangle = (2\pi)^{-3} \sum_{M_1', m_{\mathrm{s}1}} \int \mathrm{d}\boldsymbol{k}_1' |\alpha' J' M_1'; \boldsymbol{k}_1' m_{\mathrm{s}1}\rangle$$

$$\times \langle \alpha' J' M_1'; \boldsymbol{k}_1' m_{\mathrm{s}1} |\mathcal{S}|\alpha J M; \boldsymbol{k}_1 m_{\mathrm{s}}\rangle + \cdots ,$$

where $\cdots$ indicate those terms that were specified in the earlier summation but not here. Using this latter identity, we then obtain for the state-vector of the electron plus atom system in the asymptotic future the form

$$|\Phi(t)\rangle_{\mathrm{out}} = (2\pi)^{-6} \int \mathrm{d}\boldsymbol{k}_1 A(\boldsymbol{k}_1) \exp\left(-\frac{\mathrm{i}}{\hbar} E_{k_1}^{\alpha J} t\right)$$

$$\times \int \mathrm{d}\boldsymbol{k}_1' \sum_{M_1', m_{\mathrm{s}1}} \langle \boldsymbol{k}_1' m_{\mathrm{s}1}; \alpha' J' M_1' |\mathcal{S}|\alpha J M; \boldsymbol{k}_1 m_{\mathrm{s}}\rangle |\alpha' J' M_1'; \boldsymbol{k}_1', m_{\mathrm{s}1}\rangle$$

$$+ \cdots , \qquad (5.7)$$

where

$$|\alpha' J' M_1'; \boldsymbol{k}_1' m_{\mathrm{s}1}\rangle = |\alpha' J' M_1'\rangle |\boldsymbol{k}_1' m_{\mathrm{s}1}\rangle \qquad (5.8)$$

is a state-vector of the electron plus target system. The scattering operator "takes" the initial state $|\alpha J M; \boldsymbol{k}_1 m_{\mathrm{s}}\rangle$ at $t = -\infty$ to the final state $\mathcal{S}|\alpha J M; \boldsymbol{k}_1 m_{\mathrm{s}}\rangle$ at $t = \infty$ (see, e.g., Rodberg and Thaler [18], p. 224).

In (5.7) there is no double prime, only single-prime and no summation over $\alpha''$ and $J''$ because we only wrote out explicitly those terms that refer to the specific final-state $\alpha' J'$ manifold of the target; the other terms in the summation were only indicated by $\cdots$ since they will not be needed in the later development.

Let us assume that the initial atomic state is described by the density operator,[2]

$$\rho^{\mathrm{in}} = |\alpha J M'\rangle \langle \alpha J M| . \qquad (5.9)$$

---

[2] The reader is advised to have some familiarity with density matrices and density operators, like the subject that can be found in Chaps. 1 and 2 of Blum's book [19]. See also Appendix C.

By a slight generalization of the method described by Csanak et al. [17], one can obtain the final atomic reduced density operator within the atomic states-manifold characterized by $\alpha'$ and $J'$ and which has evolved from the initial state given by (5.9) following electron scattering. (In the following we shall use the notation and conceptual framework of Csanak et al. [17] with the obvious modifications and generalizations.) If we denote by $|\Phi(t)\rangle_{\text{out}}$ the final state that evolved from the $|\alpha J M'\rangle|\boldsymbol{k}_1 m_{s_1}\rangle$ initial state of the atom plus electron system, then

$$\boldsymbol{\rho}_{\text{tot}}^{\text{out}} = |\Phi(t)\rangle_{\text{out out}}\langle\Phi(t)|$$

describes the final state of the electron plus atom system that evolved from the atomic state given by the density operator in (5.9), and the free-electron state described by the free-electron density operator

$$\boldsymbol{\rho}^{\text{free}} = |\boldsymbol{k}_1 m_{s1}\rangle\langle\boldsymbol{k}_2 m_{s2}| \ .$$

The reduced density operator of the final state of the atomic system is obtained by taking the trace over the electron states $|\boldsymbol{k}_0 m_{s0}\rangle$ in the form, (see, e.g., Blum [19], p. 66)

$$\boldsymbol{\rho}^{\text{out}} = \frac{1}{2}(2\pi)^{-3} \sum_{m_s, m_{s0}} \int \mathrm{d}\boldsymbol{k}_0 \langle\boldsymbol{k}_0 m_{s0}|\boldsymbol{\rho}_{\text{tot}}^{\text{out}}|\boldsymbol{k}_0 m_{s0}\rangle,$$

where we have also introduced an averaging over the spin of the incident electron assuming an unpolarized incident electron beam. Using (5.7) in the above equation we obtain,

$$\boldsymbol{\rho}^{\text{out}} = \frac{1}{2}(2\pi)^{-9} \sum_{m_s, m_{s0}} \int \mathrm{d}\boldsymbol{k}_0 \int \mathrm{d}\boldsymbol{k}_1 A(\boldsymbol{k}_1) \exp\left(-\frac{\mathrm{i}}{\hbar}E_{k_1}^{\alpha J}t\right)$$

$$\times \int \mathrm{d}\boldsymbol{k}_2 A^*(\boldsymbol{k}_2) \exp\left(\frac{\mathrm{i}}{\hbar}E_{k_2}^{\alpha J}t\right) \sum_{M_1', M_2'} |\alpha' J' M_1'\rangle\langle\alpha' J' M_2'|$$

$$\times \langle\boldsymbol{k}_0 m_{s0}; \alpha' J' M_1'|\mathcal{S}|\boldsymbol{k}_1 m_s; \alpha J M'\rangle\langle\boldsymbol{k}_2 m_s; \alpha J M|\mathcal{S}^\dagger|\boldsymbol{k}_0 m_{s0}; \alpha' J' M_2'\rangle$$

$$+ \cdots . \tag{5.10}$$

For the matrix element (within the $\alpha' J'$ manifold) of this reduced density operator we obtain,

$$\langle\alpha' J' M_1'|\boldsymbol{\rho}^{\text{out}}|\alpha' J' M_2'\rangle \equiv \rho_{M_1' M_2'}^{\text{out}}$$

$$= \frac{1}{2}(2\pi)^{-9} \sum_{m_s, m_{s0}} \int \mathrm{d}\boldsymbol{k}_0 \int \mathrm{d}\boldsymbol{k}_1 A(\boldsymbol{k}_1) \exp\left(-\frac{\mathrm{i}}{\hbar}E_{k_1}^{\alpha J}t\right)$$

$$\times \int \mathrm{d}\boldsymbol{k}_2 A^*(\boldsymbol{k}_2) \exp\left(\frac{\mathrm{i}}{\hbar}E_{k_2}^{\alpha J}t\right)$$

$$\times \langle\boldsymbol{k}_0 m_{s0}; \alpha' J' M_1'|\mathcal{S}|\boldsymbol{k}_1 m_s; \alpha J M'\rangle$$

$$\times \langle\boldsymbol{k}_2 m_s; \alpha J M|\mathcal{S}^\dagger|\boldsymbol{k}_0 m_{s0}; \alpha' J' M_2'\rangle \ . \tag{5.11}$$

If we denote by $a, b, \dots$ all the quantum numbers of a state of the non-interacting electron plus atom system and by $E_a, E_b, \dots$ the corresponding energies, then we can define the $\mathcal{T}$ operator by its matrix elements, $\mathcal{T}_{ab}$, via the equation, (see, e.g., Bransden [20], p. 129)

$$\mathcal{S}_{ab} = \langle a|b \rangle - 2\pi i \delta(E_a - E_b)\mathcal{T}_{ab} \,. \tag{5.12}$$

The first term on the right hand side in the above equation is just the matrix element of the unit operator: $\mathbf{1}$. $\mathcal{T}$ is called the the transition operator (it enters the formula for the differential cross-section, see, e.g., Bransden [20], p. 130). If we use the above equation for the matrix elements of the $\mathcal{S}$ operator in (5.11), then we shall obtain *four terms* for the reduced density matrix of the final atomic state.

The *first term* is obtained when the $\mathcal{S} \to \mathbf{1}, \mathcal{S}^\dagger \to \mathbf{1}$ substitutions are made. The result can be given in the form

$$\rho_{M_1' M_2'}^{\text{out}(1)} = \delta_{\alpha\alpha'} \delta_{JJ'} \delta_{M_1' M'} \delta_{MM_2'} \,, \tag{5.13}$$

which is equal to the density matrix element of the initial state density operator, given by (5.9).

The *second term* is obtained when the $\mathcal{S} \to \mathbf{1}, \mathcal{S}_{ab}^\dagger \to 2\pi i \delta(E_a - E_b)\mathcal{T}_{ab}^\dagger$ substitutions are made in (5.11). This gives the result

$$\rho_{M_1' M_2'}^{\text{out}(2)} = (2\pi)^{-5} i \delta_{\alpha\alpha'} \delta_{JJ'} \delta_{M_1' M'} \frac{1}{2} \sum_{m_s} \langle \boldsymbol{k}_{\text{in}} m_s; \alpha JM | \mathcal{T}^\dagger | \boldsymbol{k}_{\text{in}} m_s; \alpha J M_2' \rangle$$

$$\times \int d\boldsymbol{k}_1 \int d\boldsymbol{k}_2 A(\boldsymbol{k}_1) A^*(\boldsymbol{k}_2) \delta(E_{k_1}^{\alpha J} - E_{k_2}^{\alpha J}) \,, \tag{5.14}$$

where we have used the fact that the $A(\boldsymbol{k})$ function is strongly peaked at $\boldsymbol{k}_{\text{in}}$ in the following manner ([18], p. 194): if $f(\boldsymbol{k})$ is an arbitrary function of $\boldsymbol{k}$ which is smooth in $\boldsymbol{k}$ around $\boldsymbol{k} = \boldsymbol{k}_{\text{in}}$ then we can use the approximation $\int d\boldsymbol{k} f(\boldsymbol{k}) A(\boldsymbol{k}) \approx f(\boldsymbol{k}_{\text{in}}) \int d\boldsymbol{k} A(\boldsymbol{k})$. We used this approximation in obtaining (5.14) with the matrix element of the $\mathcal{T}$ operator in place of the $f(\boldsymbol{k})$ function,

$$\int d\boldsymbol{k}_1 \int d\boldsymbol{k}_2 A(\boldsymbol{k}_1) A^*(\boldsymbol{k}_2) \langle \boldsymbol{k}_2 m_s; \alpha JM | \mathcal{T} | \boldsymbol{k}_1 m_s; \alpha JM \rangle \delta(E_{k_1}^{\alpha J} - E_{k_2}^{\alpha J})$$

$$\approx \langle \boldsymbol{k}_{\text{in}} m_s; \alpha JM | \mathcal{T} | \boldsymbol{k}_{\text{in}} m_s; \alpha JM \rangle \int d\boldsymbol{k}_1 \int d\boldsymbol{k}_2 A(\boldsymbol{k}_1) A^*(\boldsymbol{k}_2) \delta(E_{k_2}^{\alpha J} - E_{k_1}^{\alpha J}) \,.$$

We define now the quantity $dP/dS$, (following Rodberg and Thaler [18], p.195), by the formula,

$$\frac{dP}{dS} = 2\pi\hbar v_{\text{in}} \int \frac{d\boldsymbol{k}_1}{(2\pi)^3} \int \frac{d\boldsymbol{k}_2}{(2\pi)^3} A(\boldsymbol{k}_1) A^*(\boldsymbol{k}_2) \delta(E_{k_1}^{\alpha J} - E_{k_2}^{\alpha J}) \,, \tag{5.15}$$

where $v_{in}$ is the initial speed of the center of the wave-packet,

$$v_{in} = \frac{\hbar k_{in}}{m} .$$

It can be easily shown that $dP/dS$ is the probability that the incident wave packet (with a given spin-projection) crosses a unit area perpendicular to the direction of propagation (the $z$-axis). Since we have assumed that the incident wave packet does not spread, its behavior will be the same at any point in space. Let us choose for the examination of the wave packet the $r = 0$ position in space. The total probability at that position in space the electron passed through an area of $1$ a.u. across the $z$-axis (the direction of propagation) in the time interval $(t, t + dt)$ with a given spin-projection $m_s$ is equal to the probability that the electron (with the given spin-projection) could be found in the volume element of $v_0\,dt$ (in a.u.) which is equal to

$$v_0\,dt|\Phi(r = 0, \sigma = m_s, t)|^2 .$$

Thus, the total probability that the selected incident wave packet passed through at any time an area of $1$ a.u. that is perpendicular to the $z$-axis and located at $r = 0$ in space is equal to

$$v_0 \int_{-\infty}^{\infty} dt|\Phi(r = 0, \sigma = m_s, t)|^2 .$$

If (5.6) is now used for $\Phi(r = 0, \sigma = m_s, t)$ in the above expression, along with the assumption that $A(k)$ function has a strong peek at $k = k_{in}$, we obtain that $dP/dS$, as defined by (5.15), is equal to the above quantity, i.e., we obtain ($v_0 = v_{in}$)

$$\frac{dP}{dS} = v_0 \int_{-\infty}^{\infty} dt|\Phi(r = 0, \sigma = m_s, t)|^2 .$$

Let us now introduce the scattering amplitude by the definition

$$f_{\alpha'J'M_2',\alpha JM_1}^{m_s'm_s}(\theta, \phi) = -\frac{m}{2\pi\hbar^2}\langle k_2 m_s'; \alpha'J'M_2'|T|k_1 m_s; \alpha JM_1\rangle , \qquad (5.16)$$

where $\theta$ and $\phi$ refer to the polar and azimuthal angles of $k_2$ relative to $k_1$. Magnitudes $k_1 = |k_1|$ and $k_2 = |k_2|$ are related by the energy conservation relation

$$\frac{\hbar^2 k_1^2}{2m} + E_{\alpha J} = \frac{\hbar^2 k_2^2}{2m} + E_{\alpha'J'}, \qquad (5.17)$$

i.e., the $T$ matrix element is calculated on the energy shell between the initial state $|k_1 m_s; \alpha JM_1\rangle$ and of the final state $\langle k_2 m_s'; \alpha'J'M'|$. Then we obtain for the second term

$$\rho_{M_1'M_2'}^{out(2)} = -\frac{2\pi}{k_{in}}i\frac{dP}{dS}\delta_{\alpha\alpha'}\delta_{JJ'}\delta_{M_1'M'}\frac{1}{2}\sum_{m_s} f_{\alpha JM_2',\alpha JM}^{m_s m_s^*}(\theta = 0, \phi = 0) . \qquad (5.18)$$

The *third term* is obtained if the $\mathcal{S}_{ab} \rightarrow -2\pi i \delta(E_a - E_b)\mathcal{T}_{ab}$ and $\mathcal{S}^\dagger \rightarrow \mathbf{1}$ substitutions are made in (5.11). Again using the fact that the $A(\mathbf{k})$ function is strongly peaked around $\mathbf{k} = \mathbf{k}_{\text{in}}$, and using the definitions (5.15) for $\mathrm{d}P/\mathrm{d}S$ and (5.16) for the scattering amplitude, respectively, we obtain for the third term,

$$\rho_{M_1' M_2'}^{\text{out}(3)} = \frac{2\pi}{k_{\text{in}}} i \frac{\mathrm{d}P}{\mathrm{d}S} \delta_{\alpha\alpha'} \delta_{JJ'} \delta_{M_2' M'} \frac{1}{2} \sum_{m_s} f_{\alpha J M_1', \alpha J M'}^{m_s m_s} (\theta = 0, \phi = 0) . \quad (5.19)$$

And finally using the $\mathcal{S}_{ab} \rightarrow -2\pi i \delta(E_a - E_b)\mathcal{T}_{ab}$ and $\mathcal{S}_{ab}^\dagger \rightarrow 2\pi i \delta(E_a - E_b)\mathcal{T}_{ab}^\dagger$ substitutions in (5.11) along with the strongly peaked behavior of $A(\mathbf{k})$, we obtain for the *fourth term*,

$$\rho_{M_1' M_2'}^{\text{out}(4)} = \frac{\mathrm{d}P}{\mathrm{d}S} \frac{1}{2} \frac{k_{\text{out}}}{k_{\text{in}}} \sum_{m_s m_{s0}} \int \mathrm{d}\Omega f_{\alpha' J' M_1', \alpha J M'}^{m_{s0} m_s} (\theta, \phi) f_{\alpha' J' M_2', \alpha J M}^{m_{s0} m_s *} (\theta, \phi), \quad (5.20)$$

where $\Omega = (\theta, \phi)$, $\mathrm{d}\Omega = \sin\theta \, \mathrm{d}\theta \mathrm{d}\phi$ is the solid angle element, and $k_{\text{out}}$ is the $k_2$ value in (5.17) in the case when $k_1 = k_{\text{in}}$.

Adding all terms, (5.13), (5.18), (5.19), and (5.20), we obtain for the asymptotic form of the density matrix the result

$$\begin{aligned}
\rho_{M_2' M_1'}^{\text{out}} = &\, \delta_{\alpha\alpha'} \delta_{JJ'} \delta_{M_1' M'} \delta_{M M_2'} \\
&- \frac{\mathrm{d}P}{\mathrm{d}S} \frac{2\pi}{k_{\text{in}}} i \delta_{\alpha\alpha'} \delta_{JJ'} \frac{1}{2} \sum_{m_s} \Big[ f_{\alpha J M_2', \alpha J M}^{m_s m_s *} (\theta = 0, \phi = 0) \delta_{M_1' M'} \\
&\hspace{6cm} - f_{\alpha J M_1', \alpha J M'}^{m_s m_s} (\theta = 0, \phi = 0) \delta_{M M_2'} \Big] \\
&+ \frac{\mathrm{d}P}{\mathrm{d}S} \frac{1}{2} \frac{k_{\text{out}}}{k_{\text{in}}} \sum_{m_s m_{s0}} \int \mathrm{d}\Omega f_{\alpha' J' M_1', \alpha J M'}^{m_{s0} m_s} (\theta, \phi) f_{\alpha' J' M_2', \alpha J M}^{m_{s0} m_s *} (\theta, \phi) .
\end{aligned}$$

$$(5.21)$$

We note here that due to the cylindrical symmetry along the $z$-axis (the direction of incidence) and to the nonrelativistic nature of the incident electron, the scattering amplitude, introduced by (5.16), factorizes in the form [20, 21]

$$f_{\alpha' J' M', \alpha J M}^{m_{s0} m_s} (\theta, \phi) = f_{\alpha' J' M', \alpha J M}^{m_{s0} m_s} (\theta) \exp[i(M + m_s - M' - m_{s0})\phi] . \quad (5.22)$$

Let us assume now a completely arbitrary form for the density matrix of level $\alpha J$. The density operator representing that state can be expanded in terms of state multipoles (see, e.g., Blum [19], p. 95) in the form[3]

$$\rho_{\alpha J}^{\text{in}} = \sum_{kq} \langle T_q^{(k)} (\alpha J)^\dagger \rangle^{\text{in}} T_q^{(k)} (\alpha J), \quad (5.23)$$

---

[3] The reader is advised to have some familiarity with the material from Chap. 4 of Blum's book [19].

where the $T_q^{(k)}(\alpha J)$ operator is defined as

$$T_q^{(k)}(\alpha J) = \sum_{M'M} (-)^{J-M'} (2k+1)^{\frac{1}{2}} \begin{pmatrix} J & J & k \\ M' & -M & -q \end{pmatrix} |\alpha JM'\rangle\langle\alpha JM| \quad (5.24)$$

with

$$\begin{pmatrix} J & J & k \\ M' & -M & -q \end{pmatrix}$$

referring to the 3-$j$ symbol, and $\langle T_q^{(k)}(\alpha J)^\dagger\rangle^{\text{in}}$ is given by the formula,

$$\langle T_q^{(k)}(\alpha J)^\dagger\rangle^{\text{in}} = \text{Tr}(\rho_{\alpha J}^{\text{in}} T_q^{(k)}(\alpha J)^\dagger) . \quad (5.25)$$

## 5.2 Inelastic Scattering

In this section we shall consider in detail only *inelastic* scattering processes, i.e., processes for which $\alpha \neq \alpha'$ or $J \neq J'$. The next section will discuss the special case of alignment creation by *elastic* scattering. Other elastic processes, e.g., alignment transfer by elastic scattering can be discussed analogously.

For *inelastic* scattering processes we can write the final density matrix in the form

$$\rho_{M_1'M_2'}^{\text{out}} = \frac{dP}{dS}\frac{1}{2}\frac{k_{\text{out}}}{k_{\text{in}}} \sum_{m_s m_{s0}} \int d\Omega f_{\alpha'J'M_1',\alpha JM'}^{m_{s0}m_s}(\theta,\phi) f_{\alpha'J'M_2',\alpha JM}^{m_{s0}m_s *}(\theta,\phi) \quad (5.26)$$

for the case when the initial density matrix was given by (5.9).

In general, the initial density matrix will be given by (5.23) and (5.24) where the multipole moments $\langle T_q^{(k)}(\alpha J)^\dagger\rangle$ can take arbitrary values. Because of the linearity of the relevant quantum-mechanical equations we then obtain in this latter case for the final density matrix the formula

$$\rho_{M_1'M_2'}^{\text{out}} = \frac{dP}{dS} \sum_{kq} \langle T_q^{(k)}(\alpha J)^\dagger\rangle^{\text{in}}$$

$$\times \sum_{M'M} (-)^{J-M'} (2k+1)^{\frac{1}{2}} \begin{pmatrix} J & J & k \\ M' & -M & -q \end{pmatrix}$$

$$\times \frac{1}{2}\frac{k_{\text{out}}}{k_{\text{in}}} \sum_{m_s m_{s0}} \int d\Omega f_{\alpha'J'M_1',\alpha JM'}^{m_{s0}m_s}(\theta,\phi) f_{\alpha'J'M_2',\alpha JM}^{m_{s0}m_s *}(\theta,\phi)$$

$$(5.27)$$

and for the state-multipoles of this final density matrix we obtain

$$\langle T_{q'}^{(k')}(\alpha'J')^{\dagger}\rangle = \mathrm{Tr}(\rho_{\alpha'J'}^{\mathrm{out}}, T_q^{(k')}(\alpha'J')^{\dagger})$$

$$= \sum_{M_1'M_2'} (-)^{J'-M_1'}(2k'+1)^{\frac{1}{2}} \begin{pmatrix} J' & J' & k' \\ M_1' & -M_2' & -q' \end{pmatrix} \rho_{M_1'M_2'}^{\mathrm{out}} \, .$$

(5.28)

Using (5.27) in (5.28) we obtain the final-state state-multipoles in terms of the initial-state state-multipoles, the scattering amplitudes, and $\mathrm{d}P/\mathrm{d}S$, the probability that the incident wave packet with a given spin-projection crosses a unit area perpendicular to the direction of propagation,

$$\langle T_{q'}^{(k')}(\alpha'J')^{\dagger}\rangle^{\mathrm{out}} = \frac{\mathrm{d}P}{\mathrm{d}S} \sum_{kq} \langle T_q^{(k)}(\alpha J)^{\dagger}\rangle^{\mathrm{in}}$$

$$\times \sum_{M_1'M_2'} (-)^{J'-M_1'}(2k'+1)^{\frac{1}{2}} \begin{pmatrix} J' & J' & k' \\ M_1' & -M_2' & -q' \end{pmatrix}$$

$$\times \sum_{M'M} (-)^{J-M'}(2k+1)^{\frac{1}{2}} \begin{pmatrix} J & J & k \\ M' & -M & -q \end{pmatrix}$$

$$\times \frac{1}{2}\frac{k_{\mathrm{out}}}{k_{\mathrm{in}}} \sum_{m_s m_{s0}} \int \mathrm{d}\Omega f_{\alpha'J'M_1',\alpha JM'}^{m_{s0}m_s}(\theta,\phi) f_{\alpha'J'M_2',\alpha JM}^{m_{s0}m_s^*}(\theta,\phi) \, .$$

(5.29)

This is the principal result of this section. In the following we shall specialize this general formula to certain particular cases.

Let us consider the special case of the connecting term (denoted by the $\Leftrightarrow$ symbol) of $k' = 0$, $q' = 0$ with $k = 0$, $q = 0$. Equation (5.29) gives for this case

$$\langle T_0^{(0)}(\alpha'J')^{\dagger}\rangle^{\mathrm{out}} \Leftrightarrow \frac{\mathrm{d}P}{\mathrm{d}S}\langle T_0^{(0)}(\alpha J)^{\dagger}\rangle^{\mathrm{in}}(2J'+1)^{-\frac{1}{2}}(2J+1)^{-\frac{1}{2}}\frac{1}{2}\sum_{MM_1'}\sigma_{\alpha JM,\alpha'J'M_1'},$$

(5.30)

where we have introduced the magnetic sublevel excitation cross-section for the $\alpha JM \to \alpha'J'M'$ electron-impact induced transition by the definition

$$\sigma_{\alpha JM,\alpha'J'M_1'}^{m_s m_{s0}} = \frac{k_{\mathrm{out}}}{k_{\mathrm{in}}} \int \mathrm{d}\Omega |f_{\alpha'J'M_1',\alpha JM}^{m_{s0}m_s}(\theta,\phi)|^2$$

(5.31)

and the spin-averaged-summed magnetic sublevel excitation cross-section by the formula

$$\sigma_{\alpha JM,\alpha'J'M_1'} = \frac{1}{2}\sum_{m_s,m_{s0}} \sigma_{\alpha JM,\alpha'J'M_1'}^{m_s,m_{s0}} \, .$$

(5.32)

If we now use the relationships (see, e.g., Blum [19], p. 97)

$$\langle T_0^{(0)}(\alpha'J')^{\dagger}\rangle^{\mathrm{out}} = \frac{\mathrm{Tr}\rho_{\alpha'J'}^{\mathrm{out}}}{(2J'+1)^{\frac{1}{2}}}$$

(5.33)

and

$$\langle T_0^{(0)}(\alpha J)^+\rangle^{\text{in}} = \frac{\text{Tr}\rho_{\alpha J}^{\text{in}}}{(2J+1)^{\frac{1}{2}}}, \tag{5.34}$$

then we obtain

$$\text{Tr}\rho_{\alpha' J'}^{\text{out}} \Leftrightarrow \frac{\mathrm{d}P}{\mathrm{d}S}\text{Tr}\rho_{\alpha J}^{\text{in}}\sigma_{\alpha J,\alpha' J'}, \tag{5.35}$$

where

$$\sigma_{\alpha J,\alpha' J'} = \frac{1}{2J+1}\sum_{MM_1'}\sigma_{\alpha JM,\alpha' J'M_1'}. \tag{5.36}$$

From (5.16), (5.31), (5.32), (5.35), and (5.36) it can be seen that $\sigma_{\alpha J,\alpha' J'}$ is the integrated cross section for the $\alpha J \rightarrow \alpha' J'$ electron impact induced transition. This cross section corresponds to (4.40a) in the preceding Chap. 4.

On the other hand, according to density matrix theory (see, e.g., Blum [19]) $\text{Tr}\rho$ gives the relative number of particles in the state described by the density operator $\rho$ for a selected volume. Thus we can write

$$n(\alpha J) = \text{Tr}\rho_{\alpha J}^{\text{in}} \tag{5.37}$$

and

$$n(\alpha' J') = \text{Tr}\rho_{\alpha' J'}^{\text{out}}, \tag{5.38}$$

where $n(\alpha J)$ and $n(\alpha' J')$ refer to the relative number of atoms in a selected volume in the initial $\alpha J$ level and the final $\alpha' J'$ level, respectively. Thus (5.35) can be written in the form

$$n(\alpha' J') \Leftrightarrow n(\alpha J)\frac{\mathrm{d}P}{\mathrm{d}S}\sigma_{\alpha J,\alpha' J'}, \tag{5.39}$$

where it expresses the relationship between the number of atoms produced by the incident electron with a given $\mathrm{d}P/\mathrm{d}S$ value with cross section $\sigma_{\alpha J,\alpha' J'}$ from the initial level and the number of electrons occupying the initial level. This is just the traditional definition of the integrated electron impact excitation cross-section.

As a second example we shall consider the case when $k' = 2, q' = 0$ and $k = 0, q = 0$. This gives the relationship

$$\langle T_0^{(2)}(\alpha' J')^\dagger\rangle \Leftrightarrow \frac{\mathrm{d}P}{\mathrm{d}S}n(\alpha J)Q_0^{0,2}(\alpha J, \alpha' J'), \tag{5.40}$$

where the alignment creation cross section $Q_0^{0,2}(\alpha J, \alpha' J')$ was defined following Kazantsev et al. [15,16] by the formula

$$Q_0^{0,2}(\alpha J, \alpha' J') = \sqrt{5}\sum_{M'}(-)^{J'-M'}\begin{pmatrix} J' & J' & 2 \\ M' & -M' & 0 \end{pmatrix}\frac{1}{2J+1}\sum_{M}\sigma_{\alpha JM,\alpha' J'M'}$$

$$= \sum_{M'}(-)^{J'-M'}\langle J'J'M'-M'|20\rangle\frac{1}{2J+1}\sum_{M}\sigma_{\alpha JM,\alpha' J'M'}$$

$$\tag{5.41}$$

in which $\langle J'J'M' -M'|20\rangle$ refers to a Clebsch–Gordan coefficient. The expression given by (5.41) for the alignment creation cross-section is also formally identical to the semiclassical expression given by Fujimoto et al. [13] and by Fujimoto and Kazantsev [14], except here the cross sections are calculated quantum-mechanically. Equation (5.41) corresponds to (4.40b) in Chap. 4.

For our final example we consider the $k' = 2$ and $k = 2$ case. For the connecting relation we obtain

$$
\langle T_{q'}^{(2)}(\alpha'J')^\dagger\rangle \Leftrightarrow 5 \frac{dP}{dS} \sum_q \langle T_q^{(2)}(\alpha J)^\dagger\rangle \sum_{M_1'M_2'} (-)^{J'-M_1'} \begin{pmatrix} J' & J' & 2 \\ M_1' & -M_2' & -q' \end{pmatrix}
$$

$$
\times \sum_{M'M} (-)^{J-M'} \begin{pmatrix} J & J & 2 \\ M' & -M & -q \end{pmatrix}
$$

$$
\times \frac{1}{2} \frac{k_{out}}{k_{in}} \sum_{m_s m_{s0}} \int d\Omega f_{\alpha'J'M_1',\alpha JM'}^{m_{s0}m_s}(\theta,\phi) f_{\alpha'J'M_2',\alpha JM}^{m_{s0}m_s^*}(\theta,\phi)
$$

$$(5.42)$$

If we now use the factorization of the scattering amplitude, given by (5.22), in the integral in (5.42), and perform the integration over the azimuthal angle, then we obtain that the integral is zero unless

$$
M_1' - M_2' = M' - M . \tag{5.43}
$$

This in turn implies that we must have

$$
q' = q . \tag{5.44}
$$

The possible values of $q' = q$ are $q = 0, 1, 2$. Thus, we obtain from (5.42)

$$
\langle T_q^{(2)}(\alpha'J')^\dagger\rangle \Leftrightarrow 5 \frac{dP}{dS} \langle T_q^{(2)}(\alpha J)^\dagger\rangle \sum_{M_1'M_2'} (-)^{J'-M_1'} \begin{pmatrix} J' & J' & 2 \\ M_1' & -M_2' & -q \end{pmatrix}
$$

$$
\times \sum_{M'M} (-)^{J-M'} \begin{pmatrix} J & J & 2 \\ M' & -M & -q \end{pmatrix}
$$

$$
\times \frac{1}{2} \frac{k_{out}}{k_{in}} \sum_{m_s m_{s0}} \int d\Omega f_{\alpha'J'M_1',\alpha JM'}^{m_{s0}m_s}(\theta,\phi) f_{\alpha'J'M_2',\alpha JM}^{m_{s0}m_s^*}(\theta,\phi).
$$

$$(5.45)$$

For the special case of $q = 0$ we obtain

$$
\langle T_0^{(2)}(\alpha'J')^\dagger\rangle \Leftrightarrow 5 \frac{dP}{dS} \langle T_0^{(2)}(\alpha J)^\dagger\rangle \sum_{M_1'} (-)^{J'-M_1'} \begin{pmatrix} J' & J' & 2 \\ M_1' & -M_1' & 0 \end{pmatrix}
$$

$$\times \sum_M (-)^{J-M} \begin{pmatrix} J & J & 2 \\ M & -M & 0 \end{pmatrix}$$

$$\times \frac{1}{2} \frac{k_{\text{out}}}{k_{\text{in}}} \sum_{m_s m_{s0}} \int d\Omega f^{m_{s0} m_s}_{\alpha' J' M'_1, \alpha J M}(\theta, \phi) f^{m_{s0} m_s*}_{\alpha' J' M'_1, \alpha J M}(\theta, \phi)$$

$$(5.46)$$

which, in view of the definition of the excitation cross-section via (5.31), can also be written in the form

$$\langle T_0^{(2)}(\alpha' J')^\dagger \rangle \Leftrightarrow 5 \frac{dP}{dS} \langle T_0^{(2)}(\alpha J)^\dagger \rangle \sum_{M'} (-)^{J'-M'} \begin{pmatrix} J' & J' & 2 \\ M' & -M' & 0 \end{pmatrix}$$

$$\times \sum_M (-)^{J-M} \begin{pmatrix} J & J & 2 \\ M & -M & 0 \end{pmatrix} \sigma_{\alpha J M, \alpha' J' M'}. \qquad (5.47)$$

If we now define the alignment-transfer cross section, $Q_0^{2,2}(\alpha J, \alpha' J')$, by the formula

$$\langle T_0^{(2)}(\alpha' J')^\dagger \rangle \Leftrightarrow \frac{dP}{dS} Q_0^{2,2}(\alpha J, \alpha' J') \langle T_0^{(2)}(\alpha J)^\dagger \rangle \qquad (5.48)$$

then we obtain for it from (5.47),

$$Q_0^{2,2}(\alpha J, \alpha' J') = 5 \sum_{M'} (-)^{J'-M'} \begin{pmatrix} J' & J' & 2 \\ M' & -M' & 0 \end{pmatrix}$$

$$\times \sum_M (-)^{J-M} \begin{pmatrix} J & J & 2 \\ M & -M & 0 \end{pmatrix} \sigma_{\alpha J M, \alpha' J' M'}$$

$$= \sum_{M'} (-)^{J'-M'} \langle J'J'M' -M'|20 \rangle$$

$$\times \sum_M (-)^{J-M} \langle JJM -M|20 \rangle \sigma_{\alpha J M, \alpha' J' M'}. \qquad (5.49)$$

This latter *form* is identical to that given by Fujimoto et al. [13] and Fujimoto and Kazantsev [14], except that here it is formulated fully quantum-mechanically. Equation (5.49) corresponds to (4.40d) or (4.20d) in Chap. 4.

Analogously to (5.48) we can define the coherence-transfer cross-section, $Q_q^{2,2}(\alpha J, \alpha' J')$, $(q = 1, 2)$, by the formula

$$\langle T_q^{(2)}(\alpha' J')^\dagger \rangle \Leftrightarrow \frac{dP}{dS} Q_q^{2,2}(\alpha J, \alpha' J') \langle T_q^{(2)}(\alpha J)^\dagger \rangle \qquad (q = 1, 2), \qquad (5.50)$$

then we obtain from (5.45),

$$Q_q^{2,2}(\alpha J, \alpha' J') = 5 \sum_{M'_1 M'_2} (-)^{J'-M'_1} \begin{pmatrix} J' & J' & 2 \\ M'_1 & -M'_2 & -q \end{pmatrix}$$

$$\times \sum_{M'M} (-)^{J-M'} \begin{pmatrix} J & J & 2 \\ M' & -M & -q \end{pmatrix}$$

$$\times \frac{1}{2} \frac{k_{\text{out}}}{k_{\text{in}}} \sum_{m_s m_{s0}} \int \mathrm{d}\Omega f^{m_{s0}m_s}_{\alpha'J'M'_1,\alpha J M'}(\theta,\phi) f^{m_{s0}m_s^*}_{\alpha'J'M'_2,\alpha J M}(\theta,\phi)$$

$$= \sum_{M'_1 M'_2} (-)^{J'-M'_2} \langle J'J'M'_1 - M'_2 | 2q \rangle$$

$$\times \sum_{M'M} (-)^{J-M'} \langle JJM - M'|2q \rangle$$

$$\times \frac{1}{2} \frac{k_{\text{out}}}{k_{\text{in}}} \sum_{m_s m_{s0}} \int \mathrm{d}\Omega f^{m_{s0}m_s}_{\alpha'J'M'_1,\alpha J M'}(\theta,\phi) f^{m_{s0}m_s^*}_{\alpha'J'M'_2,\alpha J M}(\theta,\phi) . \quad (5.51)$$

This expression is *the quantum-mechanical version* of the expression for the coherence transfer cross-section given by Fujimoto et al. [13] and Fujimoto and Kazantsev [14] who have defined it within the semiclassical impact-parameter formalism. Equation (5.51) corresponds to (4.20e). In the quantum-mechanical case, just as in the semiclassical one, the coherence transfer cross-section is expressed directly in terms of the scattering amplitudes and the formula cannot be simplified to one expressed in terms of cross-sections.

So far we have presented a fully quantum-mechanical formalism for the description of creation, destruction, and transfer of atomic multipole moments by electron scattering. We have treated the scattering electron nonrelativistically and as a distinguishable particle, thus the exchange effect of the scattering electron with the target electrons was not taken into consideration. (Both exchange and spin-orbit coupling effects were fully taken into account for the target electrons.) The exchange of the scattering electron with the target electrons and the spin-orbit coupling effect in the scattering electron can be taken into account as described by Goldberger and Watson [22] and by Kelly [23], and our fundamental results given by (5.21) and (5.29) will remain intact, except that those physical effects have to be taken into account in the calculation of the scattering amplitudes entering those formulas.

## 5.3 Alignment Creation by Elastic Electron Scattering

Alignment creation by elastic heavy particle scattering has been studied for many years by Omont [1], by D'yakonov and Perel [2], by Petrashen, Rebane, and Rebane [3–11], and by Kazantsev et al. [12]. The technique has been adapted for arbitrary perturbers (including electrons) by Fujimoto et al. [13] and by Fujimoto and Kazantsev [14]. In the case of heavy particle perturbers (e.g., ions) there was an argument by Petrashen et al. [7] that under certain conditions (namely only elastic scattering is possible and the semi-classical straight-line trajectory assumption holds) in the case of an isolated level, alignment can not be created by elastic scattering. On the other hand, Dashevskaya and Nikitin [24] argued that the above conclusion of Petrashen et al. [7] is due to an extra symmetry introduced into the problem by the straight-line trajectory approximation (which introduces detailed balance for magnetic-sublevel-to-magnetic-sublevel transitions) and if a more accurate approximation is made alignment creation can be obtained by elastic scattering. (See the discussion in Fujimoto et al. [13].) In the case of inelastic scattering, Kazantsev et al. [15, 16] gave a quantum-mechanical definition of the alignment creation cross-section. In earlier works, Trajmar et al. [25] and Csanak et al. [26] adopted the inelastic alignment creation cross-section definition of Kazantsev et al. [15, 16] for elastic electron scattering and reported results for Ba [25] and OV ions [26] based on that formula. However, a closer inspection of the semi-classical formula of Fujimoto et al. [13] and Fujimoto and Kazantsev [14] as well as the quantum-mechanical rate equations of Ben-Reuven [27], Nienhuis [28], and Bommier and Sahal-Brechot [29] also indicated that the inelastic scattering formula might not hold for elastic scattering. In what follows we investigate this problem and show that indeed the alignment creation cross-section formula is different for elastic scattering, as compared to the inelastic scattering formula. Indeed, we will show that alignment creation by elastic electron scattering is possible.

### 5.3.1 Semi-Classical Background

Fujimoto et al. [13] and Fujimoto and Kazantsev [14] gave the following formula for the alignment creation cross-section by elastic scattering ((3.15b) in Fujimoto et al. [13]), which is equivalent to (4.22b) in the preceding Chap. 4,

$$
\sigma_0(20) = (2F+1)^{-1/2} \sum_{M'} (-1)^{F-M'} \langle FFM' - M'|20\rangle D_{\alpha FM'}
$$
$$
+ (2F+1)^{-1/2} \sum_{M \neq M'} \Big[ (-1)^{F-M'} \langle FFM' - M'|20\rangle
$$
$$
- (-1)^{F-M} \langle FFM - M|20\rangle \Big] Q_{\alpha FM, \alpha FM'}, \tag{5.52}
$$

where $D_{\alpha FM'}$ and $Q_{\alpha FM, \alpha FM'}$ were defined as ((31.4) in Fujimoto et al. [13] and also (4.21) and (4.19) in Chap. 4. Note the slight difference in the notations in Chap. 4 from [13] and [14].)

$$
D_{\alpha FM'} = 2\pi \int b\, db \sum_M \big[ \delta_{MM'} - |\langle \alpha FM|\mathcal{S}_c|\alpha FM'\rangle|^2 \big] \tag{5.53}
$$

and

$$
Q_{\alpha FM, \alpha FM'} = 2\pi \int b\, db |\langle \alpha FM|\mathcal{S}_c|\alpha FM'\rangle|^2 \tag{5.54}
$$

(we have adopted their [13,14] notation). These equations are somewhat involved and it is difficult to see their relationship to the inelastic scattering formulas. By simple algebraic transformation it can be shown that $\sigma_0(20)$ can be written in the form

$$\sigma_0(20) = (2F+1)^{-1/2} \sum_M (-1)^{F+M} \langle FFM - M|20\rangle$$

$$\times 2\pi \int b\, db \sum_{M'} \left[ \delta_{MM'} - |\langle \alpha FM|S_c|\alpha FM'\rangle|^2 \right] \qquad (5.55)$$

where (as in Fujimoto et al. [13]) $S_c$ refers to the $S$-matrix (actually an operator) in the "collision frame" whose axis is parallel to the incident electron beam. If we now introduce the $T_c$ operator by the definition

$$S_c = 1 + T_c \qquad (5.56)$$

we obtain

$$\sigma_0(20) = (2F+1)^{-1/2} \sum_M (-1)^{F+M+1} \langle FFM - M|20\rangle$$

$$\times \left\{ 2\pi \int b\, db \left[ \langle \alpha FM|T_c|\alpha FM\rangle + \langle \alpha FM|T_c^\dagger|\alpha FM\rangle \right] \right.$$

$$\left. + \sum_{M'} Q_{\alpha FM, \alpha FM'} \right\} . \qquad (5.57)$$

The second term on the right-hand side is identical in form to the expression for the alignment creation cross-section in the case of inelastic processes. The first term is a linear term in the $T$-operator and it is an additional term for elastic scattering. Its physical meaning in this semiclassical treatment is not clear. It will become clear in the quantum-mechanical analysis presented in Sect. 5.3.2. Similar additional linear terms appear in the relaxation rate equations of Ben-Reuven [27], Nienhuis [28], and Bommier and Sahal-Brechot [29].

## 5.3.2 Wave-Packet Formulation of Alignment Creation by Elastic Scattering

The above uncertainties prompted us to reinvestigate the quantum-mechanical problem of alignment creation by elastic scattering. Application of the wave-packet formulation of Sect. 5.1 to the case of elastic scattering leads to some simplifications in notations as initial and final states of the target atom are the same. The indexes $\alpha$ and $J$ specifying the target state will be dropped; for example, the elastic scattering amplitude is defined as

$$f_{M_1,M_2}^{m_s m_s'}(\theta,\phi) = -\frac{m}{2\pi\hbar^2} \langle \mathbf{k}_1 m_s; \alpha J M_1|T|\mathbf{k}_2 m_s'; \alpha J M_2\rangle , \qquad (5.58)$$

where $|\mathbf{k}_1| = |\mathbf{k}_2|$ and $\theta$ and $\phi$ refer to the polar angles of $\mathbf{k}_2$ relative to $\mathbf{k}_1$.

In this section we are interested in the change of the reduced atomic density matrix by elastic scattering. Thus we define the quantity

$$\Delta \rho_{M_1 M_2} = \rho^{\text{out}}_{M_1 M_2} - \rho^{\text{in}}_{M_1 M_2} \, , \tag{5.59}$$

where

$$\rho^{\text{in}}_{M_1 M_2} = \langle \alpha J M_1 | \rho^{\text{in}} | \alpha J M_2 \rangle = \delta_{M_1 M_2} \, . \tag{5.60}$$

Using (5.58), (5.59), and (5.21), we obtain

$$
\begin{aligned}
\Delta \rho^{\text{out}}_{M_1 M_2} = {} & \frac{2\pi}{k_{\text{in}}} \mathrm{i} \frac{\mathrm{d}P}{\mathrm{d}S} \frac{1}{2} \sum_{m_s} \Big[ f^{m_s m_s}_{M_1 M}(\theta = 0, \phi = 0) \delta_{MM_2} \\
& - f^{m_s m_s *}_{M_2 M}(\theta = 0, \phi = 0) \delta_{MM_1} \Big] \\
& + \frac{\mathrm{d}P}{\mathrm{d}S} \frac{1}{2} \sum_{m_s m_{s0}} \int \mathrm{d}\Omega_0 f^{m_{s0} m_s}_{M_1 M}(\theta_0, \phi_0) f^{m_{s0} m_s *}_{M_2 M}(\theta_0, \phi_0) \, . \tag{5.61}
\end{aligned}
$$

This is the fully quantum-mechanically obtained expression, which shows the linear terms in the scattering amplitude and corresponds to the semi-classical expression given by (5.57). Here we shall make an argument which is based on the assumption that the scattering electron is considered distinguishable (i.e., we neglect exchange) and is described nonrelativistically. Under these assumptions the angular momentum projection quantum number of the target state and the spin-projection of the incident electron are conserved upon elastic scattering in the forward direction and they are independent of the spin projection of the incident electron. We can therefore write

$$f^{m_s m_s}_{M_1 M}(\theta = 0, \phi = 0) = f^{m_s m_s}_{MM}(\theta = 0, \phi = 0) \delta_{MM_1} \tag{5.62}$$

with $f^{m_s m_s}_{MM}(\theta = 0, \phi = 0)$ being independent of $m_s$. Then we obtain for $\Delta \rho_{M_1 M_2}$, the formula

$$
\begin{aligned}
\Delta \rho^{\text{out}}_{M_1 M_2} = {} & \frac{2\pi}{k_{\text{in}}} \mathrm{i} \frac{\mathrm{d}P}{\mathrm{d}S} \frac{1}{2} \sum_{m_s} \Big[ f^{m_s m_s}_{MM}(\theta = 0, \phi = 0) \\
& - f^{m_s m_s *}_{M_2, M}(\theta = 0, \phi = 0) \Big] \delta_{MM_1} \delta_{MM_2} \\
& + \frac{\mathrm{d}P}{\mathrm{d}S} \frac{1}{2} \sum_{m_s m_{s0}} \int \mathrm{d}\Omega_0 f^{m_{s0} m_s}_{M_1 M}(\theta_0, \phi_0) f^{m_{s0} m_s *}_{M_2 M}(\theta_0, \phi_0) \, . \tag{5.63}
\end{aligned}
$$

If we now use the mathematical identity

$$f^{m_s m_s}_{MM}(\theta = 0, \phi = 0) - f^{m_s m_s *}_{MM}(\theta = 0, \phi = 0) = 2\mathrm{i}\,\mathrm{Im} f^{m_s m_s}_{MM}(\theta = 0, \phi = 0) \tag{5.64}$$

along with the optical theorem (see, e.g., Bransden [20] p. 131),

$$\mathrm{Im} f^{m_s m_s}_{MM}(\theta = 0, \phi = 0) = \frac{k_{\text{in}}}{4\pi} \sigma^{\text{tot}}_{\alpha J M, m_s} \, , \tag{5.65}$$

where $\sigma^{\mathrm{tot}}_{\alpha JM, m_{\mathrm{s}}}$ refers to the total electron scattering cross-section by the $|\alpha JM\rangle$ state with incident electron spin $m_{\mathrm{s}}$,

$$\sigma^{\mathrm{tot}}_{\alpha JM, m_{\mathrm{s}}} = \sum_{m_{\mathrm{s}0}} \sum_{\alpha' J' M'} \int d\Omega_0 |f^{m_{\mathrm{s}0} m_{\mathrm{s}}}_{\alpha' J' M', \alpha J M_1}(\theta_0, \phi_0)|^2 , \qquad (5.66)$$

where $|\alpha' J' M'\rangle$ refers to an arbitrary state of the target. We obtain for $\Delta\rho^{\mathrm{out}}_{M_1 M_2}$ the expression

$$\Delta\rho^{\mathrm{out}}_{M_1 M_2} = -\frac{dP}{dS}\frac{1}{2} \sum_{m_{\mathrm{s}}} \sigma^{\mathrm{tot}}_{\alpha JM, m_{\mathrm{s}}} \delta_{MM_1}$$

$$+ \frac{dP}{dS}\frac{1}{2} \sum_{m_{\mathrm{s}} m_{\mathrm{s}0}} \int d\Omega_0 f^{m_{\mathrm{s}0} m_{\mathrm{s}}}_{M_1 M}(\theta_0, \phi_0) f^{m_{\mathrm{s}0} m_{\mathrm{s}}*}_{M_2, M}(\theta_0, \phi_0) . \quad (5.67)$$

These transformations are of great physical significance as we will see below.

This expression was obtained with the assumption that the initial state was described by the density operator

$$\rho^{\mathrm{in}} = |\alpha JM\rangle\langle\alpha JM| .$$

Here we are interested in alignment creation by elastic scattering from an initial isotropic state (with no alignment). The isotropic state will be described by the density operator

$$\rho^{\mathrm{in}}_{\mathrm{iso}} = \frac{1}{2J+1} \sum_M |\alpha JM\rangle\langle\alpha JM| \qquad (5.68)$$

giving the initial density matrix as

$$\rho^{\mathrm{in, iso}}_{M_1 M_2} = \frac{1}{2J+1} \delta_{M_1 M_2} . \qquad (5.69)$$

Since $\rho^{\mathrm{in}}_{\mathrm{iso}}$ is additive from the "$\rho^{\mathrm{in}}_{\mathrm{iso}}$ elements" we can simply sum (5.67) over $M$ and divide by $(2J+1)$ to obtain

$$\Delta\rho^{\mathrm{iso}}_{M_1 M_2} = -\frac{dP}{dS}\frac{1}{2J+1} \left\{ \frac{1}{2} \sum_{m_{\mathrm{s}}} \sigma^{\mathrm{tot}}_{\alpha J M_1, m_{\mathrm{s}}} \delta_{M_1 M_2} \right.$$

$$\left. - \frac{1}{2} \sum_{m_{\mathrm{s}} m_{\mathrm{s}0}} \sum_M \int d\Omega_0 f^{m_{\mathrm{s}0} m_{\mathrm{s}}}_{M_1 M}(\theta_0, \phi_0) f^{m_{\mathrm{s}0} m_{\mathrm{s}}*}_{M_2, M}(\theta_0, \phi_0) \right\} ,$$

$$(5.70)$$

where $\Delta\rho^{\mathrm{iso}}_{M_1 M_2}$ is the change of the density matrix element of the initially isotropic state by the elastically scattered electron. The alignment created in the scattering process from the isotropic state can be given in the form (see, e.g., Blum [19], p. 98)

$$\langle T_0^{(2)}(\alpha J)\rangle = \frac{5^{1/2}}{[(2J+3)(2J+1)J(2J-1)(J+1)]^{1/2}}$$
$$\times \sum_M \left[3M^2 - J(J+1)\right] Q(M) , \qquad (5.71)$$

where

$$Q(M) = \rho_{MM}^{\text{iso}} . \qquad (5.72)$$

From (5.70) we obtain

$$\Delta\rho_{MM}^{\text{iso}} = -\frac{dP}{dS}\frac{1}{2J+1}\left\{\frac{1}{2}\sum_{m_s}\sigma_{\alpha JM,m_s}^{\text{tot}} - \frac{1}{2}\sum_{m_s m_{s0}}\sum_{M'}\int d\Omega_0 |f_{MM'}^{m_{s0}m_s}(\theta_0,\phi_0)|^2\right\}$$

$$\equiv \frac{dP}{dS}\frac{1}{2J+1}\left\{\frac{1}{2}\sum_{m_s m_{s0}}\sum_{M'}\sigma_{MM'}^{m_{s0}m_s} - \frac{1}{2}\sum_{m_s}\sigma_{\alpha JM,m_s}^{\text{tot}}\right\}$$

$$\equiv \frac{dP}{dS}\frac{1}{2J+1}\left\{\sum_{M'}\sigma_{MM'} - \frac{1}{2}\sum_{m_s}\sigma_{\alpha JM,m_s}^{\text{tot}}\right\} , \qquad (5.73)$$

where we have defined the $\sigma_{MM'}$ cross-section by the formula,

$$\sigma_{MM'} \equiv \frac{1}{2}\sum_{m_s m_{s0}}\sigma_{MM'}^{m_{s0}m_s} \equiv \frac{1}{2}\sum_{m_s m_{s0}}\int d\Omega_0 |f_{MM'}^{m_{s0}m_s}(\theta_0,\phi_0)|^2 . \qquad (5.74)$$

We can define the alignment creation cross-section by the formula

$$Q_0^{0,2}(\alpha J, \alpha J) = \left(\frac{dP}{dS}\right)^{-1}\langle T_0^{(2)}(\alpha J)\rangle , \qquad (5.75)$$

which gives

$$Q_0^{0,2}(\alpha J, \alpha J) = \frac{5^{1/2}}{[(2J+3)(2J+1)J(2J-1)(J+1)]^{1/2}}$$
$$\times \sum_M \left[3M^2 - J(J+1)\right] q(M) , \qquad (5.76)$$

where

$$q(M) = \frac{1}{2J+1}\left\{\frac{1}{2}\sum_{m_s m_{s0}}\sum_{M'}(\sigma_{MM'}^{m_s m_{s0}} - \sigma_{M'M}^{m_s m_{s0}}) - \sum_{m_s}\sigma_{\alpha JM,m_s}^{\text{inel}}\right\}$$

$$= \frac{1}{2J+1}\left\{\sum_{M'}(\sigma_{MM'} - \sigma_{M'M}) - \sum_{m_s}\sigma_{\alpha JM,m_s}^{\text{inel}}\right\} , \qquad (5.77)$$

where $\sigma_{\alpha JM,m_s}^{\text{inel}}$ refers to the sum of all inelastic cross-sections of excitations and de-excitations out of the state $|\alpha JM\rangle$ with incident electron of spin $m_s$, given by the formula

**Table 5.1.** Magnetic sublevel quantities and alignment creation cross-section

| $E_k^{in}$ (eV) | 2.74 | 20.0 | 97.7 |
|---|---|---|---|
| $q(1)$ | $-136.0$ | $-114.9$ | $-52.81$ |
| $q(0)$ | $-140.9$ | $-113.3$ | $-43.50$ |
| $Q_0^{0,2}$ | 4.03 | $-1.33$ | $-7.60$ |
| $Q$ | 385.1 | 139.7 | 62.23 |

The magnetic sublevel quantities $q(m)$, alignment creation cross section $Q_0^{0,2}$, and the total cross section $Q$ (all in atomic units) for differing incident electron energies.

$$\sigma_{\alpha JM,m_s}^{inel} = \sum_{m_{s0}} \sum_{\alpha' J' M' (\alpha' \neq \alpha J' \neq J)} \int d\Omega_0 |f_{\alpha' J' M',\alpha JM}^{m_{s0} m_s}(\theta_0, \phi_0)|^2. \qquad (5.78)$$

It is noted that (5.76) corresponds to (4.41b). For the $J = 1$ case, which is relevant to the experiment of Trajmar et al. [25],

$$Q_0^{0,2}(\alpha J = 1, \alpha J = 1) = \left(\frac{2}{3}\right)^{1/2} (q(1) - q(0))$$

$$= \left(\frac{2}{3}\right)^{1/2} \left( (\sigma_{10} - \sigma_{01}) \right.$$

$$\left. + \frac{1}{6} \sum_{m_s} (\sigma_{\alpha J=1,M=0,m_s}^{inel} - \sigma_{\alpha J=1,M=1,m_s}^{inel}) \right).$$

$$(5.79)$$

If we ignore the contribution to $Q_0^{0,2}$ from inelastic processes, then we obtain $Q_0^{0,2}((\alpha J = 1, \alpha J = 1) \approx (2/3)^{1/2}(\sigma_{10} - \sigma_{01})$, a result essentially identical to the one used by Dashevskaya et al. [31].

For electron scattering from laser excited neutral Barium, results from converged close coupling (CCC) calculations (Trajmar et al. [25]) are given in Table 5.1.

The earlier calculation by Trajmar et al. that used the inelastic formula for $Q_0^{0,2}$ gave the respective values of 8.70, 2.33, and 1.00 for the above energies.

### 5.3.3 Discussion and Conclusions

Here we have obtained a formula by quantum-mechanical methods for the alignment creation cross-section by elastic electron scattering. The formula obtained differs from the analogous formula relevant for inelastic electron scattering. In the case of a $J = 1$ to $J = 1$ transition according to the inelastic formula the alignment created is proportional to the quantity $\sigma(1) - \sigma(0)$, where $\sigma(M)$ is the excitation cross section of the $M$ magnetic sublevel and thus $\sigma(1) = (\sigma_{1-1} + \sigma_{10} + \sigma_{11})/3$ and $\sigma(0) = (\sigma_{0-1} + \sigma_{00} + \sigma_{01})/3$ where $\sigma_{MM'}$ refers to the cross section of the electron impact induced $M'$ to $M$ transition.

In the elastic scattering alignment creation formula obtained by us in the case of a $J = 1$ to $J = 1$ elastic scattering, the alignment created is proportional to the quantity $q(1) - q(0)$ and is given by (5.79). Thus, we can conclude that the "linear terms" that were missing in the earlier expression have clear physical meaning in the quantum mechanical case while their meaning in the semi-classical case was not clear [30]. Our numerical results given in Table 1 clearly show that, even in the case of an isolated level, elastic electron scattering can create alignment. Our derivation considered only direct scattering, i.e., the incident electron was considered distinguishable from the target electrons. The wave packet treatment of exchange was discussed by Goldberger and Watson [22], by Rodberg and Thaler [18], and by Kelly [23]. These works seem to indicate that our results carry over when exchange scattering is also considered if exchange is included in the calculation of the $\mathcal{T}$ matrix elements. Future work will be directed toward the treatment of exchange and spin-orbit coupling effects.

**Acknowledgement**

The authors want to thank Professors T. Fujimoto, Al Stauffer and Klaus Bartschat and Drs. Jon Weisheit, Suxing Hu, and especially Peter Hakel for reading and criticizing the original manuscript, and Chris Fontes for assistance with the preparation of the manuscript. This work was partially conducted under the auspices of the U.S. Department of Energy. Support of the Australian Research Council is also acknowledged.

# References

1. A. Omont: Jour. de Phys. **26**, 26 (1965)
2. M.I. D'yakonov, V.I. Perel: JETP (USSR) **48**, 345 (1965) (English translation: Soviet Physics JETP **21**, 227 (1965))
3. V.N. Rebane, T.K. Rebane: Opt. i Spektrosk. **20**, 185 (1966) (English translation: Optics and Spectroscopy **20**, 101 (1966))
4. V.N. Rebane: Opt. i Spektrosk. **21**, 405 (1966) (English translation: Optics and Spectroscopy **21**, 229 (1966))
5. V.N. Rebane: Opt. i Spektrosk. **24**, 296 (1968) (English translation: Optics and Spectroscopy **24**, 155 (1968))
6. V.N. Rebane: Opt. i. Spektrosk. **24**, 309 (1968) (English translation: Optics and Spectroscopy **24**, 163 (1968))
7. A.G. Petrashen, V.N. Rebane, T.K. Rebane: Opt. i Spektrosk. **55**, 819 (1983) (English translation: Optics and Spectroscopy **55**, 492 (1984))
8. A.G. Petrashen, V.N. Rebane, T.K. Rebane: Opt. i Spektrosk. **57**, 376 (1984) (English translation: Optics and Spectroscopy **57**, 230 (1984))
9. A.G. Petrashen, V.N. Rebane, T.K. Rebane: Zh. Eksp. Teor. Fiz. **87**, 147 (1984) (English translation: Sov. Phys. JETP **60**, 84 (1984))
10. A.G. Petrashen, V.N. Rebane, T.K. Rebane: Zh. Eksp. Teor. Fiz. **94**, 46 (1988) (English translation: Sov. Phys. JETP **67**, 2202 (1989))

11. A.G. Petrashen, V.N. Rebane: Opt. i Spektrosk. **67**, 6 (1989) (English translation: Optics and Spectroscopy **67**, 3 (1990))
12. S.A. Kazantsev, A.G. Petrashen, V.N. Rebane: Zh. Eksp. Teor. Fiz **106**, 698 (1994) (English translation: Sov. Phys. JETP **79**, 384 (1994))
13. T. Fujimoto, H. Sahara, G. Csanak, S. Grabbe: *Atomic States and Collisional Relaxation in Plasma Polarization Spectroscopy* (NIFS-DATA-38, Nagoya, 1996)
14. T. Fujimoto, S.A. Kazantsev: Plasma Phys. Control. Fusion **39**, 1267 (1997)
15. S.A. Kazantsev, N.Ya. Plynovskaya, L.N. Pyatnitskii, S.A. Edel'man: Usp. Fiz. Nauk **156**, 3 (1988) (English translation: Sov. Phys. Usp. **31**, 785 (1988))
16. S.A. Kazantsev, J.-C. Hénoux: *Polarization Spectroscopy of Ionized Gases* (Kluwer Academic Publishers, Dordrecht, 1995), p. 33
17. G. Csanak, D.P. Kilcrease, D.V. Fursa, I. Bray: *Alignment Creation by Elastic Scattering. A Quantum Treatment*, in Proc. Japan-US Workshop on Plasma Polarization Spectroscopy, NIFS-PROC-57 (T. Fujimoto and P. Beiersdorfer, eds., 2004) p. 133
18. L.S. Rodberg, R.M. Thaler: *Introduction to the Quantum Theory of Scattering* (Academic Press, New York, 1967)
19. K. Blum: *Density Matrix Theory and Applications* (Plenum Press, New York, 1981)
20. B.H. Bransden: *Atomic Collision Theory* (W.A. Benjamin Inc., New York, 1970)
21. J.M. Blatt, L.C. Biedenharn: Revs. of Mod. Phys. **24**, 258 (1952)
22. M.L. Goldberger, K.M. Watson: *Collision Theory* (Krieger Publishing Co., New York, 1975)
23. R.L. Kelly: Phys. Rev. **147**, 376 (1966)
24. E.I. Dashevskaya, E.E. Nikitin: Sov. J. Cem. Phys. **4**, 1934 (1987)
25. S. Trajmar, I. Kanik, M.A. Khakoo, L.R. LeClair, I. Bray, D. Fursa, G. Csanak: J. Phys. B: At. Mol. Opt. Phys. **32**, 2801 (1999)
26. G. Csanak, D.P. Kilcrease, H. Zhang, D.V. Fursa, I. Bray, T. Fujimoto, A. Iwamae: *Plasma Polarization Spectroscopy for the OV Ion: Relevant Collision Cross Sections for Kinetic Modeling*, in Proc. the 3rd US-Japan Plasma Polarization Spectroscopy Workshop, Report UCRL-ID-146907 (P. Beiersdorfer and T. Fujimoto, eds., 2001)
27. A. Ben-Reuven: Phys. Rev. **145**, 7 (1966)
28. G. Nienhuis: J. Phys. B: At. Mol. Phys. **9**, 167 (1976)
29. V. Bommier, S. Sahal-Bréchot: Ann. Phys. Fr. **16**, 555 (1991)
30. P. Paradoksov: Usp. Fiz. Nauk **89**, 707 (1966) (English translation: Sov. Phys. Uspekhi 9, 618 (1967))
31. E.I. Dashevskaya, E.E. Nikitin, S.Ya. Umanskii: Opt. Spectrosc. (USSR), **55**, 664 (1983)

# Collision Processes

T. Fujimoto

In the preceding two chapters, we introduced the cross sections and rate coefficients for various collision processes involving alignment. In this chapter, we examine the physical meaning of each of these processes. We also review the situation about our knowledge of these cross-sections, or about the cross-section data available at present.

In the following two sections, we put some emphasis on electron impact. Atom collisions will be discussed in Sect. 6.3.

## 6.1 Inelastic and Elastic Collisions

### 6.1.1 Excitation/Deexcitation and Ionization, $Q_0^{0,0}(r,p)$, and $Q_0^{0,0}(p,p)$

As is understood from (4.40a) with (4.19), $Q_0^{0,0}(r,p)$ is the excitation or deexcitation cross-section in the usual sense. Therefore, the corresponding rate coefficient was written simply as $C(r,\ p)$ in the later part of Chap. 4 after (4.42). The underlying assumption in using $C(r,\ p)$ is that the magnetic sublevels are evenly populated, i.e., there is no alignment, in the initial level. In general, an alignment is present and its effect is included as a correction. This point will be discussed later in this section. As Fig. 4.2 shows $Q_0^{0,0}(r,p)$ describes the creation of population $n(p)$ from the population $n(r)$. This is also understood from the following consideration about the method of determining the excitation cross-section or the upper-level population in a beam collision experiment as introduced in the beginning of Chap. 1.

Suppose the atoms are in level $r$ as the initial state and these atoms have no alignment. These atoms are excited to level $p$ by a monoenergetic beam of electrons having velocity $v_0$ in the $z$-direction. From (4.32), we have the expansion coefficients of the velocity distribution function $f(v,\theta)$,

$$f_0(v) = \delta(v - v_0)/4\pi v_0^2$$
$$f_2(v) = 5\delta(v - v_0)/4\pi v_0^2 \tag{6.1}$$
$$f_4(v) = 9\delta(v - v_0)/4\pi v_0^2.$$

This excitation process is expressed by (4.50) as

$$\frac{d}{dt}n(p) = C(r,p)n(r)n_e = Q_0^{0,0}(r,p)v_0 n(r)n_e, \tag{6.2}$$

where depopulation has been omitted and we have used (4.42a) with (6.1). This equation clearly shows the meaning of this cross-section.

As suggested from (4.5), determination of the upper-level population, or the excitation cross-section in (6.2), from the observed line intensity is less straightforward than it may be imagined. The contribution from the second term on the right-hand side of (4.5), or that from the alignment, should be eliminated. Appendix E discusses various methods to accomplish this objective.

The excitation and deexcitation cross-sections are related by the principle of detailed balance, or the Klein–Rosseland relationship [1]. The cross-section for deexcitation $p \to r$ is given by

$$Q_0^{0,0}(p,r;E'') = \frac{2J_r + 1}{2J_p + 1}\frac{E'}{E''}Q_0^{0,0}(r,p;E'), \tag{6.3}$$

where the incident electron energy $E''$ in deexcitation is equal to $(E' - \Delta E)$, i.e., the incident electron energy for excitation $E'$ subtracted by the threshold energy $\Delta E$.

Equation (4.41a) with (4.21) indicates that $Q_0^{0,0}(p,p)$ represents the collisional depopulation from level $p$, so that this cross-section should be the sum of all the cross-sections for excitation, deexcitation, and ionization originating from this level. See (4.43). Again, the implicit assumption is that the initial level has no alignment.

With regard to the excitation and, therefore, deexcitation processes by electron impact, we have a substantial body of cross-section data for atoms as well as for ions. This is because these cross-sections are the essential ingredient in interpreting quantitatively the experimental data in the conventional intensity spectroscopy, and enormous efforts to produce cross-sections continued for a long time. Just one example of the cross-sections, the excitation cross-section of neutral helium for transition $1^1S_0 \to 2^1P_1$, is shown in Fig. 6.1 [2]. Only the results of the most sophisticated calculations and the most recent experiments are given. Agreement among the data is excellent except immediately above the excitation threshold.

For various transitions in many atom or ion species our knowledge on the cross-section is much more limited. Sometimes, even a cross-section cannot be found. Even if it is found, it is quantitatively uncertain, even by an order near the threshold region. For the purpose of helping researchers, existing cross-sections have been collected and, sometimes, evaluated. These collections of

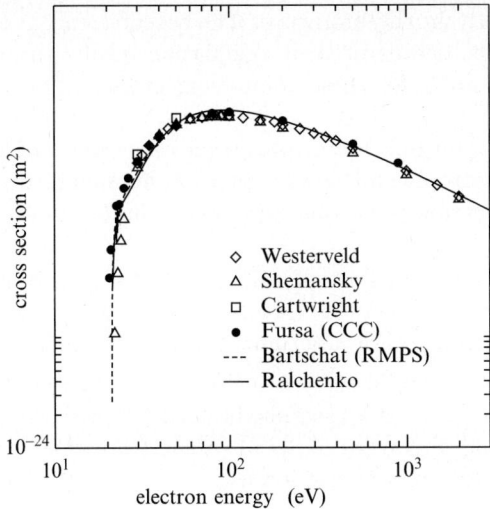

**Fig. 6.1.** Excitation cross-section of He $(1^1S_0 - 2^1P_1)$ determined by several experiments and theories. (Quoted from [2]; with permission from Elsevier.)

the data are called data bases, which are available online. One of the data bases is listed in the Reference section in this chapter. Owing to the recent progresses in experimental techniques as well as in computers, the species of atoms and ions being studied are expanding and the quality of the cross-section data is improving. Sometimes newly determined cross-sections are made available online in the internet directly by the worker who produced the data.

Ionization also contributes to $Q_0^{0,0}(p,p)$, as explicitly shown by (4.43). A large body of experimental and theoretical cross-section data has been accumulated. However, almost all of them are for ionization from the ground state atoms or ions. Since the ground state is assumed unaligned, the ionization term appearing in the second line of (4.50) has little significance in the present context. Cross-section data from excited atoms or ions are quite scanty. However, it is known that, at least for neutral hydrogen and hydrogen-like ions, the dominant depopulation mechanism from an excited level is excitation to the adjacent higher-lying level, not ionization [1]. A correction for the presence of an alignment in this level is discussed in Sect. 6.1.3.

## 6.1.2 Alignment Creation, $Q_0^{0,2}(r,p)$, and Alignment-to-Population, $Q_0^{2,0}(r,p)$

These cross-sections reduce to the magnetic-sublevel-to-magnetic-sublevel cross-sections as seen in (4.40b), (5.41), and (4.40c). Although theoretical calculations could provide these component cross-sections, they are usually not given explicitly in the literature. The recent trend in which the authors make available the details of their calculation results through his/her home page may be favorable for these component cross-sections to be provided to the public.

We look at again our beam excitation experiment of $r \to p$. Since our excitation is anisotropic and $f_2(v)$ is present, an alignment is created in level $p$ (4.42b); this creation process is expressed from (4.51) as

$$\frac{\mathrm{d}}{\mathrm{d}t} a(p) = C^{0,2}(r,p)n(r)n_e = Q_0^{0,2}(r,p)v_0 n(r)n_e, \tag{6.4}$$

where we have used (4.42b) with (6.1).

From the comparison of (6.2) with (6.4), and (4.5)–(4.8), we may express the alignment creation cross-section in terms of the experimentally determined longitudinal alignment $A_L(p,s)$ of the transition line $p \to s$, and the excitation/deexcitation cross-section as

$$Q_0^{0,2}(r,p) = (-)^{J_p+J_s} \sqrt{\frac{2}{3}} (2J_p+1)^{-1} \left\{ \begin{matrix} J_p & J_p & 2 \\ 1 & 1 & J_s \end{matrix} \right\}^{-1} A_L(p,s)Q_0^{0,0}(r,p). \tag{6.5}$$

Thus, from the experimental data such as Fig. 1.2 and the excitation cross-section such as in Fig. 6.1, we can determine the alignment creation cross-section. As can be understood from the positive and negative polarization degrees, the cross-section can be positive or negative, depending on energy. Figure 6.2 shows the alignment creation cross-section, which is constructed from Figs. 1.2 and 6.1 according to (6.5). In this figure, the cross-section calculated by other methods [3,4] are also given. The agreement among the data is quite good.

Figure 6.3 shows similar cross-section data for $2^3S$–$n^3D$ of neutral helium [3,4]. In this case, the transition is optically forbidden, and the agreement between the data is rather poor. Figure 6.4 is for helium-like iron [5]. In this case, instead of the cross-sections, the degree of polarization of the line emitted from the level excited directly from the ground state ion $1^1S_0$ is given; line $w$ is for the upper level $2^1P_1$, $x$ is for $2^3P_2$, and $y$ is for $2^3P_1$. The calculation has been substantiated experimentally for scandium on the EBIT machine [6] first and for iron [7]. For hydrogen-like, helium-like ions, and some other ions, several calculations have been reported. It is noted that, for a transition in ions in an iso-electronic sequence, the alignment creation cross-section, or the polarization degree, depends little on the charge of the ion [8]. Figure 6.5 shows an example. It is also noted that for transitions of series lines, i.e., with upper levels with different principal quantum numbers, the cross-section has similar "shape." Figure 6.6 shows an example.

For excitation of atoms and ions by heavy-particle collisions, including the electron capture process in ion–atom collisions, polarization of emission lines has been observed and theoretical calculations have been presented. See the

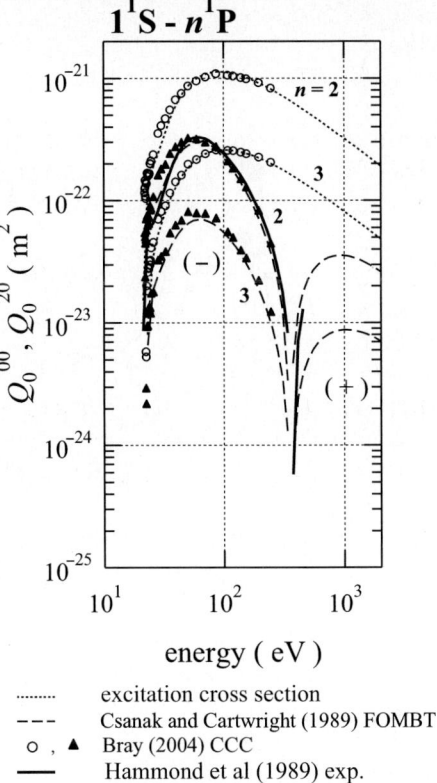

$1^1S - n^1P$

energy ( eV )

··········  excitation cross section
- - -     Csanak and Cartwright (1989) FOMBT
o , ▲    Bray (2004) CCC
———     Hammond et al (1989) exp.

**Fig. 6.2.** Population creation cross-section, $Q_0^{0,0}(1^1S, n^1P)$, of helium: *dotted line, circle.* Alignment creation cross-section, $Q_0^{0,2}(1^1S, n^1P)$, constructed from Figs. 1.2 and 6.1 according to (6.5): *straight line.* Cross-sections obtained by other methods are also shown

review papers [9, 10]. For electron capture in collisions of highly-ionized ions with atoms, it has been shown [11] that, if the quantization axis is taken to be perpendicular to the scattering plane, the captured electron tends to have the largest magnetic quantum number, or a circular orbit in the classical picture. This means that, in the system with the quantization axis taken along the beam direction, the magnetic quantum number tends to be small, or the level has a negative alignment (see Fig. 4.1). This tendency has been confirmed by a Monte Carlo calculation [12] and by experiment. This tendency seems to be valid even for a double-electron-capture process [13].

The alignment-to-population cross-section $Q_0^{2,0}(r,p)$ is, as (4.40c) and (4.50) indicate, the correction to the excitation or deexcitation process expressed by $Q_0^{0,0}(r,p)$; this is due to the presence of an alignment in the initial level $r$. Since the magnetic-sublevel-to-magnetic-sublevel cross-sections are different for different pairs of the magnetic sublevels, if the magnetic sublevels

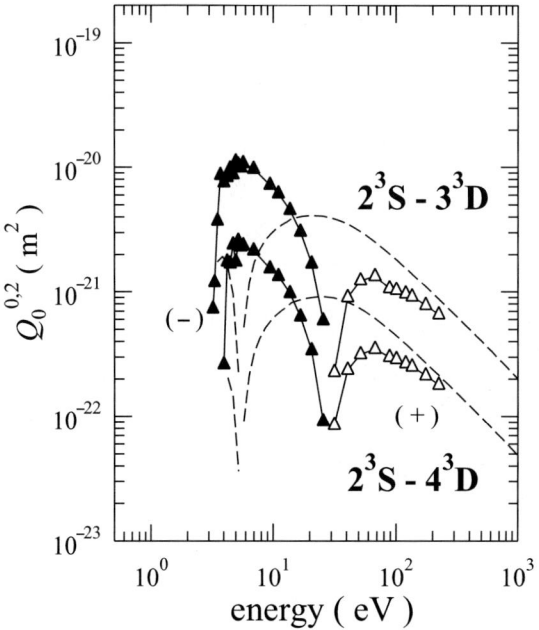

**Fig. 6.3.** Alignment creation cross-section $Q_0^{0,2}(r,p)$ for He $2^3\text{S} - n^3\text{D}$. ---: [3], ▲ △; [4]

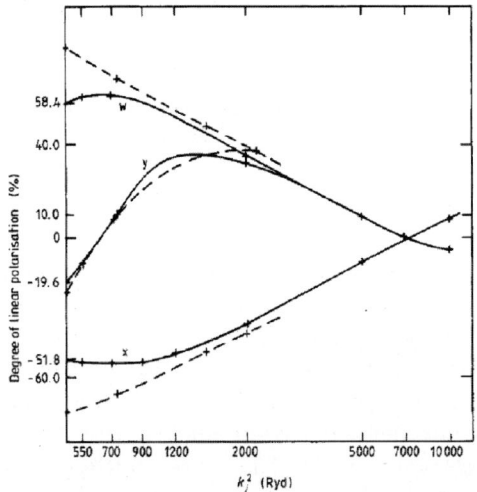

**Fig. 6.4.** Polarization degree of the helium-like iron $1s2p(^{1,3}\text{P}_J) \rightarrow 1s^2(^1\text{S}_0)$ lines against the energy of the incident electrons, which excite the upper levels from the ground state. $w$: $2^1\text{P} \rightarrow 1^1\text{S}_0$, $x$: $2^3\text{P}_2 \rightarrow 1^1\text{S}_0$, and $y$: $2^3\text{P}_1 \rightarrow 1^1\text{S}_0$. The *solid* and *dashed lines* are the result of calculation of Inal and Dubau (1987) and Schlyaptseva et al. (1981), respectively. (Quoted from [5], with permission from IOP Publishing.)

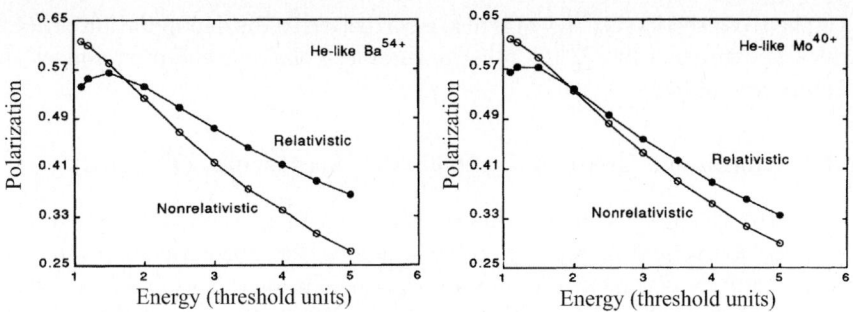

**Fig. 6.5.** Polarization degree of helium-like ion transition $1^1S - 2^1P$. These results correspond to Figs. 1.2 and 6.4. (Quoted from [8], with permission from The American Physical Society.)

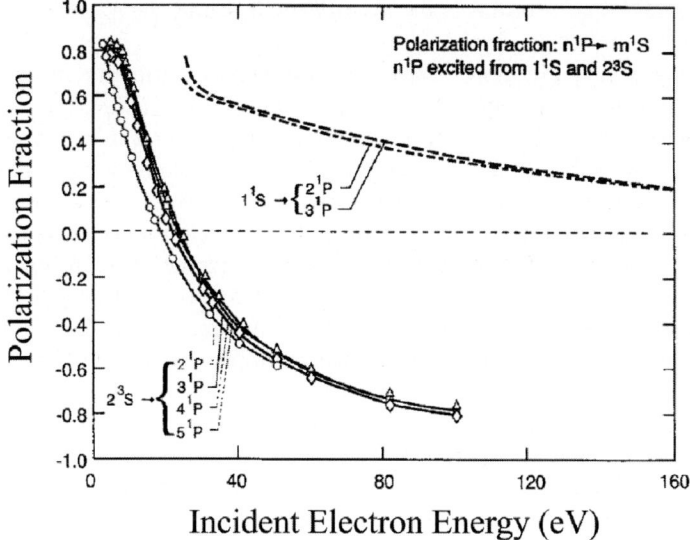

**Fig. 6.6.** Polarization degree of He I ($1^1S - n^1P$) transitions for excitation of $1^1S \rightarrow n^1P$ and $2^3S \rightarrow n^1P$. (Quoted from [3])

of the initial level have different populations, (4.40a) is insufficient to express the creation of population in the final level $p$. This is the reason why the correction term is necessary.

If we can assume a rectilinear path for the perturber, the principle of detailed balance holds for each pair of the magnetic-sublevel-to-magnetic-sublevel cross-sections, and $Q_0^{2,0}(p,r)$ is related to $Q_0^{0,2}(r,p)$. See (4.40b) and (4.40c). If this assumption is removed, the above relationship fails; this is the case for electron impact.

The cross-section $Q_0^{2,0}(p,p)$ is also a correction to the depopulation process, which is expressed by $Q_0^{0,0}(p,p)$. The reason of this correction is almost the same as the above.

### 6.1.3 Alignment Creation by "Elastic" Scattering, $Q_0^{0,2}(p,p)$

In (4.41b) the first term on the right-hand side represents the alignment creation by unequal depopulation rates among the magnetic sublevels. The second term is alignment creation by elastic collisions. If we assume a rectilinear path for collisions, this latter term vanishes, because of the principle of detailed balance [14, 15]. If this assumption is removed this principle no longer holds for the spatially anisotropic collisions, and this term may have a nonzero value [16]. An electron beam experiment confirmed that elastic scattering creates an alignment by a significant amount [17]. This problem is also examined quantum mechanically in Sect. 5.3, and the above conclusion is confirmed.

Kawakami and Fujimoto [18] perform a Monte Carlo calculation of electron impact on a rydberg hydrogen atom. The proton is fixed at the origin, and the electron motions are treated classically. The presence of the electron spin is ignored, so that the interaction is of pure Coulomb forces. The atomic electron is initially in one of the elliptic orbits corresponding to a specific $(n, l, m_l)$ quantum numbers ($n$, the principal quantum number; $l$, the orbital angular momentum quantum number; $m_l$, the magnetic quantum number); the classical orbits are quantized according to the quantization scheme that is introduced in p. 15 of [1]. The incident electron comes from the $-z$ direction, traveling in parallel with the $z$-axis with an impact parameter, an azimuthal angle, and the timing with respect to the phase of the orbit motion of the atomic electron; all these parameters are chosen from random numbers. The energy of the incident electron is 0.2–100 eV. The temporal development of the motions of the two electrons is followed, and the final state is registered according to the quantum numbers in the quantization scheme. In this way, they obtain the $(n, l, m_l)$-resolved cross-sections for deexcitation, elastic scattering, and excitation (see Fig. 6.7). They obtain these cross-sections for the direct process and the electron-exchange process. When, in the final state, both of the electrons have positive energies, this process is ionization. When possible, the results are compared with the cross-sections available from various sources, and their results are confirmed to be consistent with them. Figure 6.8 is an example of the results for the initial state of $(n, l, m_l) = (10, 6, 3)$ for the atomic electron and the incident electron energy of 0.54 eV. The magnitude of the cross-section is expressed as the volume of the sphere located at the $(n, l, m_l)$ point of the final state. This figure shows the cross-sections of deexcitation $(n = 10) \rightarrow (n = 9)$, elastic scattering $10 \rightarrow 10$, and excitation $10 \rightarrow 11$. cross-sections for other inelastic processes, including ionization, are much smaller. It is seen that excitation is more probable than deexcitation; this is consistent with the well established tendency, as seen for all the lev-

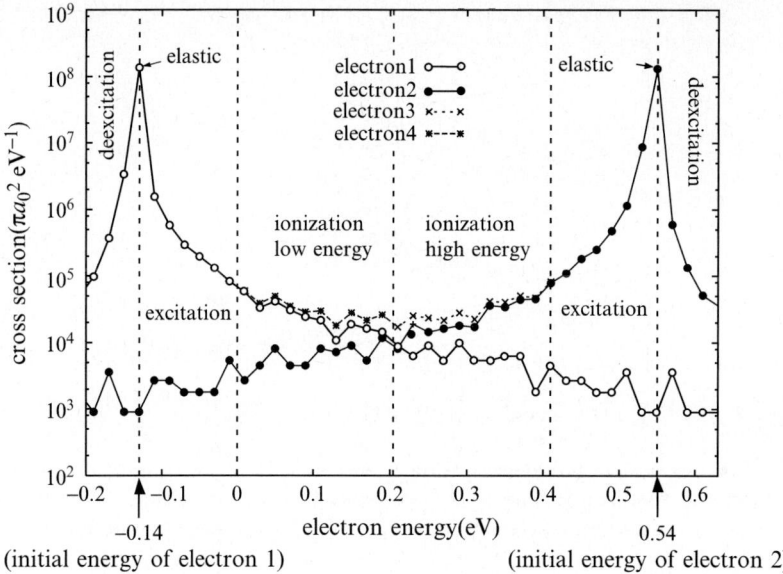

**Fig. 6.7.** "Cross-section" per unit energy interval (eV) for the initial state $(n, l, m_l) = (10, 6, 3)$ of hydrogen and $0.54\,\text{eV}$ monoenergetic electrons traveling in the $z$ direction. "electron 1" is initially the incident electron, "electron 2" is initially the atomic electron. "electron 3" and "electron 4" are the two outgoing electrons in the case of ionization event, with the higher and lower energies, respectively. The peak at $0.54\,\text{eV}$ for electron 2 corresponds to the peak at $-0.14\,\text{eV}$ for electron 2, and they indicate elastic scattering. Deexcitation and excitation are obvious. The curves for electron 1 and 2 in the energy region between 0 and $0.41\,\text{eV}$ show the "ionization cross-section." (Quoted from [18])

els in Fig. 3.7c. Another prominent feature is that elastic collisions are much more frequent than the inelastic collisions. This is also seen in Fig. 6.7 that the elastic peaks are much larger than the "cross-sections" in the inelastic regions. It is noted that these latter inelastic processes were the dominant collision processes in determining the populations (see Figs. 3.7 and 3.9).

Among the elastic collisions, cross-sections for transitions to the neighboring $(l, m_l)$ states are overwhelmingly large. This is interpreted as, in distant collisions, the orbit of the initial state is only slightly modified, with little energy and momentum transferred between the incident electron and the atomic electron. With the arguments in the last paragraph in mind, we look at Fig. 6.8. Then, we may argue as follows. Suppose that, in a plasma, the level of interest lies above Griem's boundary [1]. In Fig. 3.8 radiative transitions can be neglected in comparison with the collisional transitions. From the arguments above, we may conclude that, among the collisional transitions, elastic collisions are predominant. This situation would be independent of whether the plasma is ionizing or recombining. Thus, it should be a good approximation

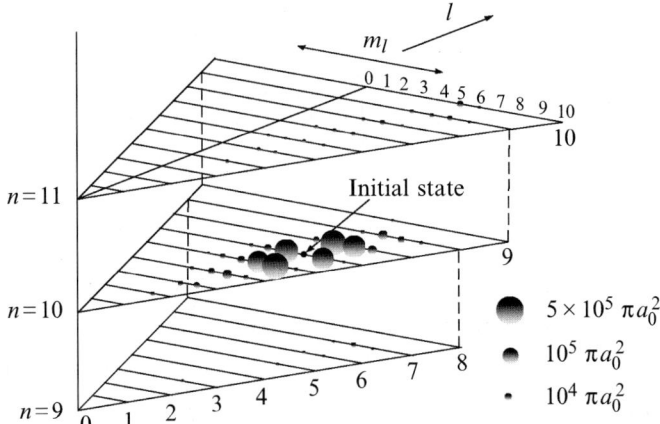

**Fig. 6.8.** Excitation, deexcitation, and elastic scattering cross-sections corresponding to Fig. 6.7. The electrons travel in the $z$ direction. Only the dominant cross-sections with the final states $9 \leqslant n \leqslant 11$ are shown. The magnitude of the cross-section is expressed by the volume of the sphere placed on the final state position. (Quoted from [18])

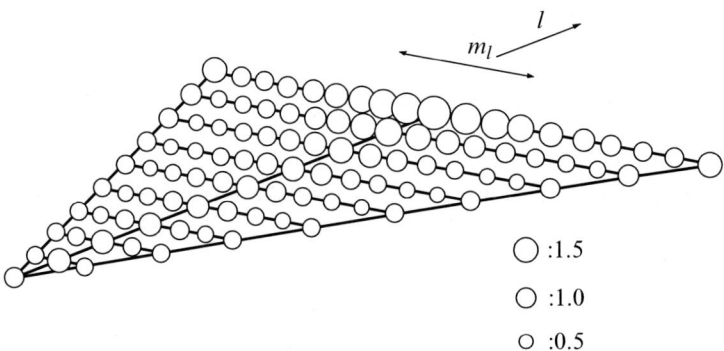

**Fig. 6.9.** Population distribution among the $n = 10$ states determined by elastic scattering. A monoenergetic (0.54 eV) electron beam traveling in the $z$ direction is incident on the atoms. (Quoted from [18])

to calculate the $(l, m_l)$ distribution within an $n$ plane only with the elastic collision cross-sections taken into account and with the inelastic collisions neglected. Figure 6.9 is the result of this approximation. The population of an $(l, m_l)$ state is expressed with the area of the circle on the $(l, m_l)$ point. We remember that the incident electrons are beam-like in the $z$ direction and the energy is 0.54 eV, rather low energy. This figure indicates that, for small $l$ states, e.g., p, d or f states, the population of the $m_l = 0$ state is about twice the populations of other $m_l$ states. A negative alignment is created.

As (4.41b) shows, ionization contributes to the alignment creation if different magnetic sublevels have different ionization cross-sections. Equations

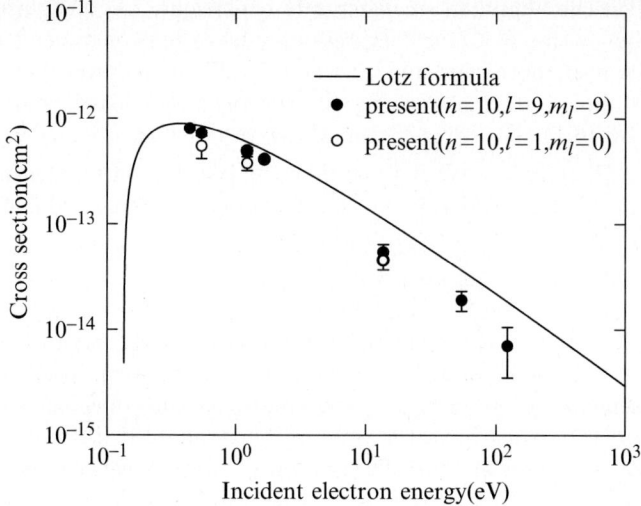

**Fig. 6.10.** The ionization cross-section of atoms with $(n, l, m_l)$ of (10, 9, 9) and (10, 1, 0). (Quoted from [18])

(4.41c) and (4.41d) (or (4.22d)) also include similar terms. However, (4.41c) is a correction term to the depopulation, (4.41a). In (4.41d), the depopulation term would be rather minor in many practical situations, as suggested later (see (4.22d)). Therefore, in this subsection we examine the ionization cross-sections of different magnetic sublevel atoms mainly in the context of (4.41b). In the Monte Carlo simulation, in Fig. 6.7, the curves in the energy range between 0 and 0.41 eV show the "ionization cross-sections." The sum of the two curves for "electron 1" and "electron 2" integrated over the energy range is the total ionization cross-section in the conventional sense. Figure 6.10 shows examples of the ionization cross-sections for two extreme cases, $l = 9, m_l = 9$ and $l = 1, m_l = 0$; the former is the circular orbits on the plane perpendicular to the direction of the incident electron, and the latter is the thin elliptic orbits with the angular momentum directed perpendicularly to the incident electron direction. It is seen that the cross-section for the latter is smaller than the former. But, the difference is quite small. Thus, we may expect that a difference in the ionization cross-sections among the magnetic sublevels is rather insignificant.

### 6.1.4 Alignment Transfer, $Q_q^{2,2}(r, p)$, and Alignment Destruction, $Q_q^{2,2}(p, p)$

In the density matrix formalism in Chap. 4, these cross sections involve $k$ and $k'$ equal to 2, so that from (4.28), $K$ can be 0, 2, or 4, and $q'$ can be 0, 1, or 2. The resulting rate coefficient, (4.42d) is thus complicated. $C^{22}(r, p)$

will be called the alignment transfer rate coefficient. The processes that the cross sections with $q = 1$ and $2$ express may be called coherence transfer. See (4.20e). See also the discussion in Sect. 5.2. When we return to our mono-energetic beam excitation experiment, the expansion coefficients of $f(v, \theta)$ are given by (6.1). By substituting these values into (4.42d) we find that $C^{2,2}(r, p) = Q_0^{2,2}(r, p)v_0$. We here assume the presence of an alignment in the initial level $r$. Then, the alignment transfer process is expressed from the third line of (4.51) as

$$\frac{\mathrm{d}}{\mathrm{d}t}a(p) = Q_0^{2,2}(r, p)v_0 a(r)n_{\mathrm{e}}. \tag{6.6}$$

It is noted that the coherence transfer cross-sections $Q_q^{2,2}(r, p)$ with $q = 1$ and $2$ have canceled out in this case. This is due to our assumption that coherence is absent in our coordinate system and that the collisions are along the quantization axis. Now the latter assumption is removed. Then, the coherence transfer cross-sections appear. The reason of this appearance is for an incident perturber approaching the atom with a finite angle with respect to the quantization axis, the population imbalance among the magnetic sublevels of the target atom is seen as a coherence. See the discussion in the following subsection. In another extreme, if the EVDF is isotorpic, we have only the first line of (4.42d). The three cross-sections with $q = 0, 1$, and $2$ altogether contribute to alignment transfer with the equal weights. Thus, in general cases, the three cross-sections constitute the alignment transfer "cross-section," with weight factors depending on the angular distribution of EVDF. We call these cross-sections altogether the alignment transfer cross-section.

It is noted here that, as (4.19) indicates, all the cross-sections are defined in the collision frame, or for collisions with the incident perturbers traveling along the quantization axis. So the alignment transfer process appears in our formulation. We may want to choose an alternative approach: in this formulation we deal only with the magnetic sublevel populations, disregarding coherence, since we assume the situation of axial symmetry and the absence of coherence in the density matrix in our formulation. If we adopt this approach what we have to do is we calculate cross-sections for collisions of perturbers with various incident angles, and average these cross-sections over the incident angles according to the actual distribution of the perturbers, or EVDF, of the plasma. These angular averaged cross-sections, or the alignment transfer cross-sections, are concerned only with the magnetic sublevels, and coherence would not appear explicitly. This approach is equivalent to the present one, but this approach is rather inconvenient, since we have to calculate the averaged cross-section according to this particular EVDF. Therefore, the present formulation including the coherence is more convenient or even universal.

The situation is almost exactly the same for the alignment destruction "cross-section" $Q_q^{2,2}(p, p)$.

We review below the cross-sections for alignment transfer and alignment destruction.

## A. Alignment Transfer

No experimental or theoretical data is found in the literature.

## B. Alignment Destruction

The quantity $C^{2,2}(p,p)$ is called the rate coefficient for alignment destruction (see (4.51) and (4.53)). This process may be understood as *disappearance of the atoms* in level $p$ that still hold the memory of their excitation to this level. In other words, as will be understood from the discussions below, this process is equivalent to the collisional relaxation of coherence of the ensemble of atoms in level $p$. Atomic coherence has been an important subject in the area of quantum electronics. In atomic physics, this process appears in, say, the level-crossing phenomena. The Hanle effect is an example. Since the half-width of the Hanle signal represents the coherence time of the excited atoms, the slope of the half-width against the number density of the ground-state atoms gives the rate coefficient for coherence relaxation. See Appendices C and D. In this way, the alignment destruction rate coefficient has been determined for excited levels of rare gas atoms and some metal atoms due to isotropic atom collisions. Substantial amount of data were produced by the self-alignment method as will be discussed in Sect. 7.1. As discussed in Sect. 6.3, in contrast to the self alignment detected by the Hanle-effect method, the pulsed-laser-excitation experiment, or the laser-induced-fluorescence spectroscopy (LIFS) experiment, gives more detailed and reliable information.

Equation (4.22d) indicates that the alignment destruction cross-section has two components: those for inelastic and elastic collisions. The former is depopulation. Note that $\sum_{(M)} \langle JJM - M|20\rangle^2 = 1$. The latter process may be called *disalignment*. This latter process is brought about by the "population" mixing among the magnetic sublevels and may be understood as *disappearance of the memory* of excitation in the atoms. Thus, the alignment destruction rate is the sum of the depopulation rate and the disalignment rate; see (C.45) in Appendix C. In the past, confusion sometimes occurred between the alignment destruction and the disalignment. For instance, an alignment destruction rate obtained by the Hanle-effect method was interpreted as a disalignment rate. This is due, at least partly, to the usage of terms like the alignment relaxation.

We now discuss the relationship between the collisional disalignment of atoms in a level, say $p$, and the collision broadening of a spectral line originating from this level. For the purpose of illustration, we consider the level having $J = 1$. We suppose that the atomic system is anisotropically excited, and in the initial state only the magnetic sublevel $M = 0$ is populated, so that a large negative alignment is created along with the population. See Fig. 4.1. Among the nine matrix elements of the density matrix $\rho_{MN}$, only $\rho_{00}$ is nonvanishing and all other elements are $\rho_{MN} = 0$. We assume $\rho_{00} = 1$. This situation is equal to that of (C.23). A disalignment process, which may be due to electron impact, is expressed as

$$
\begin{pmatrix} 0 & 0 & 0 \\ 0 & 1 & 0 \\ 0 & 0 & 0 \end{pmatrix} \xrightarrow{\text{disalignment}} \begin{pmatrix} 1/3 & 0 & 0 \\ 0 & 1/3 & 0 \\ 0 & 0 & 1/3 \end{pmatrix},
\tag{6.7}
$$

that is, the population imbalance is relaxed and all the magnetic sublevels become evenly populated. Here we assume that our density matrix is normalized. Disalignment is thus the "population" transfer among the magnetic sublevels as shown by the second term of the right-hand side of (4.22d). As shown in Sect. C.5 (Appendix C), the present disalignment rate is the decay rate of $\rho_0^2(p)$ in Table C.1, and other components of the alignment, $\rho_1^2(p)$ and $\rho_2^2(p)$, if they were present, should decay with the same rate. Now we tilt the above system by the magic angle, $54.7°$. Then the initial state is expressed from (4.27) by the rotation matrix $R$ with Euler angles $(\phi, \theta, \gamma) = (45°, 54.7°, 0°)$ to be $R\rho R^{-1}$, where $\rho$ is given by the initial state of (6.7). See Appendices B and C. The initial density matrix is different, and the process of disalignment is now expressed as

$$
\begin{pmatrix} \frac{1}{3} & -\frac{1-i}{3\sqrt{2}} & \frac{i}{3} \\ -\frac{1+i}{3\sqrt{2}} & \frac{1}{3} & \frac{1-i}{3\sqrt{2}} \\ -\frac{i}{3} & \frac{1+i}{3\sqrt{2}} & \frac{1}{3} \end{pmatrix} \xrightarrow{\text{disalignment}} \begin{pmatrix} 1/3 & 0 & 0 \\ 0 & 1/3 & 0 \\ 0 & 0 & 1/3 \end{pmatrix}.
\tag{6.8}
$$

In this case, the magnetic sublevels were already equally populated in the initial state. Instead, the off-diagonal elements appeared. It should be noted that the initial states of (6.7) and (6.8) are equivalent; in the former case, the anisitropic excitation creates the population imbalance among the magnetic sublevels, leading to $\rho_0^2(p)$. While in the latter case, it creates the off-diagonal elements, or the coherence, and therefore $\rho_0^2(p)$ disappears. Instead $\rho_1^2(p)$ and $\rho_2^2(p)$ appears. These three component of the alignment decay with the same rate. See Sect. C.5 in Appendix C.

The off-diagonal elements of the density matrix, or the coherence, represent the phase correlation between the wavefunctions of the magnetic sublevels in *an atom*; we may call this correlation *the Zeeman coherence*. Thus, disalignment is, at the same time, relaxation of the Zeeman coherence.

Now we turn to collision broadening of a spectral line. We consider the transition line $p \to s$. We assume, first, that the atomic states are stationary, i.e., the atoms in these states do not decay. The time developments of the upper- and lower-level wavefunctions are given by $\exp[-iE(p)t/\hbar]$ and $\exp[-iE(s)t/\hbar]$, respectively. The photon of this transition has frequency $\omega = [E(p) - E(s)]/\hbar$. The phase correlation between these levels is called *the optical coherence*. Since the optical coherence is perfectly conserved on the present assumption, the transition line is strictly monochromatic. In reality, however, the atomic state(s) has a finite lifetime, and the time development(s) includes the decay term. Thus, the optical coherence decays with time, and the line has a natural line width. Now, atoms are subjected to atom or electron collisions, and the optical coherence is further relaxed by the dephasing collisions. In other words, the level energies are further broadened, so that the transition line is broadened, too. Thus, the line broadening is understood as the collisional relaxation of *the optical coherence*. Dephasing may take place in the upper level or in the lower level or even in both of them. In many cases, however, the upper level is more likely to be affected by collisions, and line broadening is determined mainly by the relaxation of the phase of the upper level atoms.

We may thus expect that disalignment of level $p$ by electron impact has some correlation with Stark broadening of the transition line $s - p$. More specifically, in the case of electron scattering the disalignment rate coefficient may have some similarity with the Stark broadening half-width.

Hirabayashi et al. [19] produce an afterglow of a neon–helium mixture plasma in a thin discharge cell. The electron density and temperature are determined from the emission intensities of series lines of neutral helium to be $5.1 \times 10^{19}\,\mathrm{m}^{-3}$ and $2 \times 10^3\,\mathrm{K}$, respectively. Some of the metastable neon ($2p^5 3s$ configuration) atoms in $1s_3$ ($J_r = 0$: Paschen notation) are excited by a polarized laser pulse to one of the $2p^5 3p$ configuration levels, $2p_2$ ($J_p = 1$), and the subsequent fluorescence $1s_2$ ($J_s = 1$) – $2p_2$ is observed. See Figs. 6.19b and 7.1a later. The laser light is $\pi$ polarized, so that only the $M = 0$ magnetic sublevel is populated, i.e., the initial state of (6.7) is created. In a ($J = 1$)−($J = 1$) transition the $M = 0$ magnetic sublevel emits only the $\sigma$ light. See Figs. 6.19c and 7.1b later. Figure 6.11a shows the temporal developments of the observed intensities of the both polarized components. The $\sigma$ component is intense, but the $\pi$ component appears. This is due to the "population" transfer among the magnetic sublevels, (6.7). As noted earlier, i.e., (4.6) and (4.7), the upper-level population is given from $I_0 = (2/3)(I_\pi + 2I_\sigma)$, and the longitudinal alignment is given as $A_L = (I_\pi - I_\sigma)/(I_\pi + 2I_\sigma)$. As (4.8) shows the latter quantity is proportional to $a(p)/n(p)$, relative alignment. Figure 6.11b shows these quantities, where the sign of the longitudinal alignment has been reversed. The longitudinal alignment starts from $-(1/2)$. From the fitted lines, the depopulation rate and the disalignment rate are determined, as shown in Tables 6.1 and 6.2.

Depopulation rate is composed of four rates: depopulation due to spontaneous emission, atom collisions, electron collisions, and radiation reabsorption. The radiative decay rate is given by the lifetime as $(5.41 \pm 0.06) \times 10^7\,\mathrm{s}^{-1}$, which has been determined by a separate experiment. The depopulation rate coefficients by atom collisions were also determined to be $(0.2 \pm 0.1) \times 10^{-17}\,\mathrm{m}^3\,\mathrm{s}^{-1}$ for neon collisions and $(1.8 \pm 0.1) \times 10^{-17}\,\mathrm{m}^3\,\mathrm{s}^{-1}$ for helium collisions. The depopulation rate under the present experimental condition is calculated to be $(0.12 \pm 0.005) \times 10^7\,\mathrm{s}^{-1}$. The depopulation rate by electron collisions is estimated for the $2p_2 \rightarrow 1s_j$ ($j = 2 - 5$) optically allowed transitions to be $(2.5 \pm 0.1) \times 10^7\,\mathrm{s}^{-1}$ on the basis of Drawin's semiempirical formula. The difference between the observed depopulation rate and the sum

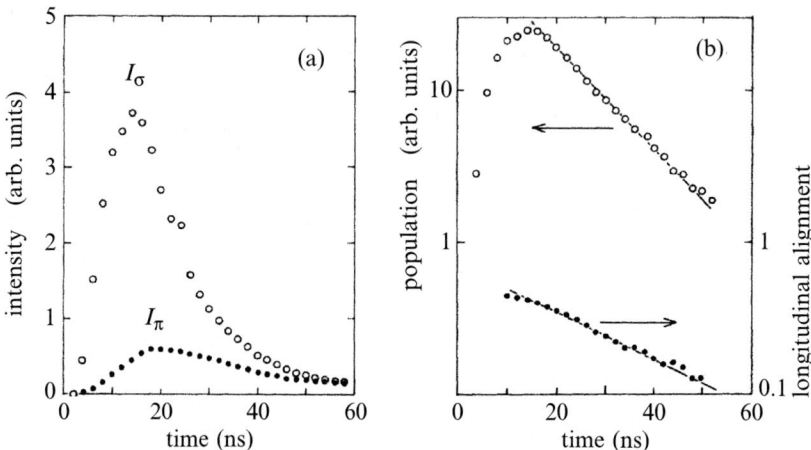

**Fig. 6.11.** (a) Time development of the intensities of the polarized components of the fluorescence Ne I ($1s_2 - 2p_2$) following pulsed excitation of ($1s_3 - 2p_2$) with the $\pi$ light. (b) Population and longitudinal alignment deduced from (a). (Quoted from [19], with permission from The American Physical Society.)

**Table 6.1.** Depopulation rate ($10^7 \, \mathrm{s}^{-1}$)

| Observed | $7.6 \pm 0.3$ |
| --- | --- |
| Spontaneous emission | $5.4 \pm 0.1$ |
| Atom collisions | $0.1 \pm 0.0$ |
| Electron collisions | $2.5 \pm 0.1$ |
| Radiation reabsorption | $-0.4 \pm 0.4$ |

**Table 6.2.** Disalignment rate ($10^7 \, \mathrm{s}^{-1}$)

| Observed | $3.7 \pm 0.5$ |
| --- | --- |
| Atomic collisions | $1.2 \pm 0.1$ |
| Ion collisions | $0.1$ |
| Trapped radiation | $0.3 \pm 0.3$ |
| *Electron collisions* | *$2.1 \pm 0.5$* |

of the above three rates, $(-0.4 \pm 0.4) \times 10^7 \, \mathrm{s}^{-1}$, is ascribed to the effect of radiation reabsorption.

The disalignment rate coefficients by atom collisions were determined in separate experiments to be $(17.0 \pm 0.3) \times 10^{-17} \, \mathrm{m^3 \, s^{-1}}$ for neon collisions and $(19.1 \pm 0.6) \times 10^{-17} \, \mathrm{m^3 \, s^{-1}}$ for helium collisions. The disalignment rate under the present condition is calculated to be $(1.22 \pm 0.05) \times 10^7 \, \mathrm{s}^{-1}$. Disalignment by ion collisions is considered in the following subsection, and its rate for the present condition is estimated to be $0.11 \times 10^7\,\mathrm{s}^{-1}$. Disalignment rate by radiation reabsorption is estimated to be $(0.3 \pm 0.3) \times 10^7\,\mathrm{s}^{-1}$. This is on the assumption that radiation reabsorption takes place on the $1s_5$ ($J{=}2$) – $2p_2$ ($J{=}1$) and $1s_3$ ($J{=}0$) – $2p_2$ ($J{=}1$) transition lines by the same degree. The reasoning for this assumption will be given in Sect. 7.2. The difference between the observed disalignment rate and the sum of the above three rates $(2.1 \pm 0.5) \times 10^7\,\mathrm{s}^{-1}$ is concluded to be the disalignment rate by electron collisions.

The disalignment rate coefficient as calculated from the above rate is $(4.1 \pm 1.0) \times 10^{-13}\,\mathrm{m^3\,s^{-1}}$. Now we compare this rate coefficient with the Stark broadening rate coefficient. Professor Griem gives an extensive list of Stark broadening of many atomic and ionic species. Unfortunately, however, Stark width of the present neon $1s_j$ – $2p_2$ transitions is not presented. We adopt for the width an averaged value over the calculated widths of some of the $1s_j$ – $2p_k$ transitions ($k = 1 - 10$); their variation is within 20%. The result is extrapolated to $T_e = 2 \times 10^3$ K. The Stark width thus obtained is $6.4 \times 10^{-6}$ nm. The corresponding rate coefficient is $5.5 \times 10^{-13}\,\mathrm{m^3\,s^{-1}}$. It is seen that this rate coefficient is rather close to the disalignment rate coefficient mentioned earlier. Thus, our expectation is supported.

Kazantsev et al. [20] applied the Hanle effect method to a self-aligned electron-beam-produced plasma, and determined the alignment destruction rate of the $4^1D_2$ state of neutral helium by charged particle collisions. The number densities of the two groups of electrons, the thermal and directional components, were determined from the probe measurement. The observed rates were converted to the disalignment rate coefficient. The result is $(1.5 \pm 1.0) \times 10^{-11}\,\mathrm{m^3\,s^{-1}}$. Electrons should be responsible for the observed alignment destruction under the plasma conditions. It is found that this value is about an order smaller than the Stark broadening rate coefficient calculated from (7.44) of [1].

Sakimoto [21] takes hydrogen atoms placed in a magnetic field, and calculates the excitation transfer cross-section between the magnetic sublevels induced by proton collisions. By using a peculiar symmetry property of hydrogen, stemming from the degeneracy of the different $l$ levels, he succeeds in reducing substantially the number of coupled levels. Figure 6.12 shows an example of the results. This figure shows the magnetic-field dependence of the excitation transfer cross-section for incident ions traveling along the magnetic field direction. He extends his calculation to isotropic collision cases. This process should be important for application of PPS to fusion plasmas.

## C. Disalignment by Ion Collisions

In many cases, the ion velocities in plasma are much smaller than electron velocities, and the electric field produced by the surrounding ions and felt by atoms or ions is well approximated as a quasi-static field. We assume here the electric field distribution at our atom is the Holtsmark field distribution

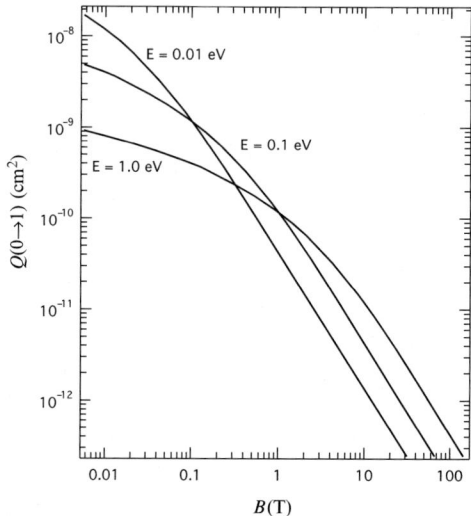

**Fig. 6.12.** Excitation transfer cross-section for $n = 3, M = 0 \rightarrow 1$ by proton collisions under a magnetic field. The parameter is the relative collision energy, and the incident direction is along the magnetic field. (Quoted from [21], with permission from IOP Publishing.)

(Fig. 7.1 with $\alpha = 0$ in [1]). We take the example of the $J = 1$ atoms, and calculate the time evolution of an ensemble of aligned atoms [19].

In an electric field, the equation of motion of the density matrix $\rho(t)$ is given as

$$\frac{\mathrm{d}}{\mathrm{d}t}\rho(t) = -\Gamma\rho(t) - \mathrm{i}\,[H_\mathrm{a}, \rho(t)] - \mathrm{i}\,[H_\mathrm{S}, \rho(t)]\,, \qquad (6.9)$$

where $\Gamma$ is the depopulation rate of the level, and $H_\mathrm{a}$ and $H_\mathrm{S}$ are the Hamiltonians of the isolated atom and the Stark effect, respectively. See Sect. C.2 in Appendix C. The matrix elements of $H_\mathrm{S}$ are given as

$$\langle M|H_\mathrm{S}|M'\rangle = -\frac{1}{4}\alpha_\mathrm{tens}E^2[M^2 - J(J+1)]\delta_{MM'} - \frac{1}{2}\alpha_\mathrm{sc}E^2\delta_{MM'}, \qquad (6.10)$$

where $E$ is the electric field strength, and $\alpha_\mathrm{tens}$ and $\alpha_\mathrm{sc}$ are the tensor and scalar polarizabilities, respectively. In the following we neglect $\Gamma$ and $H_\mathrm{a}$ for simplicity. The rate equation is solved as

$$\rho_{MM'}(t) = \rho_{MM'}(0)\,\exp[\mathrm{i}\omega(M^2 - M'^2)t] \qquad (6.11)$$

with

$$\omega = \frac{1}{4}\alpha_\mathrm{tens}E^2. \qquad (6.12)$$

In the coordinate system with the $z$-axis directed along the direction of the alignment we take the initial density matrix

$$\rho_z(0) = \begin{pmatrix} 0 & 0 & 0 \\ 0 & 1 & 0 \\ 0 & 0 & 0 \end{pmatrix}. \tag{6.13}$$

Remember (6.7). Suppose an electric field is applied in the $z'$ direction, which has the Euler angle $(\phi, \theta, 0)$ with respect to the original coordinate system. The initial density matrix in the $z'$ system is expressed by

$$\rho_{z'}(0) = R^{-1}(\phi, \theta, 0)\rho(0)R(\phi, \theta, 0). \tag{6.14}$$

The temporal evolution of the atomic system is obtained from substitution of (6.14) into (6.9). The density matrix of the atoms that have evolved in this static electric field is obtained as

$$\rho_z(t) = R(\phi, \theta, 0)\rho(t)R^{-1}(\phi, \theta, 0)$$

$$= \begin{pmatrix} p^2q^2(2 - \delta - \delta^*) & \begin{matrix} pq[2p^2(1 - \delta) \\ -q^2(1 - \delta^*)]\,e^{-i\phi} \end{matrix} & -p^2q^2(2 - \delta - \delta^*)\,e^{-2i\phi} \\ \begin{matrix} pq[2p^2(1 - \delta^*) \\ -q^2(1 - \delta)]\,e^{i\phi} \end{matrix} & 1 - 2p^2q^2(2 - \delta - \delta^*) & \begin{matrix} -pq[2p^2(1 - \delta^*) \\ -q^2(1 - \delta)]\,e^{-i\phi} \end{matrix} \\ p^2q^2(2 - \delta - \delta^*)\,e^{2i\phi} & \begin{matrix} -pq[2p^2(1 - \delta) \\ -q^2(1 - \delta^*)]\,e^{i\phi} \end{matrix} & p^2q^2(2 - \delta - \delta^*) \end{pmatrix}$$

$$\tag{6.15}$$

where

$$p = \frac{1}{\sqrt{2}}\sin\theta, \qquad q = \cos\theta, \qquad \delta = e^{i\omega t}.$$

The longitudinal alignment $A_L$ of an emission line of a transition from $J = 1$ to $J = 0$ is obtained from (6.15) as

$$A_L = 1 - 3p^2q^2(2 - \delta - \delta^*) = 1 - \frac{3}{4}\sin^2(2\theta)(1 - \cos\omega t). \tag{6.16}$$

In plasma, the directions of the electric field is random, and the second term of (6.16) is averaged over angle $\theta$

$$\overline{\sin^2 2\theta} = \frac{2}{\pi}\int_0^{2\pi}(\sin^2 2\theta)2\pi(\sin\theta)\frac{3}{2\pi}\,d\theta = \frac{16}{5\pi}.$$

Then (6.16) reduces to

$$A_L = 1 - \frac{12}{5\pi}(1 - \cos\omega t). \tag{6.17}$$

Since the electric field strength $E$ is distributed according to the Holtsmark distribution, $\omega$ is distributed.

We take the plasma discussed in the preceding subsection as an example here. The normal field strength is $F_0 = 5 \times 10^4\,\mathrm{Vm^{-1}}$, and $\alpha_{tens} =$

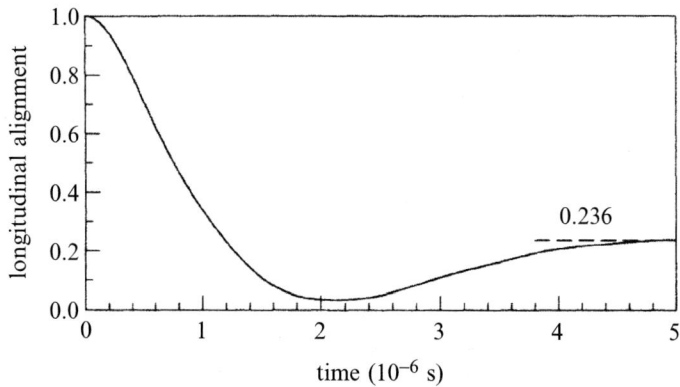

**Fig. 6.13.** Temporal development of the longitudinal alignment of the aligned 2p$_2$ atoms due to the Holtsmark field. The plasma condition is the same as for Fig. 6.11, $n_e = 5 \times 10^{19}$ m$^{-3}$. (Quoted from [19], with permission from The American Physical Society.)

$10^{-3}$ s$^{-1}$(Vm)$^{-2}$ for level 2p$_2$. Equation (6.17) is numerically averaged. The temporal development of the averaged $A_L$ is shown in Fig. 6.13. Starting from $t = 0$, the initial alignment decreases. If this decrease is represented by the $1/e$ decay time the effective rate is $1.1 \times 10^6$ s$^{-1}$. This is the value given in Table 6.2 as the disalignment rate by ion collisions. For sufficiently large $t$ the value of $A_L$ tends to $1 - 12/5\pi = 0.236$. The effective rate may be compared with the quantum beat frequency of $\omega_0 = 7 \times 10^5$ s$^{-1}$ for the present normal field strength $F_0$ and $\alpha_{\text{tens}}$.

## 6.2 Recombination

### 6.2.1 Radiative Recombination

In the context of PPS, radiative recombination is significant from the two aspects: the polarization of the recombination continuum radiation and the alignment creation in excited levels produced by recombination, leading to polarization of transition lines to still lower-lying levels.

### A. Polarization of Recombination Continuum

For recombination of fully stripped ion to hydrogen-like levels, there are several calculations. Cooper and Zare [22] give a succinct formula for recombination of beam electrons having velocity $\boldsymbol{v}$ and the observed polarization $\boldsymbol{e}$ (unit vector);

$$dQ_{nl}(\boldsymbol{e}, \boldsymbol{v}) = Q_{nl}(v)[1 + \beta_{nl}P_2(\boldsymbol{v} \cdot \boldsymbol{e}/v)]d\Omega/4\pi, \qquad (6.18)$$

where $Q_{nl}(v)$ is the total or ordinary recombination cross-section, $n$ and $l$ are the principal quantum number and the orbital quantum number of the level after recombination, respectively, $d\Omega$ is the solid angle subtended by the optics of observation and $P_2(x)$ is the second-order Legendre polynomial,

$$P_2(x) = \frac{1}{2}(3x^2 - 1). \tag{6.19}$$

In (6.18), $\beta_{nl}$ is called the asymmetry parameter. This parameter specifies the polarization characteristics of the recombination continuum. In general, $\beta_{nl}$ takes values between –1 and 2, and for s states ($l = 0$), $\beta_{ns}$ is 2. For $n \geqslant 2$, several recombination continua overlap, and the effective asymmetry parameter is defined. Table 6.3 gives examples of the asymmetry parameters [23].

Suppose we observe the recombination continuum in the same geometry as that for Fig. 1.2, i.e., from the direction perpendicular to the electron beam.

**Table 6.3.** Hydrogenic asymmetry parameters

| $\omega/\omega_2$ | $\beta_{2p}$ | $\beta_{n=2}$ |
|---|---|---|
| 1.00 | 0.48485 | 0.88889 |
| 1.02 | 0.49186 | 0.89977 |
| 1.04 | 0.49880 | 0.91047 |
| 1.06 | 0.50566 | 0.92101 |
| 1.08 | 0.51246 | 0.93138 |
| 1.10 | 0.51917 | 0.94160 |
| 1.15 | 0.53566 | 0.96646 |
| 1.20 | 0.55172 | 0.99038 |
| 1.25 | 0.56738 | 1.0134 |
| 1.30 | 0.58263 | 1.0356 |
| 1.40 | 0.61602 | 1.0777 |
| 1.50 | 0.64000 | 1.1169 |
| 1.75 | 0.70440 | 1.2039 |
| 2.00 | 0.76190 | 1.2778 |
| 2.50 | 0.86022 | 1.3960 |
| 2.90 | 0.92615 | 1.4696 |
| 3.00 | 0.94118 | 1.4857 |
| 3.10 | 0.92486 | 1.4863 |
| 3.50 | 0.86486 | 1.4909 |
| 4.00 | 0.80000 | 1.5000 |

$\hbar\omega$ is the photon energy and $\hbar\omega_2$ is the binding energy of the electron in level $n=2$. (Quoted from [23])

The longitudinal alignment is defined similarly to (4.7). It is straightforward to show that

$$A_{\mathrm{L}} = \beta/2, \tag{6.20}$$

and, for the case of "oblique" observation, the angular distribution of the intensity averaged over the polarization is

$$I(\theta) = 1 - \beta P_2(\cos\theta), \tag{6.21}$$

where $\theta$ is the angle between the beam and the observation direction.

For other ion species, calculations are scanty. Scofield [24] calculates polarization of the recombination continuum for several cases, including the relativistic effects. Figure 6.14 is an example: this is for sodium-like barium ions produced by recombination of neon-like barium. This figure shows the polarization as the ratio of the cross-sections for the $\sigma$ light and the $\pi$ light.

For an anisotropic EVDF in plasma, which is axially symmetric, the longitudinal alignment corresponding to (6.20) is given by

$$A_{\mathrm{L}} = \frac{\beta}{10} \frac{f_2(v)}{f_0(v)}. \tag{6.22}$$

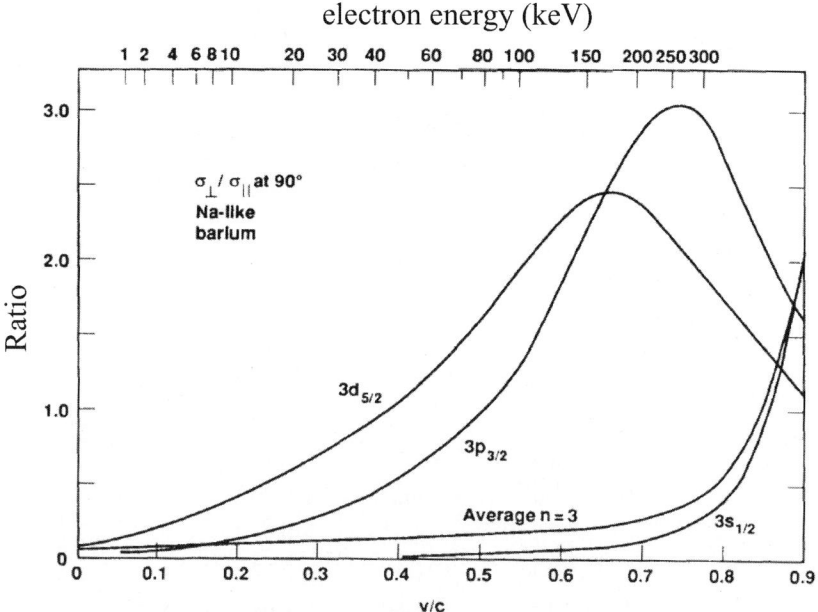

**Fig. 6.14.** Ratio of the radiative recombination cross-sections producing recombination continua polarized in the direction parallel and perpendicular to the electron beam. Sodium-like barium. (Quoted from [24], with permission from The American Physical Society.)

## B. Alignment Creation by Recombination

By recombination an alignment along with a population is created in level $p$ according to (4.45b) and (4.45a), respectively. Scofield [25] calculates these cross-sections for several cases with the relativistic effects included. Figure 6.15 is an example; this is for hydrogen-like and helium-like titanium ions, and the ordinate is $\sqrt{2J+1}Q_0^2/Q_0^0$. In this calculation cascading contributions from still higher-lying levels are included.

### 6.2.2 Dielectronic Recombination: Satellite Lines

Dielectronic recombination is characteristic of ions and is a resonant process through doubly excited levels;

$$A^{z+}(J_0) + e \rightarrow [A^{(z-1)+}(J)]^{**} \rightarrow [A^{(z-1)+}(J')]^* + h\nu, \qquad (6.23)$$

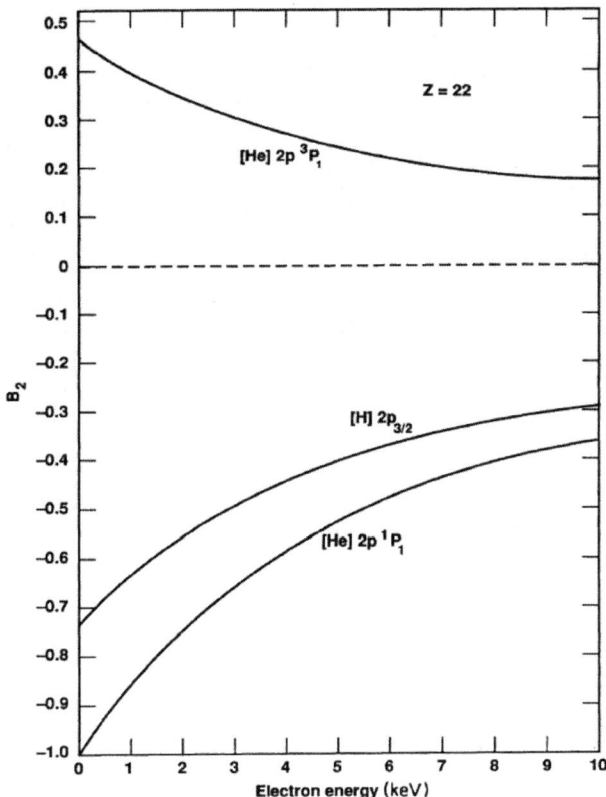

**Fig. 6.15.** The longitudinal alignment parameter $\rho_0^2/\rho_0^0$ of hydrogen- and helium-like levels of titanium produced by radiative recombination. (Quoted from [25], with permission from The American Physical Society.)

where the double asterisk means the doubly excited level and the single asterisk is a singly excited level. The emitted photon constitutes a satellite line of the $(z-1)+$ ion accompanying the parent line of the $z+$ ion. Since the doubly excited level has a definite energy position with a narrow width, the satellite line gives the information of EVDF at that particular energy.

Chen and Scofield [26] calculate polarization characteristics of helium-like satellite lines accompanying the resonance line of hydrogen-like ions. They find that, for observation from the direction perpendicular to the beam, polarization is expressed by

$$A_{\mathrm{L}} = -\beta', \tag{6.24}$$

and the intensity distribution is

$$I(\theta) = 1 + \beta' P_2(\cos\theta), \tag{6.25}$$

where the angular asymmetry parameter $\beta'$ is different from $\beta$ in Sect. 6.2.1, and takes values between $-1 \leqslant \beta' \leqslant 0.5$. They include the relativistic effects. Table 6.4 shows their results.

Inal and Dubau [27] investigate the case of $J_0 = 0$ and the electric dipole transition. No overlapping between the satellite lines is assumed. This situation is encountered in many experiments as lithium-like satellite lines accompanying the helium-like resonance line $1^1\mathrm{S} - 2^1\mathrm{P}$. The initial state of (6.23) is $1^1\mathrm{S}$. When observed from the direction perpendicular to the beam,

**Table 6.4.** Angular anisotropy parameter $\beta'$ for the helium-like satellite lines accompanying the hydrogen-like resonance line. (Quoted from [26])

| Transition | $F^{7+}$ | $Ti^{20+}$ | $Ni^{26+}$ | $Mo^{40+}$ | $U^{90+}$ |
|---|---|---|---|---|---|
| $1s2s\,^3S_1 - 2s2p\,^3P_1$ | $-0.249$ | $-0.153$ | $-0.029$ | $0.125$ | $-0.111$ |
| $1s2s\,^1S_0 - 2s2p\,^3P_1$ | $0.499$ | $0.306$ | $0.059$ | $-0.249$ | $0.221$ |
| $1s2s\,^3S_1 - 2s2p\,^3P_2$ | $-0.350$ | $-0.348$ | $-0.347$ | $-0.343$ | $-0.334$ |
| $1s2p\,^3P_0 - 2p^2\,^3P_1$ | $-0.328$ | $-0.303$ | $-0.296$ | $-0.271$ | $-0.159$ |
| $1s2p\,^3P_1 - 2p^2\,^3P_1$ | $0.164$ | $0.152$ | $0.148$ | $0.136$ | $0.080$ |
| $1s2p\,^3P_2 - 2p^2\,^3P_1$ | $-0.033$ | $-0.030$ | $-0.030$ | $-0.027$ | $-0.016$ |
| $1s2p\,^1P_1 - 2p^2\,^3P_1$ | $0.164$ | $0.152$ | $0.148$ | $0.136$ | $0.080$ |
| $1s2p\,^3P_1 - 2p^2\,^3P_2$ | $-0.500$ | $-0.500$ | $-0.500$ | $-0.500$ | $-0.499$ |
| $1s2p\,^3P_2 - 2p^2\,^3P_2$ | $0.500$ | $0.500$ | $0.500$ | $0.500$ | $0.499$ |
| $1s2p\,^3P_1 - 2p^2\,^1D_2$ | $-0.500$ | $-0.500$ | $-0.500$ | $-0.499$ | $-0.462$ |
| $1s2p\,^3P_2 - 2p^2\,^1D_2$ | $0.500$ | $0.500$ | $0.500$ | $0.499$ | $0.462$ |
| $1s2p\,^1P_1 - 2p^2\,^1D_2$ | $-0.500$ | $-0.500$ | $-0.500$ | $-0.499$ | $-0.462$ |
| $1s2s\,^3S_1 - 2s2p\,^1P_1$ | $0.500$ | $0.499$ | $0.498$ | $0.477$ | $0.338$ |
| $1s2s\,^1S_0 - 2s2p\,^1P_1$ | $-1.000$ | $-0.999$ | $-0.996$ | $-0.954$ | $-0.677$ |

the longitudinal alignment is given by the combination of angular momenta of the levels involved;

$$A_{\mathrm{L}} = (-)^{J'+1/2} \sqrt{\frac{15}{2}} (2J+1) \begin{pmatrix} 2 & J & J \\ 0 & 1/2 & -1/2 \end{pmatrix} \begin{Bmatrix} 2 & J & J \\ J' & 1 & 1 \end{Bmatrix}. \qquad (6.26)$$

Shlyaptseva et al. [28] extend the treatment to include the beryllium-like satellite lines accompanying the helium-like resonance line. The initial state is the lithium-like $1s^2 2S$. Figure 6.16 shows a theoretical spectrum for iron ions near the electron energy 4.695 keV. Line intensities of the polarized components parallel and perpendicular to the electron beam direction are shown.

For an anisotropic EVDF in plasma, it is straightforward to derive from (6.26) with (6.1) that

$$A_{\mathrm{L}} = (-)^{J'+1/2} \sqrt{\frac{3}{10}} (2J+1) \begin{pmatrix} 2 & J & J \\ 0 & 1/2 & -1/2 \end{pmatrix} \begin{Bmatrix} 2 & J & J \\ J' & 1 & 1 \end{Bmatrix} \frac{f_2(v)}{f_0(v)}. \qquad (6.27)$$

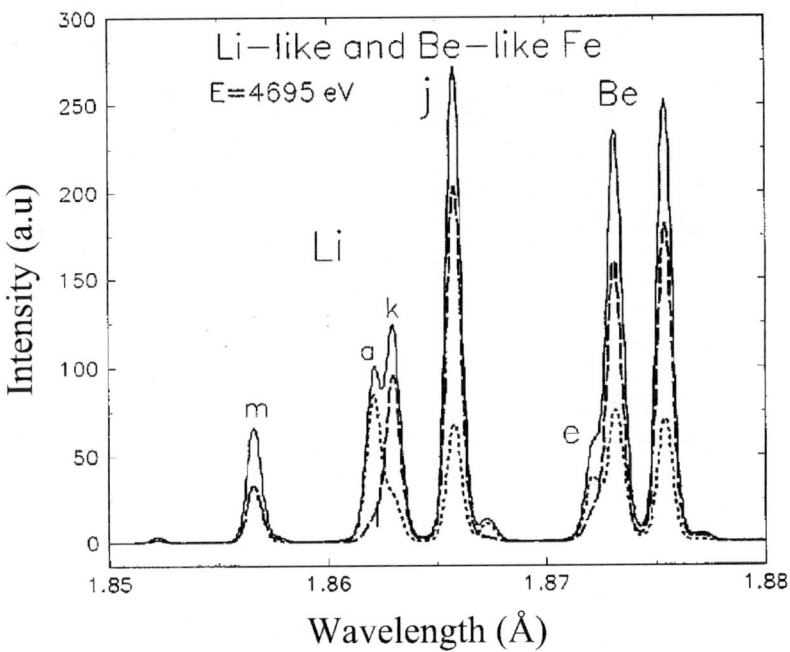

**Fig. 6.16.** Theoretical spectrum of the satellite lines due to dielectronic recombination of highly-ionized iron ions. Lines "m" through "e" are the lithium-like satellites, and the two peaks in the long wavelength are the beryllium-like satellites. *Dashed line*: parallel component and *dotted line*: perpendicular component. (Quoted from [28], with permission from The American Physical Society.)

In some cases, the doubly excited level can also be produced by inner-shell excitation;

$$A^{(z-1)+}(J'') + e \rightarrow [A^{(z-1)+}(J)]^{**} + e \rightarrow [A^{(z-1)+}(J')]^* + h\nu + e.$$

In this case the satellite line has a contribution from the inner-shell excitation, so that its polarization characteristics cannot be given by (6.24)–(6.27).

No calculation is found about the alignment created in the singly excited level $[A^{(z-1)+}(J')]^*$ in (6.23).

### 6.2.3 Ionization

Three-body recombination is the inverse process to ionization. The rate coefficient for three-body recombination for the isotropic part of EVDF is given by (4.48), while those for anisotropic part, $\alpha^{2,2}(p)$ and $\alpha^{2,0}(p)$, are not known yet. They should be related with anisotropy characteristics of the ionization process. We discuss here a remarkable feature of ionization.

The ionization process of atoms was treated by Wannier [29]; on the basis of classical mechanics, he derived several characteristics of the ionization process. Now many of them are theoretically shown correct qualitatively or even quantitatively, and confirmed by experiment. One of them is the correlation of the two outgoing electrons. They tend to repel each other and go in opposite directions in the case of collisions with low excess energy; near the ionization threshold the relative angle is close to 180°. When the excess energy is small and the two electrons share equally the excess energy, the angle should be exactly 180°.

In reality, the second condition is not satisfied in general. For example, in Fig. 6.7 the excess energy is 0.41 eV and in the energy region between 0 and 0.41 eV, the "cross-sections" (per unit energy) represent ionization. The sum of the two cross-sections integrated over this energy interval is the total ionization cross-section. It is seen that the cross-section has the minimum at the equal energies of 0.2 eV. The majority of the ionization events create two electrons with unequal energies.

Figure 6.17 shows the relative angle of the two outgoing electrons averaged over their energies. The open triangles correspond to Fig. 6.7. The relative angles close to 180° correspond to the equally shared energies in Fig. 6.7. In Fig. 6.17, if the two electrons did not have any angular correlations, the distribution should be flat. However, the curve is almost flat or slightly increasing only from 180° down to 110°, and it decreases sharply with the further decrease in the relative angle. There are almost no ionization events with relative angles smaller than 60°. Other examples show similar characteristics, some having more peaked distributions. Thus the ionization process is quite anisotropic. This is understandable from the above argument concerning the Wannier theory.

From the above observation it is concluded that, in three-body recombination, two electrons traveling in almost the same directions cannot recombine.

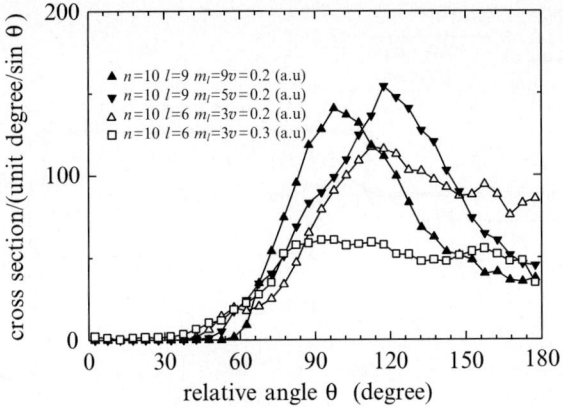

**Fig. 6.17.** Distribution of the relative angles of the two outgoing electrons in the events of ionization in the Monte Carlo simulation of electron impact on atomic hydrogen. See Sect. 6.1.3. (Quoted from [18])

This is especially true for low-energy electrons for which the three-body recombination "cross-section" is quite large. Therefore, electrons having anisotropic EVDF tend to recombine less. This is expressed by $\alpha^{2,0}(p)$. The produced atoms (ions) should also be aligned, which is expressed by $\alpha^{2,2}(p)$.

## 6.3 Alignment Relaxation by Atom Collisions

In a gas discharge plasma, an alignment created by, say, anisotropic electron impact excitation, may be relaxed by isotropic collisions with the gas atoms. As noted in Sect. 6.1.4, the term "alignment relaxation" has been accompanied by an ambiguity and should be defined carefully. As (4.22d) indicates the alignment destruction cross-section consists of the depopulation "cross-section" and the disalignment "cross-section." See also Appendix C. As discussed in Appendix D, the self alignment as observed by the Hanle effect method against atom density gives the alignment destruction rate coefficient. By this method as applied to self-aligned atoms (ions) in various plasmas, a substantial amount of data on the rate coefficients (cross-sections) is accumulated [30]. The validity of this method will be examined later in this section and in Chap. 7.

### 6.3.1 LIFS Experiment: Depopulation and Disalignment

Depopulation and disalignment can be independently examined by the pulsed laser-induced-fluorescence spectroscopy (LIFS) experiment. Figure 6.18 shows the block diagram of the experimental set-up (Nimura et al. [31]). A 5 ns pulse of dye laser light is polarized perpendicularly to the plane of this figure and

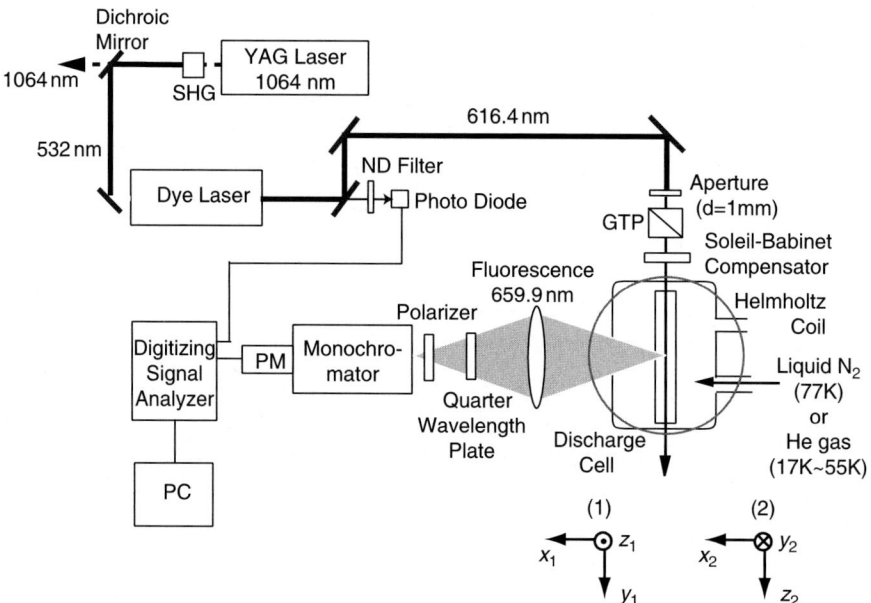

**Fig. 6.18.** The pulsed LIFS experiment. The pulsed (5 ns) laser light is polarized (linearly or circularly) and incident on the discharge plasma with variable temperature. The fluorescence light is observed by the spectrometer equipped with the linear polarizer (and a quarter-wave plate). (Quoted from [31], with permission from IOP Publishing.)

incident on the discharge plasma of a neon–helium mixture gas. With reference to the coordinate system (1), this excitation light is the $\pi$ light. Figure 6.19a shows the partial energy-level diagram of the neon $2s^53s - 2s^53p$ ($1s_j - 2p_k$: Paschen notation) transitions relevant to the experiment discussed below and that discussed in Sect. 6.1.4 [32]. The excitation–observation transitions are the same as those in Sect. 6.1.4: the laser pulse excites the $1s_3$ ($J = 0$) atoms to the $2p_2$ ($J = 1$) level. See Fig. 6.19b. By this excitation, only the ($M = 0$) magnetic sublevel is populated, so that a large negative alignment is created in the upper level. See Fig. 4.1. Fluorescence of the $1s_2(J = 1) - 2p_2(J = 1)$ transition (Fig. 6.19c) is observed from the $x$ direction with the spectrometer equipped with a linear polarizer in front of the entrance slit. Since the ($M = 0$) level atoms can emit only the $\sigma$ light, the intensity of this component ($I_\sigma$) is strong, as shown in Fig. 6.20a. However, the $\pi$ component ($I_\pi$) also appears, and the intensity difference decreases with time. This relaxation is due to the excitation transfer from the ($M = 0$) magnetic sublevel to the ($M = \pm 1$) sublevels (see Fig. 6.19c). This phenomenon is the disalignment.

The disalignment rate is three times the excitation transfer rate for the ($M = 0$) $\leftrightarrow$ ($M = \pm 1$) transition. This is understood as follows; First, we

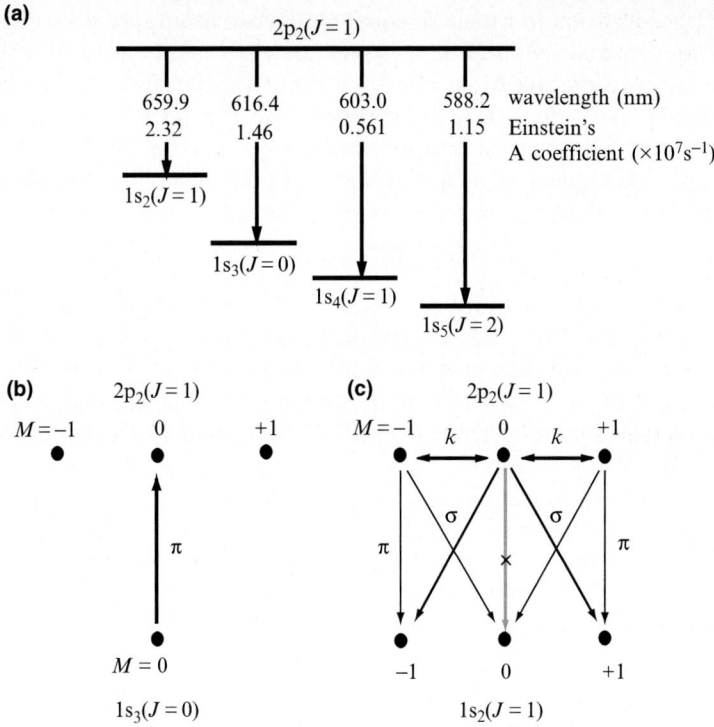

**Fig. 6.19.** (a) The partial energy level diagram of neon $2p_2$ and $1s_k$ connected by the electric-dipole transition lines with wavelength and the transition probability. (b) The $\pi$ light excitation from $1s_3$ to $2p_2$. (c) The fluorescence line $1s_2 - 2p_2$ with the polarized components resolved. $k$ denotes the excitation transfer rate between the $(M = 0)$ and $(M = \pm 1)$ magnetic sublevels. (Quoted from [32], with permission from IOP Publishing.)

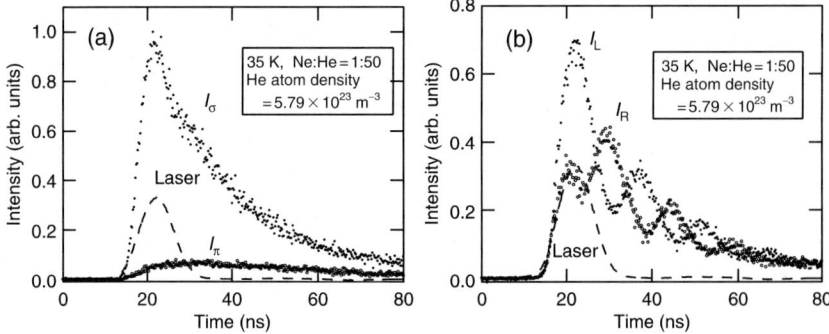

**Fig. 6.20.** (a) The intensities of the $\sigma$ and $\pi$ components of the fluorescence of $1s_2(J{=}1) - 2p_2(J{=}1)$ transition. (b) The intensities of the left- and right-circularly polarized components. (Quoted from [31], with permission from IOP Publishing.)

consider the situation in which an equal number of atoms are present in these three magnetic sublevels, and these atoms are subjected to isotropic collisions.

In a certain time duration, *one* excitation transfer event may take place from the $(M = 1)$ to $(M = 0)$ sublevel (see Fig. 6.19c). Then, from the symmetry property another transfer event should take place from $(M = -1)$ to $(M = 0)$. This means that, in the same time duration, in order to conserve the population balance among the magnetic sublevels, *two* excitation transfer events of $(M = 0)$ to $(M = 1)$ and $(M = 0)$ to $(M = -1)$ should take place. We now assume another situation in which only the $(M = 0)$ atoms are present. In the same time duration *two* excitation transfer events of $(M = 0)$ to $(M = 1)$ and $(M = 0)$ to $(M = -1)$ take place. For the purpose of simplicity, suppose that we are observing the fluorescence of the $(J = 0) - (J = 1)$ transition. See Fig. 7.1a later. Then, in this time duration $I_\pi$ decreases by two, while $I_\sigma$ increases by one. Thus, from the definition of (4.7), the longitudinal alignment decreases by three.

The sum of the intensities $(I_\pi + 2I_\sigma)$, which is proportional to the total population (4.6), and the longitudinal alignment $A_\mathrm{L} = (I_\pi - I_\sigma)/(I_\pi + 2I_\sigma)$, which is proportional to the relative alignment (4.7) and (4.8), are calculated from Fig. 6.20a. The results are shown in Fig. 6.21a, b, respectively. In the latter figure, the sign has been reversed. On the assumption of a single exponential decay for the both quantities, a least squares fitting is applied to yield the straight lines. The slope of these lines gives the depopulation rate and the disalignment rate, respectively. Figures 6.20 and 6.21 are for atom temperature of 35 K. These rates are plotted in Fig. 6.22a, b, respectively. A similar procedure is repeated for plasmas with varying helium atom densities. The depopulation rate, Fig. 6.22a, is the sum of the radiative decay rate (the inverse life time) and the population decay rate by atom collisions, predominantly by helium atoms in the present case. The slight increase with the helium atom density indicates the collisional depopulation rate. It is seen that the collisional disalignment, elastic collisions, is faster than the collisional depopulation, inelastic collisions. This means that the alignment destruction is predominantly by disalignment, with a small contribution from depopulation. Remember (4.22d).

As will be discussed in the following chapter, radiation reabsorption affects both the population decay and the disalignment; the population tends to decay slower and disalignment is brought about by radiation reabsorption. In the present case, the four lines in Fig. 6.19a suffer radiation reabsorption, although only by a small degree. By the self-absorption method (see [32]), the lower-level populations, $1s_2 - 1s_5$, are determined, and the apparent depopulation rate and disalignment rate are corrected for the effects of radiation reabsorption of the four lines. The procedure will be described in more detail in Sect. 7.2. The points in Fig. 6.22b are the results after this correction. The slope of the least squares fitted line gives the disalignment rate coefficient by helium atom collisions and plotted in Fig. 6.23 with the closed symbol.

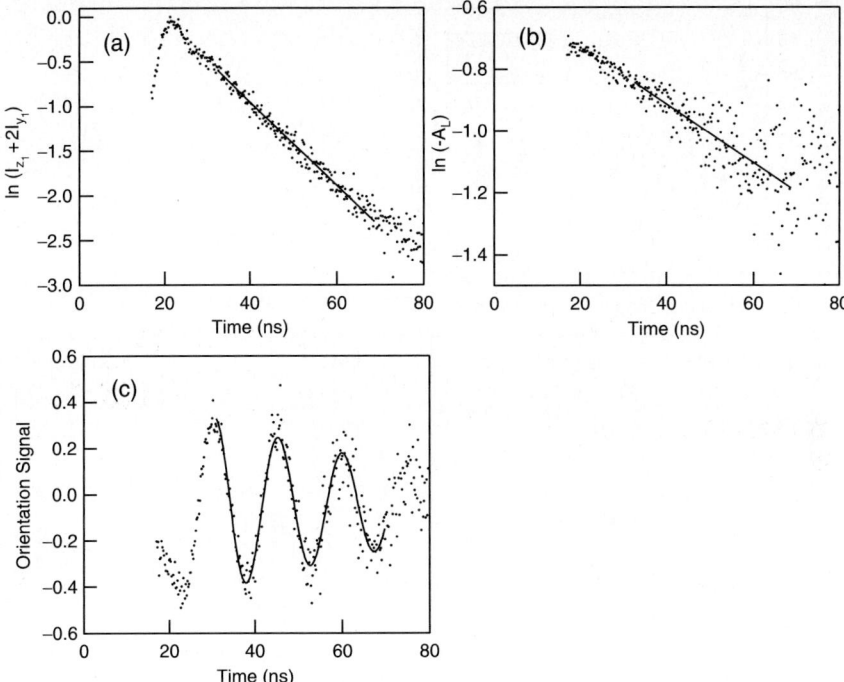

**Fig. 6.21.** (a) The "total intensity" $(I_\pi + 2I_\sigma)$ determined from Fig. 6.20a. (b) The longitudinal alignment $(I_\pi - I_\sigma)/(I_\pi + 2I_\sigma)$ determined from Fig. 6.20a. (c) The orientation signal $(I_R - I_L)/(I_\pi + 2I_\sigma)$ determined from Fig. 6.20a, and b. (Quoted from [31], with permission from IOP Publishing.)

The atom temperature is varied over 15–600 K. All the above procedures are followed at each temperature. The closed symbols in Fig. 6.23 show the temperature dependence of the disalignment rate coefficient. A rate coefficient $R(T)$ at temperature $T$ is given from the energy dependent cross-section $\sigma(E)$,

$$R(T) = \int_0^\infty \sigma(E) v f_T(E) \, \mathrm{d}E, \qquad (6.28)$$

where $v$ is the relative speed and $f_T(E)$ is the energy distribution function of the colliding atoms, respectively; $f_T(E)$ is assumed to be a Maxwell distribution and its definition is different from in Sect. 4.2. Note that the disalignment collisions are elastic process so that there is no threshold. Figure 6.24 shows $v f_T(E)$ for several temperatures; this picture shows that the rate coefficient at low temperatures reflects well the cross-section values at low energies. For example, the rate coefficient at 17 K is given mainly from the cross-section values in a couple of meV. Figure 6.23 shows that the disalignment rate coefficient is well fitted by a straight line at low temperatures. From the experimental

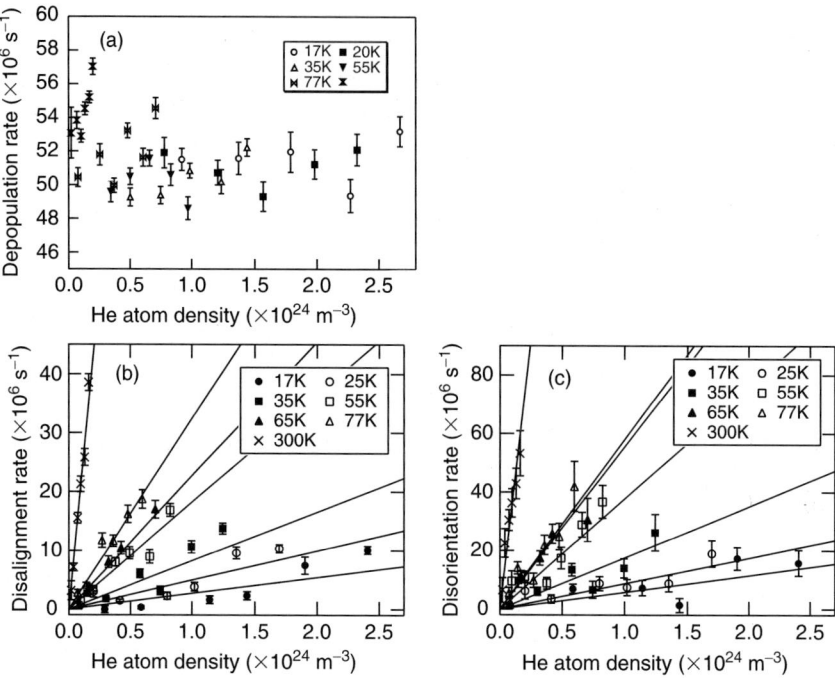

**Fig. 6.22.** (a) The depopulation rate against the helium atom density for various atom temperatures. (b) The disalignment rate. (c) The disorientation rate. (Quoted from [31], with permission from IOP Publishing.)

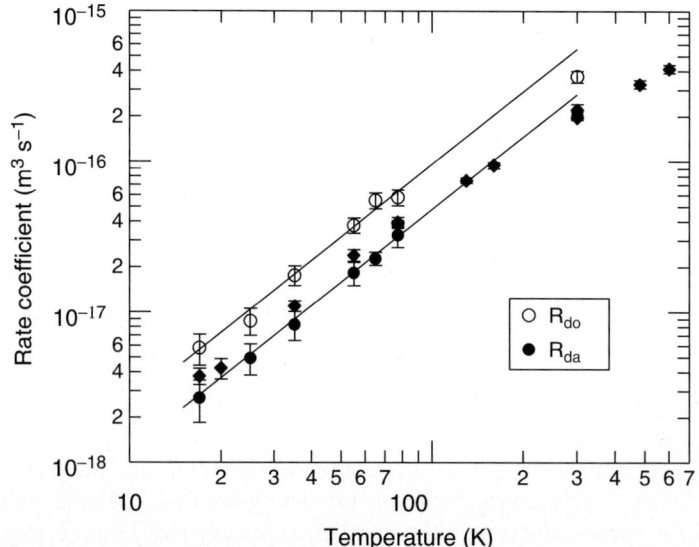

**Fig. 6.23.** The disalignment rate coefficient (*closed symbols*) of neon $2p_2$ atoms due to helium atom collisions against temperature. The open symbols are for disorientation. (Quoted from [31], with permission from IOP Publishing.)

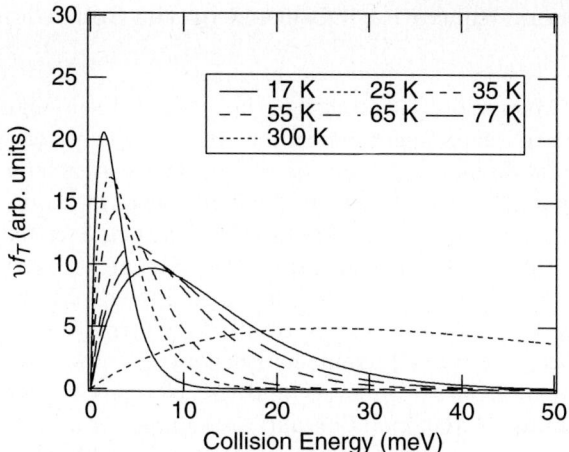

**Fig. 6.24.** The Maxwell distribution to be multiplied by the cross-section. (Quoted from [31], with permission from IOP Publishing.)

temperature dependence of $R(T) \propto T^{1.6}$, the energy dependence of the cross-section is estimated to be $\sigma(E) \propto E^{1.1}$; starting from virtually zero it increases almost linearly with energy up to 10 meV.

Figure 6.23 includes the result of another experiment: disorientation. In Fig. 6.18, the incident laser light is right-circularly polarized so that a "population" is produced only in the ($M = -1$) magnetic sublevel in the coordinate system (2). A magnetic field of 3.64 mT is applied in the direction perpendicular to the plane of the figure, and the produced atomic magnetic dipole precesses around the magnetic field, i.e., the Larmor precession. In front of the spectrometer, a quarter-wave plate and a linear polarizer are placed, and the left-circularly and right-circularly polarized light is observed separately. Figure 6.20b is the result. The oscillations with the opposite phase show the Larmor precession. The orientation signal is defined as $(I_R - I_L)/(I_\pi + 2I_\sigma)$, where the subscripts R and L indicate the right- and left-circularly polarized light. By a similar procedure to the determination of the disalignment rate coefficient, Figs. 6.21c and 6.22c, the disorientation rate coefficient is obtained against temperature. The open circles in Fig. 6.23 show the results. The disorientation rate is the sum of the excitation transfer rate of ($M = 0$) $\leftrightarrow$ ($M = \pm 1$) and two times the rate of ($M = \pm 1$) $\leftrightarrow$ ($M = \mp 1$). The experiment shows that, at low temperatures, the disorientation rate coefficient is about twice the disalignment rate coefficient. This means that the ($M = \pm 1$) $\leftrightarrow$ ($M = \mp 1$) rate coefficient is 2.5 times the ($M = 0$) $\leftrightarrow$ ($M = \pm 1$) rate coefficient. Classically speaking, the excitation transfer of ($M = 0$) $\leftrightarrow$ ($M = \pm 1$) corresponds to a 90° rotation of the atomic angular momentum after a collision, while the excitation transfer of ($M = \pm 1$) $\leftrightarrow$ ($M = \mp 1$) corresponds to a 180° rotation. The above result that the latter rotation is more frequent than the former appears rather puzzling.

## 6.3.2 Alignment Relaxation Observed by the Self-Alignment Method

Fujimoto and Matsumoto [33] compare the results of self-alignment method and a pulsed LIFS experiment concerning the alignment relaxation of neon atoms in a glow discharge plasma of a neon or neon–helium mixture gas. The former experiments are based on the Hanle effect method (Appendix D), and the latter is essentially the same as described in Sect. 6.3.1. The atom temperature is $300\,\mathrm{K}$. They first determine the alignment destruction rate coefficient by the self-alignment method for neon $2\mathrm{p}_2$ and $2\mathrm{p}_7$ levels. Table 6.5 shows their results together with the results by Carrington and Corney [34]. After the discovery of the self-alignment phenomenon, these authors made an extensive study on neon $2\mathrm{p}^5 3\mathrm{p}$ levels, and this is a part of their results. This table also contains the depopulation and disalignment rate coefficients as determined by the pulsed LIFS method and the resulting alignment destruction rate coefficient. The former rate coefficient is determined by the two methods: as described in Sect. 4.1 (4.6) the decay of $(I_\pi + 2I_\sigma)$ is determined and as in Appendix E the effective magic-angle excitation is implemented. As Fig. E.2 shows, both the methods give the identical result.

If the two kinds of experiment, i.e., the self-alignment method and the LFS method, give correct results for the alignment relaxation, the alignment destruction rate as given by the former method and the sum of the collisional depopulation rate and the disalignment rate as given by the latter method should agree with each other. We now compare these rate coefficients in Table 6.5. First, we compare the alignment destruction rate coefficients for neon $2\mathrm{p}_2$ as determined by Carrington and Corney (CC) and by Fujimoto and Matsumoto (FM), both by the self-alignment method. For helium collisions the results from the two experiments agree within the experimental

**Table 6.5.** Comparison of the rate coefficients for alignment destruction with the sum of those for disalignment and collisional depopulation ($10^{-16}\,\mathrm{m}^3\,\mathrm{s}^{-1}$)

| Level | Self-alignment method | Pulsed LIFS method | | |
|---|---|---|---|---|
| | Alignment destruction | Depopulation + | disalignment = | alignment destruction |
| Ne collision | | | | |
| $2\mathrm{p}_2$ | 1.90±0.12    2.22±0.08[a] | 0.02±0.01 | 1.70±0.03 | 1.72±0.03 |
| $2\mathrm{p}_7$ | –    4.18±0.17[a] | 0.12±0.01 | 3.08±0.05 | 3.20±0.05 |
| He collision | | | | |
| $2\mathrm{p}_2$ | 3.07±0.28    2.79±0.15[a] | 0.18±0.01 | 1.91±0.05 | 2.09±0.05 |
| $2\mathrm{p}_7$ | –    5.26±0.22[a] | 0.59±0.02 | 3.17±0.11 | 3.76±0.11 |

[a] Carrington and Corney (1971). (Quoted from [33], with permission from IOP Publishing.)

uncertainties. For neon collisions they differ slightly. We now compare these results with those of the LIFS method; the last column shows the alignment destruction rate coefficient. For neon collisions the latter result is close to the self alignment result of FM and not to CC, but for helium collisions the LIFS result differs substantially from the both self-alignment results. We also compare the results for the neon $2p_7$ level atoms. Both the results, the LIFS method and CC, differ substantially from each other.

The above fact is interpreted as follows; since the pulsed LIFS method is simple and straightforward in its principle and interpretation of the results, and its validity is tested on many occasions like the experiment described above and the determination of atomic lifetimes (Fujimoto [35]), we have little reason to doubt its result. On the other hand, the self-alignment method involves a serious ambiguity. The self-alignment in a positive column discharge plasma is ascribed to radiation reabsorption in the anisotropic geometry. As will be pointed out in Sect. 7.1.2, the implicit assumption that the rate of creation of alignment in the upper-level atoms is independent of the applied magnetic field is quite questionable. Thus, we are led to question the suitability of the self-alignment method for the determination of the alignment relaxation rate. The fact that the two self-alignment experiments give inconsistent results appears to support our suspicions; the disagreement could perhaps be ascribed to the difference in the geometry of the discharge tubes used in these experiments. (The tube of FM has inner diameter of 5 mm and length of 30 mm, while that of CC is quite long with diameter of 7 mm.)

So far we have been concerned with the alignment relaxation. The natural lifetime of the upper level is also determined by the self-alignment method. For example, Carrington et al. [36] gives the lifetime of $2p_2$ to be 17.6 ns, while the LIFS method gives 18.5 ns and a delayed coincidence measurement gives 18.8 ns (Bennett and Kindlmann [37]). The self-alignment experiment of FM gives 15.4 ns. This variation, especially between the two results obtained by the self-alignment method, strengthens our suspicions. We thus conclude that the self-alignment method based on the Hanle effect is not suitable for a quantitative investigation of atomic parameters.

# References

Excitation cross-sections by electron impact are provided by the data base **AMDIS** (National Institute for Fusion Science, Toki) http://dbshino.nifs.ac.jp/

1. T. Fujimoto: *Plasma Spectroscopy* (Oxford University Press, Oxford, 2004)
2. M. Goto: J. Quant. Spectrosc. Radiat. Transfer **76**, 331 (2003)
3. G. Csanak, D.C. Cartwright, S.A. Kazantsev, I. Bray: NIFS-PROC-37 (National Institute for Fusion Science, Toki, 1998) p.136, Los Alamos Atomic Physics Codes http://aphysics2.lanl.gov/tempweb/

4. I. Bray: (private communications, 2004), CCC Data Base `http://atom.murdoch.edu.au/CCC-WWW/index.html`
5. M.K. Inal, J. Dubau: J. Phys. B **20**, 4221 (1987)
6. J.R. Henderson et al.: Phys. Rev. Lett. **65**, 705 (1990)
7. P. Beiersdorfer: NIFS-PROC-37 (National Institute for Fusion Science, Toki, 1998) p.67
8. K.J. Reed, M.H. Chen: Phys. Rev. A **48**, 3644 (1993)
9. R.K. Janev, H. Winter: Phys. Rep. **117**, 265 (1985)
10. T. Fujimoto, S.A. Kazantsev: Plasma Phys. Control. Fusion **39**, 1267 (1997)
11. M.F.V. Lundsgaard, C.D. Lin: J. Phys. B **25**, L429 (1992)
12. M.R.C. McDowell, R.K. Janev: J. Phys. B **18**, L295 (1985)
13. H. Watanabe (Ph.D. thesis, Kyoto University, 1996)
14. A. Omont: Prog. Quantum Electron. **5**, 69 (1997)
15. A.G. Petrashen, V.N. Rebane, T.K. Rebane: Sov. Phys-JETP **60**, 84 (1984)
16. E.I. Dashyvskaya and E.E. Nikitin: Sov. J. Chem. Phys. **4**, 1934 (1987)
17. S. Trajmar, I. Kanik, M.A. Khakoo, L.R. LeClair, I. Bray, D. Fursa, G. Csanak: J. Phys. B **32**, 2801 (1999)
18. K. Kawakami, T. Fujimoto: UCRL-ID-146907 (University of California Lowrence Livermore National Laboratory, 2001) p.187
19. A. Hirabayashi, Y. Nambu, M. Hasuo, T. Fujimoto: Phys. Rev. A **37**, 83 (1988)
20. S. Kazantsev, V.V. Luchinkina, A.P. Mezentsev, V.N. Rebane, A.G. Rys, Yu.L. Stepanov: Opt. Spectrosc. **76**, 809 (1994)
21. K. Sakimoto: J. Phys. B **25**, 3641 (1992)
22. J. Cooper, R.N. Zare: in *Lectures in Theoretical Physics: Atomic Collision Processes* (S. Geltmann, K.T. Mahanthappa, W.E. Brittin, eds., Gordon and Breach, New York, 1968) p.317
23. H.M. Milchberg, J.C. Weisheit: Phys. Rev. A **26**, 1023 (1982)
24. J.H. Scofield: Phys. Rev. A **40**, 3054 (1989)
25. J.H. Scofield: Phys. Rev. A **44**, 139 (1991)
26. M.H. Chen, J.H. Scofield: Phys. Rev. A **52**, 2057 (1995); **58**, 5011 (1998)
27. M.K. Inal, J. Dubau: J. Phys. B **22**, 3329 (1989)
28. A.S. Schlyaptseva, R.C. Mancini, P. Neill, P. Beiersdorfer: Rev. Sci. Instrum. **68**, 1095 (1997)
29. G.H. Wannier: Phys. Rev. **90**, 817 (1953)
30. S.A. Kazantsev: Sov. Phys.-Usp. **26**, 328 (1983)
31. M. Nimura, M. Hasuo, T. Fujimoto: J. Phys. B **37**, 4647 (2004)
32. M. Seo, M. Nimura, M. Hasuo, T. Fujimoto: J. Phys. B **36**, 1869 (2003)
33. T. Fujimoto and S. Matsumoto: J. Phys. B **21**, L267 (1988)
34. C.G. Carrington, A. Corney: J. Phys. B **4**, 849 (1971)
35. T. Fujimoto, C. Goto, K. Fukuda: Phys. Scripta **26**, 443 (1982)
36. C.G. Carrington, A. Corney, A.V. Durrant: J. Phys. B **5**, 1001 (1972)
37. W.R. Bennet Jr., P.J. Kindlmann: Phys. Rev. **149**, 38 (1966)

# 7

# Radiation Reabsorption

T. Fujimoto

In Chap. 1, we introduced the phenomenon of self-alignment, which was spontaneous polarization of emission lines first discovered in laboratory discharge plasmas. The origin of the self-alignment in a positive column of a glow discharge was identified with anisotropic radiation reabsorption of the line emitted from the level of interest. It was also found that, even an isotropic radiation field could produce an alignment, a latent alignment. On the other hand, an alignment created by, say, anisotropic electron impact excitation may also be destroyed by radiation reabsorption. In this chapter we deal with these phenomena.

## 7.1 Alignment Creation by Radiation Reabsorption: Self-Alignment

### 7.1.1 Basic Principle

Let level $r$ be the lower level and $p$ the upper level, and these levels are connected by an electric-dipole transition. We consider a plasma having an anisotropic geometry; specifically we take an example of a sufficiently long cylinder. We encounter this situation frequently, e.g., a positive column of a glow discharge plasma. Let the central axis be the $z$-axis or the quantization axis. Suppose the plasma is optically thick to the transition line $r - p$. Sometimes, level $r$ is the ground state and the transition is the resonance transition. The optical thickness at the line center is substantially high. The radiation field of the emission line at this frequency is strong in the plasma, and atoms in level $r$ may absorb this radiation and be excited to level $p$. An atom near the axis tends to be illuminated more from the axial direction than from the radial direction. This is because the number of emitting atoms $p$ surrounding this atom in $r$ is larger in the axial direction than that in the radial direction. We consider an extreme case here that the atoms are illuminated only in the $z$ direction. Since the source of this radiation field is excited atoms in the plasma, this radiation is unpolarized. The beam of unpolarized radiation is expressed as an incoherent superposition of the left-circularly polarized light and the right-circularly polarized light.

$$e_{\pm} = \mp \frac{1}{\sqrt{2}} (e_x \pm ie_y) , \tag{7.1}$$

where $e_x$ and $e_y$ are the unit polarization vectors (the electric field) in the $x$ and $y$ directions, respectively, and the upper signs are for the left-circularly polarized light (positive helicity, having a positive angular momentum in the propagation direction) and the lower signs are for the right-circularly polarized light. We assume that the ensemble of the lower level atoms are unpolarized, i.e., it has no orientation nor alignment, so that its density matrix has only the population, $n(r) = \sqrt{2J_r + 1}\rho_0^0(r)$. According to Omont [1], the atoms are excited into the state of the upper level

$$\frac{d\rho(p)}{dt} = \frac{C_b}{2J_r + 1} \frac{1}{4\pi\varepsilon_0} \sum_{kq} \phi_q^k(e)(-)^{J_r+J_p+k+1}|d_{pr}|^2$$

$$\times \left\{ \begin{matrix} 1 & 1 & k \\ J_p & J_p & J_r \end{matrix} \right\} T_q^{(k)}(p)n(r) , \tag{7.2}$$

where $J_r$ and $J_p$ are the angular-momentum quantum numbers of the lower and upper levels, respectively, $\phi_q^k(e)$ is called the standard irreducible component of the normalized electric dipole part of the density matrix of the radiation field of polarization $e$, $|d_{pr}|^2$ is the square of the reduced electric dipole matrix element which is proportional to the absorption oscillator strength, $\{:::\}$ is the 6-$j$ symbol and $\varepsilon_0$ is the dielectric constant of vacuum. The radiation field intensity is expressed as

$$C_b = \frac{2\pi}{\hbar^2} u , \tag{7.3}$$

where $u(\nu)$ is the energy density per unit frequency ($\nu$) interval. Here we assume that the radiation field is broad and covers the whole absorption line profile.

The interaction of the radiation (angular part) and the atom is expressed by

$$e \cdot d = \sum_s (-)^s e_{-s} d_s , \tag{7.4}$$

with

$$e_0 = \gamma$$
$$e_{\pm 1} = \mp \frac{1}{\sqrt{2}} (\alpha \pm i\beta) , \tag{7.5}$$

where $\alpha$, $\beta$, and $\gamma$ are the Cartesian components of $e$ onto the $x$-, $y$-, and $z$-axes, respectively. The density matrix is expressed as

$$\phi_q^k(e) = \sum_{ss'} (-)^{1+q} e_{-s}(e^*)_{-s'} \langle 11ss'|kq\rangle . \tag{7.6}$$

We note that the Cartesian components of $\boldsymbol{e}^*$ are $\alpha^*$, $\beta^*$, and $\gamma^*$, so that $(e^*)_{-s'} = (-)^{s'}(e_{s'})^*$.

We first consider excitation by left-circularly polarized light, $e_+$. In this case, $\alpha = -1/\sqrt{2}$ and $\beta = -i/\sqrt{2}$ (see (7.1)), so that $e_0 = e_1 = 0$ and $e_{-1} = -1$. Equation (7.6) leads to $k = 0$, 1, and 2, and $q = 0$. We readily obtain

$$\begin{aligned}
\phi_0^0(e_+) &= 1/\sqrt{3} \\
\phi_0^1(e_+) &= 1/\sqrt{2} \\
\phi_0^2(e_+) &= 1/\sqrt{6} .
\end{aligned} \tag{7.7}$$

For the purpose of illustration, we consider the case of $J_r = 0$ and $J_p = 1$. On the right-hand side of (7.2), we omit the factors common to all the $k$s, so that we obtain

$$\frac{d\rho(p)}{dt} \propto \sum_k \phi_0^k(e_+)(-)^k \begin{Bmatrix} 1 & 1 & k \\ 1 & 1 & 0 \end{Bmatrix} T_0^{(k)}(p) \tag{7.8}$$

$$= \frac{1}{3\sqrt{3}} T_0^{(0)}(p) + \frac{1}{3\sqrt{2}} T_0^{(1)}(p) + \frac{1}{3\sqrt{6}} T_0^{(2)}(p) . \tag{7.8a}$$

Table 7.1 shows the example of $J = 1$, the multipole moments (the *population*, orientation, and alignment) defined from the "populations" in the magnetic sublevels $\rho_{MM}$.

The conventional definition of the population is $n(p) = \sqrt{3}\rho_0^0$, as noted earlier. Since, we are now exciting the upper level with left-circularly polarized photons which have an angular momentum $+\hbar$ in the $+z$ direction, the atoms are excited only in the $M = +1$ state, so that $\rho_{11} \neq 0$ and $\rho_{00} = \rho_{-1-1} = 0$ (see Fig. 7.1a). Equation (7.8a) is consistent with this picture.

We now excite the atoms with the unpolarized light, which is expressed as an incoherent superposition of the left- and right-circularly polarized light with equal intensities. For the right-circularly polarized light, $e_-$, we have

$$\begin{aligned}
\phi_0^0(e_-) &= 1/\sqrt{3} \\
\phi_0^1(e_-) &= -1/\sqrt{2} \\
\phi_0^2(e_-) &= 1/\sqrt{6} .
\end{aligned} \tag{7.9}$$

**Table 7.1.** The multipole moments for $J = 1$

| | |
|---|---|
| $\rho_0^0$ | $\frac{1}{\sqrt{3}}(\rho_{11} + \rho_{00} + \rho_{-1-1})$ |
| $\rho_0^1$ | $\frac{1}{\sqrt{2}}(\rho_{11} - \rho_{-1-1})$ |
| $\rho_0^2$ | $\frac{1}{\sqrt{6}}(\rho_{11} - 2\rho_{00} + \rho_{-1-1})$ |

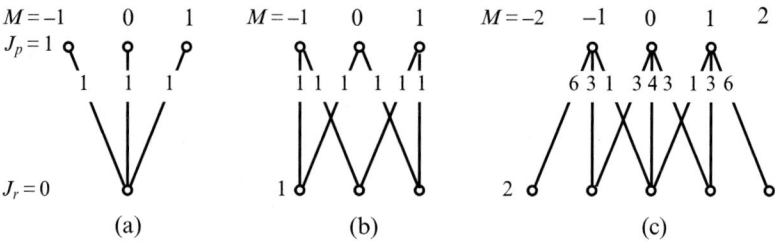

**Fig. 7.1.** Kastler diagrams for $J_p = 1$. The vertical lines are the $\pi$ light and the oblique lines are the $\sigma$ light. The numbers are proportional to the emission/absorption probabilities

It is obvious that, when the light is unpolarized, the orientation disappears. This is because of the equal populations in the $M = \pm 1$ levels. We thus arrive at the conclusion that, in the axial excitation, a positive alignment is created (see Fig. 4.1). The intensities of the polarized components of the emission line are given by (4.5). In (4.5), since $J_s = 0$, the second term in the brackets is $(-1)\frac{\sqrt{6}}{2}(1 - 3\cos^2\eta)3 \cdot \frac{1}{3} \cdot (\frac{1}{\sqrt{6}})$, the $\sigma$ component ($\eta = \pi/2$) has a finite intensity while the $\pi$ component ($\eta = 0$) is absent.

So far, we have considered the light propagating in the $+z$ direction. For the light propagating in the $-z$ direction, the above conclusion is the same for unpolarized light excitation.

We took an example of the transition with the angular-momentum pair of 0–1. This transition can be considered as if it is a classical electric dipole. For example, the above axial excitation is excitation of dipoles oscillating perpendicularly to the axis. Emission radiation is thus polarized perpendicularly to the axis, or only the $\sigma$ component is emitted.

For other angular-momentum pairs, we have to start from (7.2). We consider the cases of $J_r = 1$ and 2, besides 0, for the same upper level $J_p = 1$. The Kastler diagrams for these transitions are shown in Fig. 7.1. For $k = 0$, the value of $(-)^{J_r+k}\{:::\}$ is $(1/3)$ for all the values of $J_r$. For $k = 1$, it is $(1/2)$, $-(1/6)$, and $-(1/6)$ for $J_r = 0, 1$, and 2, respectively. For $k = 2$, it is $(1/3)$, $-(1/6)$, and $(1/30)$ for $J_r = 0, 1$, and 2, respectively. Thus, when the lower level has $J_r = 1$, the self-alignment is negative, and when the lower level has $J_r = 2$, a positive self-alignment is created but it would be very small. These features are understood from the Kastler diagrams. For $J_r = 0$, only the levels $M = \pm 1$ are populated, and for $J_r = 1$, the level $M = 0$ has a population twice those of $M = \pm 1$. For $J_r = 2$, the populations in the $M = \pm 1$ and $M = 0$ levels are 7 : 6; the difference is quite small. Of course, we have assumed here that the radiation field is weak.

We now consider an infinite slab geometry having plane parallel boundaries. In this geometry, the quantization axis is taken perpendicular to the planes. The atoms in the slab tend to be illuminated more from the directions parallel to the planes than from the quantization axis direction. In this case, therefore, the "axial" illumination is lacking as contrasted to the case of the cylindrical geometry above. Thus, for the $(J_r = 0)-(J_p = 1)$ case, for example, a negative alignment is created.

## 7.1.2 Latent Alignment

Until now, we have assumed that the spectrum of the radiation field is broad, so that atoms with any velocities are excited uniformly. In fact, in the limit that the optical thickness of the line at the line center is large, the radiation field there tends to the blackbody (Planck) radiation field, and the profile becomes very broad. In many real situations, we have a finite optical thickness. We assume an isotropic plasma and consider an ensemble of atoms near the origin of a coordinate system. The frequency profile of the radiation field is assumed something like a Gaussian profile with a certain temperature. The atom velocities have a Maxwell distribution. In the velocity space, we consider atoms having, e.g., a large $x$ component, $v_x$. Since $v_x$ is large, other components $v_y$ and $v_z$ tend to be small. This group of atoms are in the radiation field and absorb photons from the field. In this case, the atoms more likely absorb photons coming from the $y$ or $z$ directions, not from the $x$ direction. This is because, owing to the Doppler effect in the $x$ direction, these atoms interact with photons in the wing of the radiation field profile traveling in the $\pm x$ directions while they interact with the photons in the central strong field traveling in the perpendicular directions. Thus, the situation is similar to the slab geometry in the $y$-$z$ plane. In the case of the $(J_r = 0)-(J_p = 1)$ transition, for example, a negative alignment in the upper level develops with respect to the $x$ direction, or the direction of the atom velocity.

Since the plasma is isotropic and a negative alignment is created with respect to the direction of velocity, all the atoms except for those having $v_x = v_y = v_z = 0$ develop microscopic alignments in the velocity space. When integrated over the angle, there is no macroscopic alignment. This phenomenon is called *the latent* (*hidden* or *concealed*) *alignment*.

Chaika [2] discusses the effect of the latent alignment on the radiation field profiles. Depending on the angular-momentum pair, the effect of the alignment makes the profile narrower or broader.

We now apply a magnetic field in the $z$ direction. Although this magnetic field does not affect the atoms on the $v_z$ axis, it affects the alignment (more exactly, the coherence) in atoms having finite $v_x$ or $v_y$ values. In the classical picture, it rotates the component of the dipoles created by the radiation reabsorption around the $z$ direction, i.e., the Larmor precession. For atoms having $v_z = 0$ and large $v_x$, for example, the situation is similar to the slab geometry, so that the number of dipoles oscillating in the $x$ direction is twice those in the $y$ or $z$ directions. The electric dipoles produced in the $z$ direction are unaffected but those in the $x$ or $y$ directions rotate.

We now consider consequences of the application of the magnetic field. If the radiation field is strong and induced emission cannot be neglected any

more, the total population of the upper level would increase. However, this is
unlikely to occur in ordinary situations. How about the radiation field? The
wing part of the field is produced by atoms having large speeds and thus large
alignments. In the above example, the original dipoles produced in the $x$ direction rotate and emit radiation polarized in the $y$ direction. Thus, they begin
to contribute to the radiation field traveling in the $x$ direction, decreasing the
central part and increasing the wing part of the original radiation field. Thus,
the magnetic field can modify the details of the radiation field. This means
that the applied magnetic field affects the production of the latent alignment.

A similar situation to the above may occur to the anisotropic radiation
field. Thus, the self-alignment created by anisotropic radiation reabsorption
*and* observed by the application of a magnetic field suffers this problem.

Chaika [3] discusses the situation in which an excited level having $J = 0$
suffers radiation reabsorption of a line to the resonance excited level having
$J = 1$, in which a latent alignment is created by radiation reabsorption of the
resonance line terminating on the ground state with $J = 0$. She treats the
problem quantitatively and shows that the effect of the latent alignment on
the population of the excited level with $J = 0$ is substantial.

### 7.1.3 Self-Alignment

The self-alignment has been examined in detail for the infinite slab geometry
which was discussed qualitatively in Sect. 7.1.1. The atoms near the plane in
the middle of the slab are illuminated more from the directions parallel to the
plane. On the other hand, near the boundary, illumination is stronger from
the direction parallel to the quantization axis. Thus, a positive alignment is
created for the $J = 0-1$ transition. The solid line in Fig. 7.2b shows an example of an approximate analytic evaluation [4]. This curve is for the alignment
(sign reversed) normalized by the population at each position, which is shown
in Fig. 7.2a. (The actual population distribution in this approximate treatment is slightly different from that in Fig. 7.2a; see below.) This is for the
case of optical thickness of $k_0 L = 13.6$, where $k_0$ is the absorption coefficient
of the line at the line center and $L$ is the half thickness of the slab. The overall
feature is consistent with the above qualitative discussion. The procedure to
obtain this result is outlined below.

According to the formulation by D'Yakonov and Perel' [5], the kinetic
equation of the multipole moments of the upper-level atoms at the position $\boldsymbol{r}$
having momentum $\boldsymbol{p}$ under the radiation reabsorption condition is given as

$$\frac{\mathrm{d}}{\mathrm{d}t}\rho_q^k(\boldsymbol{r},\boldsymbol{p},t) = -A(p,r)\rho_q^k(\boldsymbol{r},\boldsymbol{p},t)$$

$$+A(p,r)\iint \mathrm{d}\boldsymbol{r}'\mathrm{d}\boldsymbol{p}'\sum_{k'q'}S_{kq}^{k'q'}(\boldsymbol{r}-\boldsymbol{r}',\boldsymbol{p},\boldsymbol{p}')\rho_{q'}^{k'}(\boldsymbol{r}',\boldsymbol{p}',t).$$

$$(7.10)$$

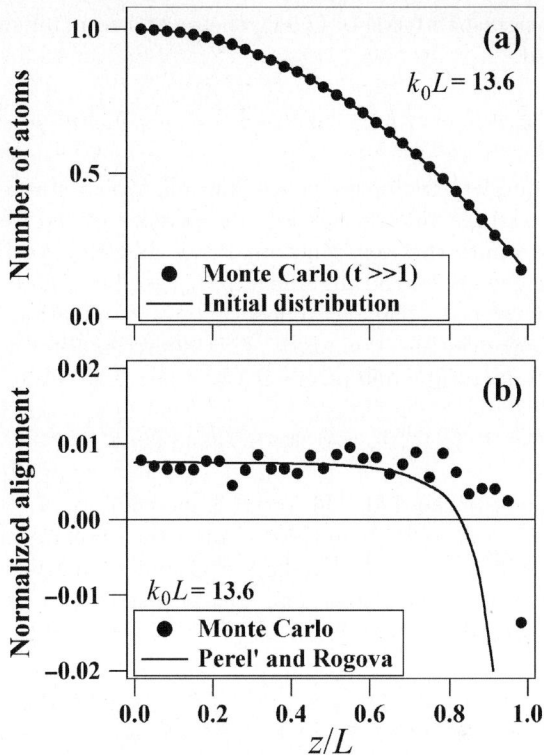

**Fig. 7.2.** The results of the Monte Carlo simulation of radiation trapping in an infinite slab. (**a**) The population distribution. (*solid line*) Initial distribution, (*filled circle*) in late times. (**b**) Distribution of alignment normalized by the population. (*filled circle*) Monte Carlo result, (*solid line*) analytical approximation by Perel' and Rogova (Quoted from [6], with permission from The American Physical Society.)

The other source of creating the multipole moments by, say, anisotropic electron impact is neglected in this formulation. The first term on the right-hand side is the radiative decay, and the second term is the radiative couplings between the multipole moments, including those between the same multiplole moment, e.g., population transfer from position $r'$ to $r$ by reabsorption of photons and creation of alignment from anisotropic distribution of the population. For the present infinite slab geometry, (7.10) reduces to two equations for the "population" and for the alignment, the latter of which is

$$\frac{\mathrm{d}}{\mathrm{d}t}\rho_0^2 = -A\rho_0^2 + A \iint \mathrm{d}r'\mathrm{d}p' S_{20}^{20}(r-r',p,p')\rho_0^2$$
$$+A \iint \mathrm{d}r'\mathrm{d}p' S_{20}^{00}(r-r',p,p')\rho_0^0. \tag{7.11}$$

We have assumed that the orientation is absent. In ordinary situations, the time derivative is absent or very small and $\rho_0^2$ is substantially smaller than $\rho_0^0$, so that, on the right-hand side of (7.11), the second term, alignment transfer, can be neglected.

$$\rho_0^2(r,p) = \iint \mathrm{d}\boldsymbol{r}'\mathrm{d}\boldsymbol{p}'S_{20}^{00}(\boldsymbol{r}-\boldsymbol{r}',\boldsymbol{p},\boldsymbol{p}')\rho_0^0(\boldsymbol{r}',\boldsymbol{p}'). \tag{7.12}$$

By assuming complete frequency redistribution, which states that an atom excited by absorbing a photon has lost its memory of the frequency of this photon when it emits the next photon, Perel' and Rogova [4] obtained an approximate expression for the alignment in this geometry and for the angular-momentum pair of 0–1. They assumed that the population distribution is parabolic, vanishing at the boundary. This approximation is obviously too crude as noted below. The line profile is Gaussian. The result is

$$\rho_0^2/\rho_0^0 = -\left[1.2k_0L\sqrt{2\pi\ln k_0L}\right]^{-1} \tag{7.13}$$

on the central plane of the slab. This result is shown in Fig. 7.2b and in Fig. 7.3 with the solid curve. Note that, in these figures, the ordinate values have been multiplied by $-\sqrt{2}/3 = -0.471$ for the purpose of comparison with a Monte Carlo calculation as given below. The spatial distribution is also obtained, which we have already seen in Fig. 7.2b. For a cylindrical plasma, a similar result is obtained;

$$\rho_0^2/\rho_0^0 = \pi\left[4.8k_0R\sqrt{2\pi\ln k_0R}\right]^{-1} \tag{7.14}$$

on the axis, where $R$ is the radius of the cylinder.

A Monte Carlo simulation of the temporal development of an anisotropically excited atoms is performed for the infinite slab geometry [6]. Again

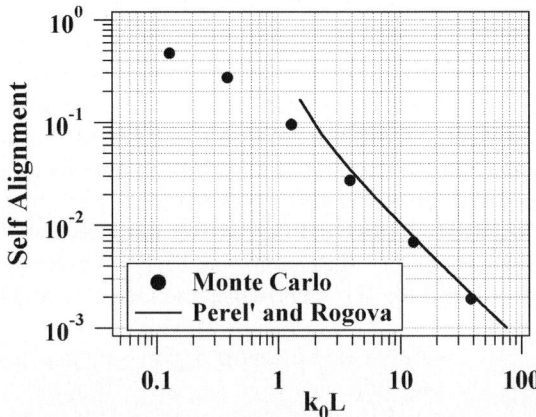

**Fig. 7.3.** Self-alignment vs. optical thickness. (*filled circle*): Monte Carlo results. (*solid line*): Perel' and Rogova [4] (Quoted from [6], with permission from The American Physical Society.)

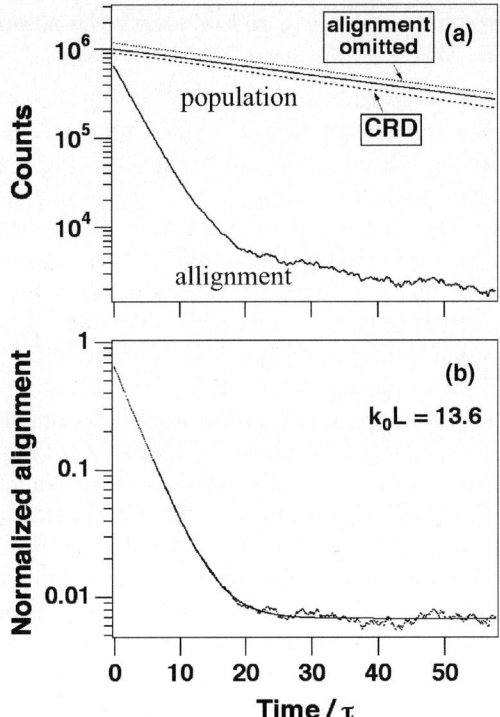

**Fig. 7.4.** Temporal developments of population and alignment. (**a**) (*dotted line*): Alignment effect omitted, (*solid line*): complete frequency redistribution removed. (**b**) Alignment normalized by the population (Quoted from [6], with permission from The American Physical Society.)

$J_r = 0$ and $J_p = 1$ are assumed. An excited atom is prepared in the magnetic sublevel $M = 0$ at a location in the slab according to the random numbers, and the subsequent photon emission–absorption process is followed until the last photon escapes the medium. Figure 7.4 shows the temporal development; (a) is for the population $n(p)$ with/without several assumptions and for the "alignment." The time scale is in units of the natural lifetime given from the transition probability. Population decays slowly with a small effective decay rate; the decay is expressed in terms of the escape factor which characterizes the effective decay rate of the fundamental decay mode for the population in radiation trapping (Chap. 8 of [7]). In order that the population distribution over the slab is strictly of the fundamental decay mode, the initial distribution of the positions of the excited atom is adjusted as given by the solid curve in Fig. 7.2a. For a sufficiently late time, the population distribution is found to be given by the dots, confirming that the distribution is strictly of the fundamental one all the time during the decay. In Fig. 7.4a, the alignment decays rapidly first, and then it decays slowly, in parallel with the population decay. The initial fast decay is the alignment relaxation which will be treated in the next subsection. Figure 7.4b shows the alignment normalized by the population. Here, the ordinate, the normalized alignment, is defined as $\sqrt{\frac{2}{3}}a(p)/n(p) = 0.816a(p)/n(p)$. The small constant background in later times is nothing but the self-alignment corresponding to (7.13). The spatial distribution of the normalized alignment is shown in Fig. 7.2b with the dots. The agreement with the analytic approximation, (7.13), is excellent except near the boundary. This disagreement may be ascribed to the population distribution assumed by Perel' and Rogova, which tends to zero at the boundary. The real distribution has a finite value as seen in Fig. 7.2a. The distribution with a finite value at the boundary is a characteristic of the populations of the decay mode by trapped radiation (Chap. 8 of [7]; especially Figs. 8.9 and 8.10). Figure 7.3 shows the normalized alignment, i.e., the self-alignment, against the optical thickness at the line center. The dots are the results of the Monte Carlo calculation and the line is (7.13), where the factor $\sqrt{2}/3$ has been multiplied to the approximation. For large optical thicknesses, agreement is satisfactory. With the decrease in the optical thickness, the self-alignment increases. However, for small optical thickness, the effect of radiation reabsorption itself is small; the population decays too rapidly to develop a substantial alignment, and processes other than radiation trapping, e.g., excitation by electron impact, would be strong. Under certain conditions, however, the self-alignment could be quite large, say 10%, at around unit optical thickness.

## 7.2 Alignment Relaxation: Alignment Destruction and Disalignment

First we consider a case in which the medium is infinitely optically thick to the transition line $r - p$, and the lower and upper levels have $J_r = 0$ and $J_p = 1$, respectively (see Fig. 7.1a). Suppose excited atoms are prepared in the magnetic sublevel $M = 0$. These atoms emit the $\pi$ polarized light. The polarization direction of the emitted photon is within the plane which includes the quantization axis and the propagation direction of this photon. Thus the angle $\eta$ in (4.5) is related with the polar angle $\theta$ of the photon propagation direction, i.e., $\eta = \pi/2 - \theta$. From $a(p)/n(p) = -2/\sqrt{6}$ (see Table 7.1) and $\{ ::: \} = 1/3$, the angular intensity distribution is readily obtained;

$$I(\theta) \propto \sin^2 \theta. \tag{7.15}$$

See also (2.4). This distribution is identical with the distribution of a classical oscillator as shown in Appendix A (Fig. A.1b). The distribution of the photon directions is peaked in the $x$–$y$ plane, $\theta = \pi/2$, with less probability in the directions with smaller or larger $\theta$ values. These photons are immediately absorbed by other atoms on the present assumption. Thus, the population is conserved. The polarization directions of these emitted photons are predominantly in the $z$ direction, but corresponding to the distribution of the

**Table 7.2.** Disalignment factors for $J_p = 1$

| $J_r$ | $\beta$ |
|-------|----------|
| 0 | 3/10 |
| 1 | 33/40 |
| 2 | 993/1000 |

propagation direction out of the $x$–$y$ plane, the $x$ or $y$ components are also present. These "oblique" photons have components of the $\sigma$ light and excite atoms in the $M = \pm 1$ levels. Of the polarizations of the emitted photons, 80% is in the $z$ direction, and 20% is in the $x$–$y$ directions. Thus the alignment is only partly conserved, or it is destroyed by radiation reabsorption. We call this phenomenon the alignment relaxation. It is readily deduced that its rate is $(3/10)A(p, r)$ in this case. We may call this $(3/10)$ factor the disalignment factor and is given in Table 7.2.

For other pairs of the angular momenta, alignment relaxation can be brought about also by another mechanism. Take an example of the 2–1 pair of the angular momenta as shown in Fig. 7.1c. A $\pi$ light photon emitted by an $M = 0$ atom may be absorbed by another lower level atom in the $M = 1$ level, producing an upper level atom in the $M = 1$ level. This process is in effect excitation transfer from the $M = 0$ level to the $M = 1$ level and contributes to disalignment.

Thus, for this pair of angular momenta, the disalignment factor is larger than that for the 0–1 pair. This factor, including both the processes is given by

$$\beta = 1 - \frac{21}{10}(2J_p + 1)\left\{ \begin{matrix} 1 & 1 & 2 \\ J_p & J_p & J_r \end{matrix} \right\}^2. \tag{7.16}$$

Table 7.2 gives the disalignment factors for three $J_r$s.

In the theory of radiation transport, the transmission probability is defined by

$$T(\rho) = \int_{\text{line}} P(\nu)\exp(-\kappa_\nu \rho)\mathrm{d}\nu, \tag{7.17}$$

where $P(\nu)$ is the emission profile of this line (normalized to unity) and $\kappa_\nu$ is the absorption coefficient. On the assumption of complete frequency redistribution, $P(\nu)$ is proportional to $\kappa_\nu$. The transmission probability is the probability that a photon of this emission line, on the average, survives absorption by other atoms in traversing distance $\rho$. Figure 8.8 in [7] schematically shows the situation. The quantity $[1 - T(\rho)]$ is thus the probability that the photon is absorbed, leading to (partial) alignment relaxation, in the present context.

For finite optical thicknesses, D'Yakonov and Perel' [5] derive an approximate expression for the alignment relaxation rate

$$\gamma_2 = A(p, r)\{1 - (1 - \beta)[1 - T(\rho)]\}, \tag{7.18}$$

where $\rho$ is a typical dimension of the medium. This rate is for the decay of $\rho_0^2$, which is the alignment destruction rate. This rate is rewritten as

$$\gamma_2 = A(p, r)T(\rho) + A(p, r)\beta[1 - T(\rho)]. \qquad (7.18a)$$

The right-hand side of this equation is interpreted as: the first term is the effective decay rate of the population, which may be expressed as $\gamma_0$. As noted above, this decay is expressed in terms of the escape factor when the population distribution is of the fundamental decay mode. $T(\rho)$ here is nothing but the escape factor. The second term is the disalignment rate. Thus, the alignment destruction rate is the sum of the depopulation rate and the disalignment rate. This is consistent with (4.22d): the alignment destruction cross section is the sum of the depopulation cross section (the first term) and the disalignment cross section (the second term). Disalignment is the decay of $\rho_0^2/\rho_0^0$ or of $a(p)/n(p)$. See Appendices C and D.

We have already seen the result of the Monte Carlo calculation in Fig. 7.4; (a) shows the temporal decays of the population and the alignment. The decay of the population corresponds to the first term in (7.18a), and the initial decay of the alignment correspond to $\gamma_2$. The initial decay of the normalized alignment in Fig. 7.4b is nothing but the disalignment and corresponds to the second term in (7.18a).

In D'Yakonov and Perel's formulation (7.18a), the decays of the population and of the alignment are expressed in terms of $T(\rho)$. For an infinite slab geometry with a half thickness $L$, the transmission probability for alignment destruction is given as

$$T(L) = \frac{75/64}{k_0 L(\pi \ln k_0 L)^{1/2}}. \qquad (7.19)$$

This is plotted in Fig. 7.5 with the dotted line. The escape factor for the fundamental decay mode of the population is given by Holstein [8] as a similar equation to (7.19) with the factor 75/64 replaced by 1.06. This is also plotted in this figure with the solid line. Also shown are the results of the Monte Carlo calculation for the population decay expressed as the escape factor (filled circle) and for the alignment decay expressed as the transmission probability (open circle).

Disalignment due to radiation reabsorption is also investigated experimentally [9]. In the positive column of a neon discharge, a substantial amount of atoms are in the four excited levels of configuration $2p^5 3s$: $1s_2(J_r = 1)$, $1s_3(J_r = 0)$, $1s_4(J_r = 1)$, and $1s_5(J_r = 2)$ (Paschen notation) (see Fig. 6.19). The populations of these levels are determined by means of the self-absorption method. Figure 7.6 shows these populations against the discharge current; all the populations increase with an increase in the discharge current. Atoms in $1s_3$ ($J_r{=}0$) is excited to $2p_2(J_p{=}1)$ by a laser pulse (Fig. 7.7a) which is $\pi$ polarized, and only the $M = 0$ sublevel is populated (see Figs. 6.19b and 7.1a). The subsequent fluorescence emission line of $1s_2{-}2p_2$ is observed with its polarized

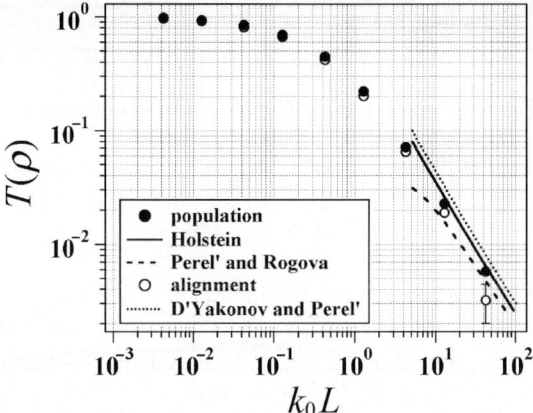

**Fig. 7.5.** Escape factor for population and transmission probability for alignment. Escape factor; (*filled circle*): Monte Carlo results, (*solid line*): results of eigenmode analysis by Holstein [8], (*dashed line*): Perel' and Rogova [4]. Transmission probability; (*open circle*): Monte Carlo results, (*dotted line*): D'Yakonof and Perel' [5] (Quoted from [6], with permission from The American Physical Society.)

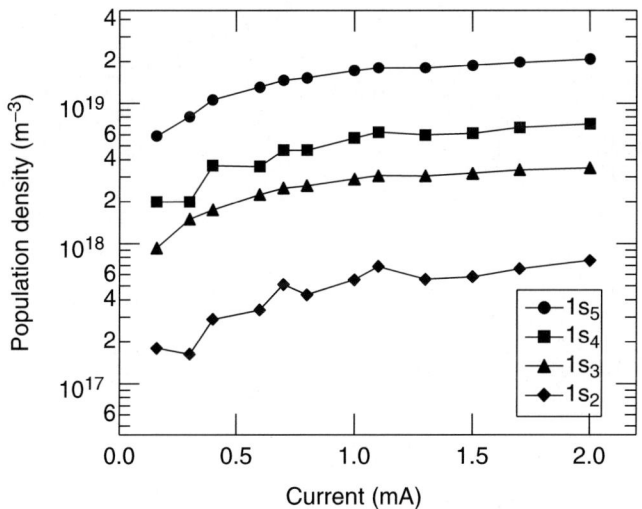

**Fig. 7.6.** Populations of the lower levels as determined by the self-absorption method (Quoted from [9], with permission from IOP Publishing.)

components resolved. The original $M = 0$ atoms emit only the $\sigma$ component as shown in Fig. 7.7a. The $\pi$ component also appears as a result of excitation transfer from the $M = 0$ level to the $M = \pm 1$ levels in Fig. 7.1b, or the disalignment. The longitudinal alignment $A_L = (I_\pi - I_\sigma)/(I_\pi + 2I_\sigma)$, which is proportional to $a(p)/n(p)$, is calculated from the intensities in Fig. 7.7a,

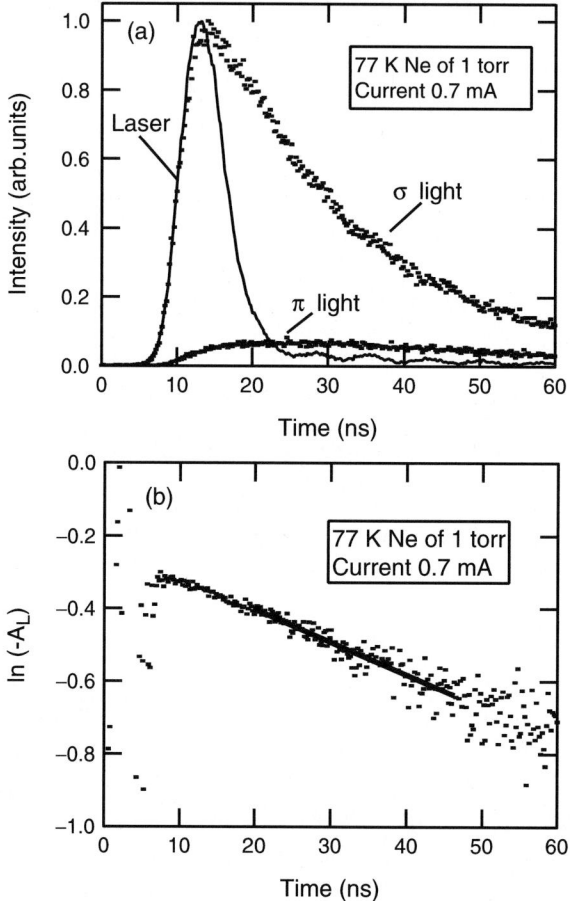

**Fig. 7.7.** Temporal development of the fluorescence intensities after laser excitation. (**a**) $\pi$ and $\sigma$ components are resolved. (**b**) The longitudinal alignment (sign reversed) is obtained from (**a**) (Quoted from [9], with permission from IOP Publishing.)

and plotted in Fig. 7.7b. The slope is the disalignment rate. By changing the discharge current, the disalignment rate is determined and plotted in Fig. 7.8 with the closed circles. It increases with the increase in the discharge current.

Under the present experimental conditions, disalignment can be brought about by collisions with ground-state atoms as well as by radiation reabsorption. Disalignment rates by electron and ion collisions are smaller by more than three orders of magnitude in Fig. 7.8. A Monte Carlo simulation is performed for disalignment of the upper level atoms by reabsorption of emission/absorption lines terminating on the four lower levels (see Figs. 6.19a and 7.1a–c). The optical thickness for these lines are given from Fig. 7.6. An example of the simulation results is shown in Fig. 7.9, which corresponds to Fig. 7.7.

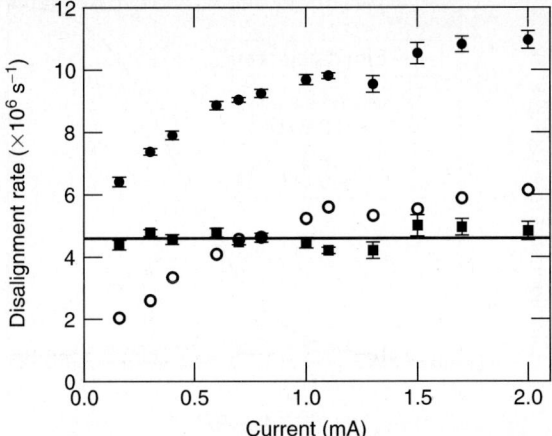

**Fig. 7.8.** Disalignment rate vs. the discharge current. (*filled circle*): Experimentally determined rates, (*open circle*): Monte Carlo simulation results, (*filled square*): Experimental rates subtracted by the simulation results (Quoted from [9], with permission from IOP Publishing.)

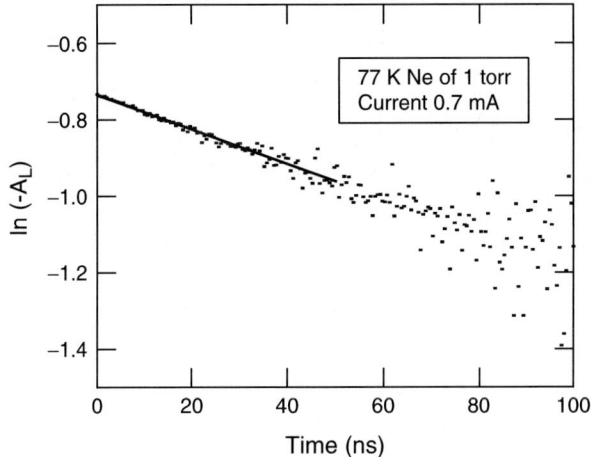

**Fig. 7.9.** Decay of the longitudinal alignment by the Monte Carlo simulation (Quoted from [9], with permission from IOP Publishing.)

The disalignment rate by reabsorption of these four lines are thus obtained, and the rate against the discharge current is plotted in Fig. 7.8 with the open circles. It increases with the current, being consistent with the populations in Fig. 7.6. On the assumption that disalignment by atom collisions and that by radiation reabsorption are additive, the experimental disalignment rate is subtracted by the simulation result. The result is shown with the closed squares. Since atom density and temperature (77 K) are constant over the discharge

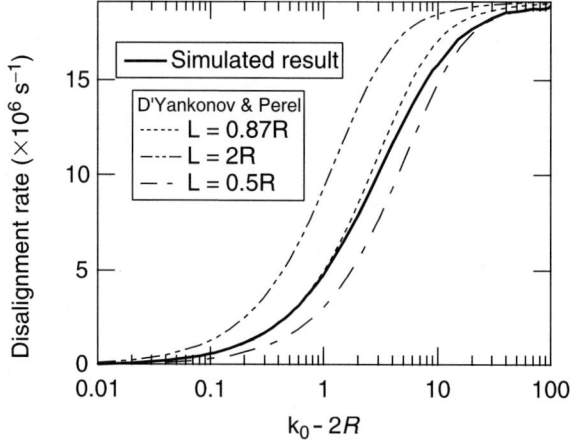

**Fig. 7.10.** Disalignment rate by Monte Carlo simulation for the transition with the 1–1 angular momentum pair (1s$_2$–2p$_2$) vs. the optical thickness. Equation (7.20) with (7.17) is compared for several values of $\rho$ (Quoted from [9], with permission from IOP Publishing.)

currents, the disalignment rate by atom collisions should be constant, which is acturally confirmed in Fig. 7.8. In this experiment, the disalignment rate by radiation reabsorption is of similar magnitude to that by atom collisions. Thus, disalignment by radiation reabsorption is quantitatively understood.

As noted earlier, D'Yakonov and Perel' [5] propose an approximate disalignment rate, the second term in (7.18a),

$$\gamma_2 - \gamma_0 = A(p, r)\beta[1 - T(\rho)]. \tag{7.20}$$

However, $\rho$ is specified only to be "a typical dimension" of the medium. For a transition of a specific angular-momentum pair of 1–1, the disalignment rates calculated by the Monte Carlo simulation for an infinite cylinder geometry with radius $R$ are compared with (7.20) and (7.17) for $\rho$ between $0.5R$ and $2R$. Figure 7.10 shows the comparison; the abscissa is the optical thickness of the line at the line center over the plasma diameter and the ordinate is the disalignment rate for this specific transition 1s$_2$–2p$_2$. For small optical thickness, $\rho = 0.87R$ gives the best fit of (7.20) to the simulation result. For large optical thickness, $\rho$ tends to $R/2$. For other pairs of angular momenta, the conclusion is almost the same.

# References

1. A. Omont, Prog. Quant. Electron. **5**, 69 (1977)
2. M.P. Chaika, Opt. Spectrosc. **31**, 274 (1971)
3. M.P. Chaika, Opt. Spectrosc. **30**, 443 (1971)

4. V.I. Perel', I.V. Rogova, Sov. Phys. – JETP **34**, 965 (1972)
5. M.I. D'Yakonov, V.I. Perel', Sov. Phys. – JETP **20**, 997 (1965)
6. A. Hishikawa, T. Fujimoto, P. Erman, Phys. Rev. A **52**, 189 (1995)
7. T. Fujimoto, *Plasma Spectroscopy* (Oxford University Press, Oxford, 2004)
8. T. Holstein, Phys. Rev. **72**, 1212 (1947)
9. M. Seo, M. Nimura, M. Hasuo, T. Fujimoto, J. Phys. B **36**, 1869 (2003)

# 8

# Experiments: Ionizing Plasma

T. Fujimoto, E.O. Baronova, and A. Iwamae

In this and next chapters, we review laboratory PPS experiments performed so far. According to the classification in Chaps. 3 and 4, we classify the plasmas into ionizing plasma and recombining plasma. In the present chapter, we review experiments on plasmas belonging to the former class. Low-pressure gas discharge plasmas naturally fall into this class, since, in the stationary state, the ionization flux of atoms in the plasma is balanced with the recombination flux of ions due to the diffusion motion of the ions to the cell wall, not with the CR (collisional-radiative) or volume recombination flux. Another group would be $z$-pinch plasmas which are formed by strong plasma contraction leading to high temperature and density in a short time. Some laser-produced plasmas have been interpreted as ionizing plasma in the original papers so that they are introduced in this chapter. The magnetically confined plasmas belong to this class and are now becoming the subject of PPS experiments.

## 8.1 Gas Discharge Plasmas

This group of plasmas has a long history of PPS research. In fact, the phenomenon of *self-alignment* was discovered in experiments on discharge plasmas. We further divide these plasmas into (1) direct current discharges including positive columns of glow discharge, hollow-cathode discharges, and atmospheric arcs, (2) high-frequency (*rf*) discharges, and (3) other sorts of plasmas like those heated by a gas–plasma collision. The atomic species investigated in these experiments are mainly rare gas atoms, mostly neutral atoms and, in a few cases, singly ionized ions. The method of polarization detection is based on the Hanle effect in most cases (see Appendix D for the Hanle effect). In some other cases, polarized components in various directions, including circular polarization, are measured to yield the Stokes parameters (see Appendix A). The PPS experiments on discharge plasmas were extensively performed in the former Soviet Union.

Since the Hanle signal reflects the coherence time of an atomic ensemble in the upper level of the transition, the lifetime and the collisional depolarization rate of this level can be deduced from a self-alignment experiment. Under the assumption that the alignment creation is unaffected by the applied magnetic field, the half width of the Hanle signal against atom density gives the depolarization rate; from the slope of the fitted line, the depolarization (actually alignment destruction) rate coefficient is obtained, and the intercept at the vanishing density gives the lifetime. Numerous experiments were performed in which these atomic parameters were determined. Several review papers appeared [1, 2]. It is noted, however, that the depolarization rate coefficient obtained from the Hanle signal is the alignment destruction rate coefficient, which is the sum of the depopulation rate coefficient and the disalignment rate coefficient, as discussed in Sect. 6.1.4. As noted there, sometimes confusion occurred between the alignment destruction rate and the disalignment rate. This is partly due to the usages of terms like the alignment relaxation (see also Appendices C and D).

As noted at the close of Sect. 7.1.2, the spectral profile of the radiation field can be modified by the applied magnetic field, so that the assumption that the creation of alignment is independent of magnetic field is violated. Therefore, in the case that the self-alignment is created by radiation reabsorption, and its relaxation is detected by the Hanle effect method, the experimental results cannot be relied on quantitatively.

### 8.1.1 Direct Current Discharge

**Positive Column of Glow Discharge**

After the discovery of the self-alignment phenomenon in a high-frequency discharge, as will be introduced in the next subsection, self-alignment was also discovered on neutral neon excited atoms in a positive column of neon or neon–helium glow discharges. Carrington and Corney [3] applied a magnetic field in the direction perpendicular to the axis of a thin and long discharge tube. They observed a line out of the 30 red neon lines of the $2p^5 3s$–$2p^5 3p$ transitions from the direction of the applied magnetic field. The difference of intensities of the polarized components of the line in one direction and in its perpendicular direction was measured against a change in the magnetic field strength and sign. The polarization degree was of the order of 1%. When the first polarization direction was parallel to the discharge tube axis, the signal was of a Lorentzian profile, and when it was at an angle $\pi/4$, a dispersion profile was obtained. Figure 8.1 is an example of the observed signals. These magnetic-field-dependent signals are obviously due to an alignment in the upper level atoms of the transition. Concerning the origin of this alignment, radiation reabsorption of the $2p^5 3s$–$2p^5 3p$ lines in the anisotropic geometry was strongly suggested.

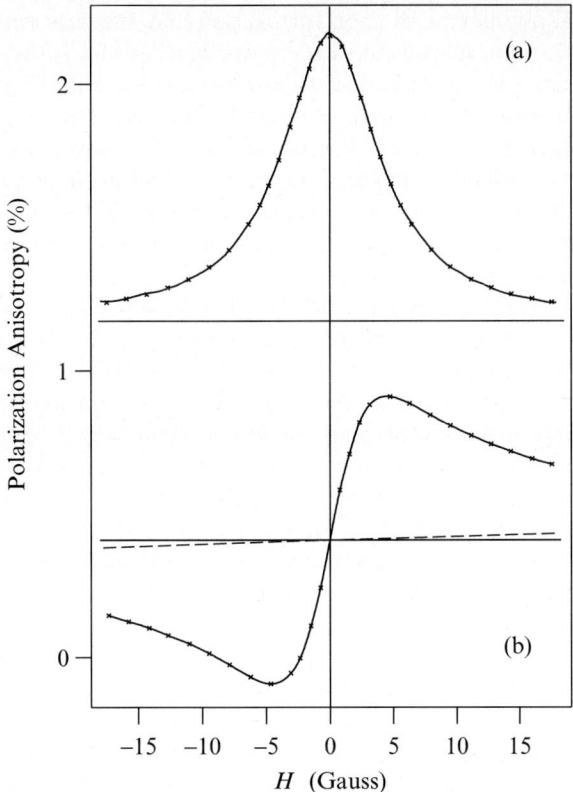

**Fig. 8.1.** Hanle signal of NeI λ626.6 nm (1s₂–2p₅) (Paschen notation) line in a positive column of a neon–helium glow discharge. (**a**) The first angle parallel to the discharge tube axis, and (**b**) The angle $\pi/4$ with respect to the axis (Quoted from [3], with permission from Elsevier.)

Slightly earlier than this paper, Kallas and Chaika [4] reported a similar experiment. They applied a magnetic field perpendicular or parallel to the discharge tube axis and measured the intensity of a line (a polarized component) against the magnetic field. A slight change in the intensity about 0.5% with a Lorentzian profile was observed. They found one symmetry axis in the axial direction in the axial part of the plasma and the radial and axial directions in the peripheral part of the plasma. By placing light reflecting prisms on the both sides of the discharge tube, they found a strong effect of the reflected light on the signal. They concluded that an alignment was produced in the upper level of the transitions. The radiation reabsorption was again suggested as the origin of this alignment. They also observed a similar signal on lines which originated from levels having total angular momentum $J = 0$, which emit only unpolarized lines. Thus this signal came from a change in the upper-level population. It soon turned out that this was due to the latent (hidden or concealed) alignment in the lower-level atoms $2p^5 3s$ having $J = 1$, i.e., the resonance levels; this new alignment was produced by *isotropic* radiation reabsorption of the resonance lines $2p^6$–$2p^5 3s$. This phenomenon was discussed already in Sect. 7.1.2. The latent alignment developed in the lower-level atoms was reflected somehow in the $2p^5 3p$ level populations through collisional excitation or radiation reabsorption from $2p^5 3s$, and when the latent alignment was modified by the magnetic field, the upper-level populations were accordingly modified.

Although the self-alignment in the low-lying level atoms were interpreted as due to radiation reabsorption, indirectly through the $2p^5 3s$ levels in the case of neon as discussed above, or directly from the ground state, it was recognized that transition lines from high-lying levels that were not connected radiatively with the ground state also showed the Hanle signal [5]. Alignment production by anisotropic electron collisions was suggested. It is known that, in a positive column plasma, EVDF is almost isotropic. It is possible, however, that high-energy electrons have some directional distributions. Kazantsev et al. [6], by using probes, determined the axial and radial electric field strengths to find that, with a decrease in pressure, the radial field strength increased dramatically. This tendency is understood from the free-fall model of a glow discharge at low pressures [7]. On the assumption that the EVDF had a symmetry axis in the direction of the electric field, they explained the observed Hanle signal from directional electron collisions. By assuming this mechanism for the alignment production, Kazantsev determined the quadrupole moment of EVDF for the first time [8]. Drachev et al. [9] deduced the radial electric field strength from the observed polarization degree against the radial distance from the axis.

### Hollow-Cathode Discharge

It is well known that a hollow-cathode discharge plasma has a cathode fall region near the cathode surface and, in this region, the electric field is strong and electrons are accelerated by this field. The field strength is highest toward the surface and it decreases linearly toward the central axial region where the plasma is almost isotropic. The first PPS experiment on a hollow-cathode discharge plasma was performed by Zhechev and Cahika [10], and a succeeding experiment gave a proof that the self-alignment produced in excited atoms was actually due to the directional electron impact excitation [11].

Self-alignment produced in atoms by drifting motion of ions was proposed by Petrashen et al. [12–14]. The idea was that, different populations (per unit statistical weight) among the fine-structure or hyperfine-structure levels of atoms would be converted by directional collisions with drifting ions into an alignment in one of the levels. Since the energy difference among these levels was quite small, low-energy collisions with ions were much more effective than electron collisions. Experimentally, an alignment was observed in argon ion levels, e.g., $4p\,^2P_{3/2}$ [15–17]. In this case, the alignment in this level was produced from the population difference in the $4p\,^2P_{1/2}$ and $4p\,^2P_{3/2}$ levels by the directional collisions of these ions against the background atoms.

### Atmospheric Arc

A sophisticated system for determining the Stokes parameters (see Apendix A) of a spectral line was developed, and it was applied to an atmospheric arc discharge plasma [18]. The stabilized electric arc in argon, which was produced with the current up to 42 A, had diameter of 2 mm and length of 10 mm. The plasma was well diagnosed, and the electron temperature was about 1 eV and density about $10^{23}\,\mathrm{m}^{-3}$. Several argon ion lines of the 4s–4p transitions were observed, and polarization with polarization degrees of several percent was found for some lines originating from levels having $J = 3/2$ or $5/2$, and virtually no polarization for lines from $J = 1/2$. Figure 8.2 is taken from this paper.

EVDF was calculated from the balance between the force exerted on electrons by the axial electric field, which was about $1.5\,\mathrm{kV\,m}^{-1}$, and the braking force by collisions of electrons with ions. In the latter collisions, owing to the Coulomb interaction, collision efficiency is lower for higher-energy electrons, so that the mean electron drift speed due to the axial electric field is higher for higher-energy electrons. On the basis of this idea, the second moment of EVDF, $f_2(v)$ of (4.32) was expressed analytically. The alignment production cross section $Q_0^{0,2}(E)$ from the ground state ion $3p^5\,^2P_{3/2}$ to excited levels was assumed to be proportional to the excitation cross section $Q_0^{0,0}(E)$ in the energy range just above the excitation threshold; the ratio of these cross

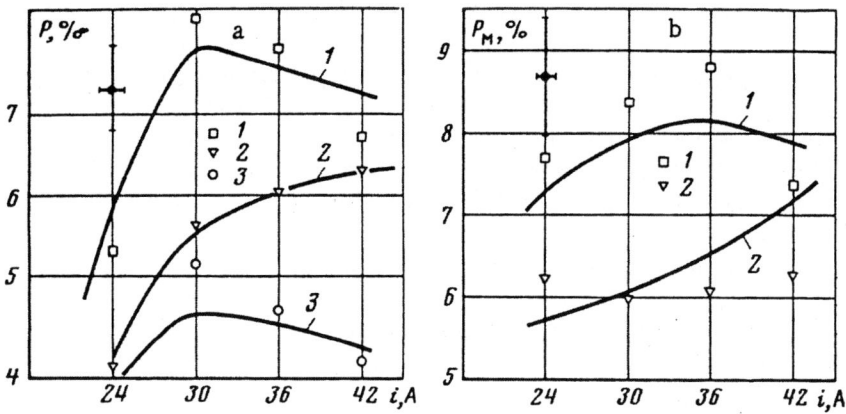

**Fig. 8.2.** Polarization degree of several argon ion lines near the axis of the arc plasma. (1) ArII $\lambda$488.0 nm ($4s\,^2P_{3/2}$–$4p\,^2D_{5/2}$). (2) ArII $\lambda$480.6 nm ($4s\,^4P_{5/2}$–$4p\,^4P_{5/2}$). (3) ArII $\lambda$454.5 nm ($4s\,^2P_{3/2}$–$4p\,^2P_{3/2}$) (Quoted from [18].)

sections was assumed to be that at the threshold; i.e., excitation is without a change in the magnetic quantum number of the magnetic sublevels. In the calculation of alignment in the upper levels, they included alignment destruction by electron collisions and by atom collisions. In fitting the calculated polarization of several lines to the experiment (Fig. 8.2), they estimated the cross sections for these processes. The former cross section was compared with the inelastic part of the Stark broadening cross section which was determined theoretically, and a surprisingly good agreement was obtained. It should be noted here that the relationship between the Stark broadening and disalignment has been discussed in Sect. 6.1.4. Since the species is an ion, elastic (Coulomb) collisions are quite substantial, but they do not contribute to alignment destruction as well as to Stark broadening. Alignment destruction by atom collisions was assumed to be by elastic collisions, i.e., disalignment, and its cross section was found to be much larger than that by electron collisions.

In this experiment, the degree and the direction of polarization were measured against the distance from the axis, and larger polarization degrees and substantial deviations of the polarization direction from the axial direction, e.g., 22% and 35%, respectively, were found. They ascribed these changes in polarization to different EVDFs which were caused mainly by the steep radial gradient of electron temperature.

### 8.1.2 High-Frequency Discharge

In 1965, Lombardi and Pebay-Peyroula [19] first discovered the self-alignment on a capacitive discharge plasma. A glass cell containing 0.01 torr helium was placed between the capacitor-type electrodes and an *rf* field of 250 MHz was applied. By the Hanle effect method, they observed substantial degree of polarization on some singlet lines of neutral helium; more than 10% for the $\lambda 492.2$ nm ($2^1$P–$4^1$D) line. Directional excitation by electrons which were accelerated by the *rf* electric field was identified as the origin of this polarization.

Kazantsev et al. [20] observed argon plasmas in a cell containing 0.08–0.16 torr argon excited by a 100 MHz *rf* field. They found polarization on neutral argon lines with polarization degree less than 1%. A theoretical EVDF for an *rf* discharge plasma was expressed as the ratio of the second to zeroth moments of the Legendre expansion (see (4.31))

$$f_2(E) = AEf_0(E) \tag{8.1}$$

with

$$A = \frac{2(eF)^2}{m(kT_e)^2(\omega_0^2 + \nu^2)}, \tag{8.2}$$

where $F$ is the electric field strength, $E$ is the electron kinetic energy, $m$ is the electron mass, $\omega_0$ is the driving frequency, and $\nu$ is the collision frequency. They assumed the zeroth moment of EVDF to be Maxwellian with temperature of $5\,\mathrm{eV}$. On the other hand, on the assumption that the alignment production cross section $Q_0^{0,2}(E)$ is proportional to the excitation cross section $Q_0^{0,0}(E)$ in the energy range just above the excitation threshold, $A$ is calculated from the experimental polarization degree;

$$
A = \varphi(P) \frac{\int\limits_{E_{\mathrm{th}}}^{\infty} E Q_0^{0,0}(E) f_0(E) \mathrm{d}E}{\int\limits_{E_{\mathrm{th}}}^{\infty} E^2 Q_0^{0,2}(E) f_0(E) \mathrm{d}E}, \tag{8.3}
$$

where $E_{\mathrm{th}}$ is the excitation threshold energy and $\varphi(P)$ is a function relating the ratio of the multipole moments of the density matrix of the upper level, $\rho_0^2(p)/\rho_0^0(p)$, to the observed polarization degree, $P$. They obtained $A$ values of $(5-10) \times 10^{-3}\mathrm{J}^{-1}$, and deduced $F$ to be about $600\,\mathrm{V/m}$.

Another experiment was reported by Kazantsev and Subbotenko [21] on a similar plasma with a wider range of experimental parameters, i.e., atom species and gas pressure. The spatial distributions of polarization and intensity of emission lines were measured. When observed from the direction parallel to the electrode plates along the cell axis, prominent peaks of polarization degree were observed close to the electrodes. The maximum polarization degree was more than 10% in the case of helium $\lambda492.2\,\mathrm{nm}$ ($2^1\mathrm{P}-4^1\mathrm{D}$) line, and of the order of 1% for argon lines. The peak position was near the boundary between the central luminous plasma and the dark space in front of the electrode. Since the electron plasma frequency was much higher than the driving frequency, which in turn was much higher than the ion plasma frequency, it may be imagined that, while ions were at rest, electrons traveling from the central region to the electrode region were reflected by the potential wall which was oscillating in amplitude. This idea is known as the Fermi acceleration. They adopted the rigid wall model: an incoming electron was reflected elastically by the moving wall, resulting in an increase in the axial component of velocity by twice the velocity of the wall (besides the thermal component). From an analytical approximate calculation, they found a beam-like component of EVDFs for the reflected electrons. Directional excitation of atoms by this component was identified as the origin of the peak of the polarization degree. They also performed a particle simulation and the results were consistent with the analytical approximation. They also measured the intensity and polarization degree along the direction perpendicular to the axis on the mid-plane between the electrodes. Again, polarization peaks were observed in the peripheral region which was the outer boundary of the central luminous plasma. However, some lines originating from low-lying levels did not show this feature: the polarization degree was highest near the axis and it monotonically decreased toward the periphery. They adopted a picture that electrons travel-ing toward the wall experience the potential wall. Low-energy electrons were reflected back by this potential wall, maintaining almost isotropic EVDF, but high-energy electrons could reach the wall and disappeared there. Therefore, a kind of "loss cone" was formed. As its result, in the periphery region, high-energy electrons lacked the radial component, but low-energy electrons were more or less isotropic. This is their interpretation of the different features of the polarization degrees of the lines.

As discussed above, the *rf*-discharge plasma shows various interesting features of polarization in the context of gas discharge physics and plasma physics. In all the cases, the Hanle effect is adopted to determine the polar-ization degree. A paper discussed above reports that when a weak magnetic field of the order of $0.1\,\mathrm{mT}$ is applied to the plasma, it affects the geome-try of the plasma itself; the beam-like electrons obviously suffer the Lorentz force and their trajectories, thus the luminous plasma region, deviate from the original position when the field is absent. Thus, the application of a magnetic field on the plasma changes the structure of the plasma and makes the situa-tion much complicated: it affects not only the coherence properties of excited atoms, i.e., the Larmor precession, but also the direction and magnitude of the initial coherence or alignment of the excited atoms. Furthermore, the spatial profile of the plasma itself may change. Even a residual earth magnetic field could affect the polarization features. Thus, an interpretation of the Hanle signal becomes less straightforward. This difficulty seems to be a drawback of *rf* discharge plasmas as applied to PPS. Although this point was not ob-vious for other sort of discharge plasmas, this difficulty may be common to any experiments of the self-alignment produced by directional excitation by electron collisions *and* observed by the Hanle-effect method.

### 8.1.3 Neutral Gas Plasma Collision

For the purpose of studying the heating mechanism of electrons in collisions of a plasma with a neutral gas, Danielsson and Brenning [22] made a se-ries of experiments. A hydrogen plasma was produced by a plasma gun and driven along the magnetic field of $0.18\,\mathrm{T}$ (In the original paper, the magnetic field is given as "$0.18\,\mathrm{V\,m^{-2}}$." However, the units should be something like $\mathrm{Vs\,m^{-2}}$. Then, that is equal to T. We adopt this correction here.) with speed of $(4\text{--}5) \times 10^5\,\mathrm{ms^{-1}}$ in the tube toward the interaction region 2 m downstream, where the direction of the magnetic field was perpendicular to the tube axis. A small cloud of helium gas with density of $10^{20}\,\mathrm{m^{-3}}$ was formed in the inter-action region with a high-speed gas valve. The plasma ($n_\mathrm{e} = 10^{17}\text{--}10^{18}\,\mathrm{m^{-3}}$ and $T_\mathrm{e} = 5\text{--}10\,\mathrm{eV}$) collided with this gas cloud. Emission lines were observed from the direction perpendicular to the magnetic field and to the tube axis. Examples of the observed signals are shown in Fig. 8.3. The emission dura-tion is about $5\,\mu\mathrm{s}$. They performed a very careful observation of polarization of helium lines. The polarization degree of the 492.2 nm ($2^1\mathrm{P}\text{--}4^1\mathrm{D}$) line was

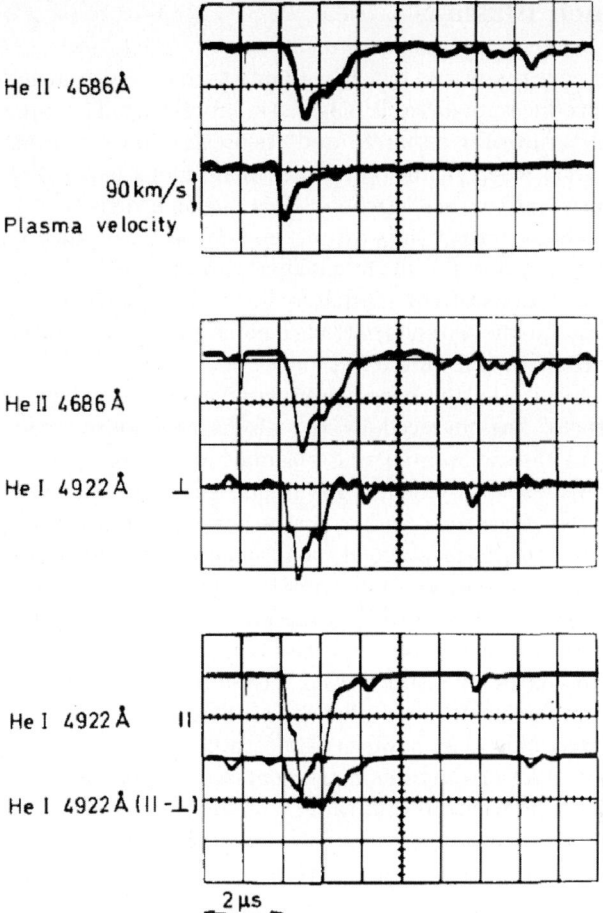

**Fig. 8.3.** Ionized helium line and the perpendicular and parallel components of HeI $\lambda\,492.2\,\text{nm}$ $(2\text{p}\ ^1\text{P}_1\text{--}4\text{d}\ ^1\text{D}_2)$ line observed simultaneously. Difference signal is also shown (Quoted from [22], with permission from The American Institute of Physics.)

found to be $(6.3\pm1.4)\%$ (Fig. 8.3) and that of the 501.6 nm $(2^1\text{S--}3^1\text{P})$ line was $(2.5\pm0.7)\%$, where the polarization direction was referred to the direction of the magnetic field. From other information it was known that the electrons were heated by the collision to about 100 eV, so that the observed positive polarization indicated that the electron motion was predominantly in the direction of the magnetic field. They interpreted this characteristic as due to the two-stream instability. The interaction of a magnetized plasma with neutral gas particles lead to energy transfer from plasma streaming protons, through helium ions, to the perpendicular (parallel to the magnetic field) motion of electrons in a short time of nanoseconds.

(T. Fujimoto)

## 8.2 *Z*-Pinch Plasmas

The term *pinch* was adopted by early scientists to express the high current discharge that produces remarkable plasmas in laboratory. The *z*-pinch has a long history of experimental research, and theoretical investigation also started even before 1934 when Bennet published his famous paper [23]. The Bennet equation $16\pi N k_{\mathrm{B}} T_{\mathrm{e}} = \mu_0 I^2$ (where $N$ is the line density, $k_{\mathrm{B}}$ Boltzmann's constant, $T_{\mathrm{e}}$ the electron temperature, and $I$ the total pinch current) is the rule of thumb even now for all the pinch plasmas irrespective of the temporal and spatial variations of the discharge state. The pinch phenomena appear quite universal to the extent that, e.g., some phenomena in solar flares are interpreted from the pinching effect.

The *z*-pinch is the discharge producing a plasma which is compressed by the magnetic field induced by the discharge current itself. This definition applies to the vacuum sparks, plasma focuses (so called noncylindrical *z*-pinches), cylindrical *z*-pinches, gas liners, gas puff machines, wire arrays, exploding wires, X-pinches, and so forth. For some kinds of exploding wires and wire arrays, the initial stage of the discharge should be evaporation of the wire material and expansion. Only in the later stage, compression becomes the dominant dynamics and the plasma falls into the category of *z*-pinch.

*Z*-pinches are investigated vigorously, since this scheme is considered as one of the candidates for future nuclear fusion reactors. In this context in particular, it is important to determine quantitatively various parameters of the *z*-pinch plasma. This task, however, is not well established yet; this is due to the difficulties in developing the instruments for observation which can resolve the strongly inhomogeneous and rapidly changing plasma, which is also irreproducible. This is also due to the lack of a theoretical framework adequate to interpret the observation. In this regard, a unique characteristic of the *z*-pinch plasma is its anisotropy which stems from the presence of a definite axis defined by the electrodes. The anisotropy may be an anisotropic electron velocity distribution function (EVDF) including the presence of electron beams, anisotropic neutrons, intense hard and soft X-rays, fast ions, and/or directional and turbulent electric and/or magnetic fields [24–26]. From the standpoint of application, such anisotropic characteristics may be regarded as even a "privilege" of pinch plasmas.

Before discussing the details below, we first look at typical parameters of the *z*-pinch plasmas:

- dimension of the hot, dense plasma region ($r$): $10^{-3}$–$10^0$ mm
- plasma lifetime: 1–50 ns or less
- electron temperature ($T_{\mathrm{e}}$): 0.5–10 keV
- electron density ($n_{\mathrm{e}}$): $10^{24}$–$10^{29}$ m$^{-3}$
- beam current ($I_{\mathrm{beam}}$): a few tenth kA in the time scale of a few ns
- beam-electron energy: up to 500 keV
- electric field ($E$): up to $10^{12}$ V/m (theoretical prediction)

- energy of X-rays: up to 300–500 keV
- energy of fast ions: up to a few MeV

Among various diagnostic techniques, X-ray spectroscopy is a versatile method. It provides information on the electron temperature and density, the energy of the electron beams and fast ions [27], the dimension of the emitting regions and their locations [28], and so forth. It is found in many $z$-pinch experiments that the hot, dense plasma radiates helium-like ion lines of various elements. It is thus natural that the helium-like lines are intensively studied theoretically [29].

In classical plasma diagnostics, it is a common practice to derive $T_e$ and $n_e$ from the observed relative intensities of the helium-like lines. These lines are assumed to be unpolarized [30]. This assumption, however, has been found unjustified: over the last few decades there appeared several publications that reported polarized radiation. These polarizations are ascribed to the anisotropic EVDF of energetic electrons or the strong electric and/or magnetic fields [31, 32]. As mentioned above, however, it is not yet possible for us to interpret quantitatively the observed X-ray polarization in $z$-pinches in terms of these anisotropic characteristics.

As mentioned already, enormous difficulties accompany the X-ray polarization spectroscopy of $z$-pinch plasmas. One of them stems from the fact that the investigated radiation comes from a small region, which is called the hot spot, the bright spot, or the plasma point [33]. The position (X-pinches excluded), the size and the plasma parameters of this spot, and the number of the spots (especially with plasma focuses) are irreproducible [34]. In addition, the intensity of the investigated radiation is sometimes too low to obtain enough information from one shot (especially for vacuum sparks). The plasma also changes too rapidly to be resolved by a conventional instrument. As a result, the measurements are the result of an average, or an integration, over space, time, and shots. Another difficulty, which is not even well recognized yet, is that the observed polarization may be modified during the passage of the photons through the hot, dense plasma itself, i.e., the opacity effect.

From the above discussions it is clear that, for a quantitative interpretation of the observations, especially of polarization, a reliable set of theoretical models is necessary which reproduces the temporal and spatial development of the plasma. This set consists of an MHD model, which describes the overall behaviors of the plasma, and a collisional-radiative (CR) model, which describes the ionization-recombination of the ions and (polarized) line emissions from the plasma. To be practical, both the models should be simple, still reliable. Complicated models do not necessarily lead to correct results.

A steady-state CR model tends to overestimate the role of various time-dependent factors; this is because the real state of the ionization-recombination may lag behind the rapidly changing plasma state. It is thus essential to use a time-dependent CR model. As will be shown later, a combination of a simple and reliable MHD model and a CR model reveals that, for instance, the occurrence of the electron beam and the electric fields may not coincide with the emission of the helium-like lines [35]. This coincidence is especially important in polarization analysis: how long the helium-like ions coexist in time and space with these anisotropy factors.

In the following, we review the polarization measurements on $z$-pinches and give a preliminary interpretation of the results by using a set of the models. In these measurements, we use two spectrometers/crystals which are arranged perpendicular to each other, or one spectrometer which registers the lines in different orders of reflection. The difficulties are [36, 37]: (1) It is generally difficult to find a crystal that has the interlayer spacing $d$ that exactly matches the Bragg angle 45° (Brewster angle) for the particular line of interest, (2) in most cases, polarization calibration of crystal is necessary, which is, however, quite difficult, (3) the characteristics of the two crystals may not be identical, and (4) the field of view for the two crystals may not be identical.

### 8.2.1 Vacuum Spark and X-Pinch

The first polarization measurement of the helium-like FeXXV lines was performed on a 150 kA vacuum spark by Veretennikov et al. [38]. Flat crystal spectrometers were used with different orientations of the dispersion planes with respect to the discharge axis, and the recorded spectra were found different for the different spectrometers. By the time of this experiment, polarization of a helium-like line due to anisotropy of EVDF was already known in solar flares, and it was well known that, in a vacuum spark plasma, electron beams are always generated in the axial direction. It was thus quite natural to expect polarization of helium-like lines and to interpret it from an anisotropic EVDF.

Polarization measurements were repeated on the same vacuum spark machine by Baronova et al. [32]. This time, two focusing Johann spectrometers were used; these spectrometers were equipped with quartz crystals ($2d = 0.851$ nm) which were of a special design. These crystals were fitted on cylindrical glass substrates by optical contact. The dispersion planes were perpendicular to the discharge axis. The distance of the crystals from the plasma was about 140 mm. Time integrated spectra of the FeXXV lines were recorded in the second ($\theta_B = 25°$) and third ($\theta_B = 41°$) orders of reflection by these two spectrometers, respectively. In the third order, the Bragg angle is close to the Brewster angle 45°, so that virtually only the polarized component perpendicular to the dispersion plane, the $\sigma$ light, was reflected. To obtain enough exposure 5,000 shots were necessary. The location of the hot spot(s), the dimension of which was found to be 0.1–0.5 mm, fluctuated shot-to-shot. The radial (1 mm) and axial (1.5 mm) displacements were, however, within the field of view of the spectrometers. The result is shown in Fig. 8.4. Both the spectra show the FeXXV ($\lambda = 0.185$ nm) $1s^2\,^1S_0$–$1s2p\,^1P_1$ transition, the resonance line ($w$), and the $1s^2\,^1S_0$–$1s2p\,^3P_1$ intercombination

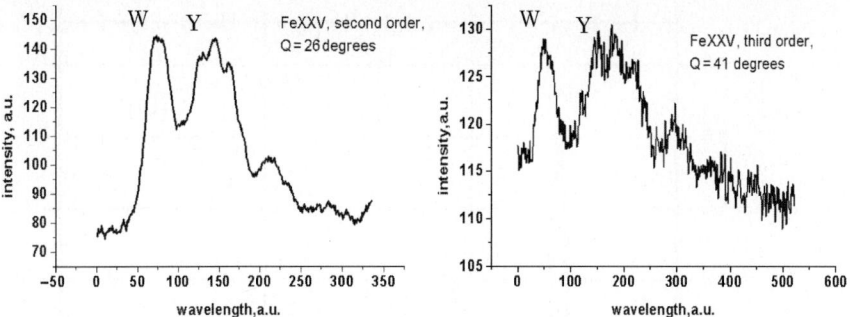

**Fig. 8.4.** Spectra of helium-like FeXXV lines produced by a vacuum spark of $I =$ 150 kA. (**a**) The second and (**b**) third orders. $\lambda_w = 1.8500$ Å, $\lambda_y = 1.8591$ Å

line ($y$), although the latter line is not well resolved. All the lines appear broadened; this broadening is mainly due to the combination of the effects of the plasma motion, opacity, Stark effect and, to a less extent, to the presence of lithium-like satellite lines. The two spectra in Fig. 8.4a, b are substantially different.

The spatial averaging done by the instrument should tend to reduce, in effect, the effects of the temperature and density gradients in and around the hot spots, so that the recorded spectra are expected to result in similar patterns. The different azimuthal directions for the spectrometers should produce no difference: it was confirmed that two almost identical spectrometers oriented in the same plane showed almost the same spectra. The crystal reflectivity can be assumed to be the same for all the lines in the helium-like group in a same reflection order. Therefore, the difference in the two spectra is explained by the presence of polarization of the observed lines. For polarization of these lines, see Fig. 6.4: the $w$-line can be substantially polarized but the $y$-line is little polarized. In reality, hyperfine structure further reduces the polarization of the latter line.

Walden et al. [39] carried out polarization measurements of the helium-like lines including AlXII ($\lambda = 0.776$ nm) $1s^2\,^1S_0$–$1s2p\,^1P_1$ on a vacuum spark. Two Johann spectrometers of identical design equipped with ADP crystals ($2d = 1.0648$ nm), so that the Bragg angle was almost exactly 45°, were arranged with their optical axes directed perpendicular to each other. One spectrometer was fixed while another was rotated around it's optical axis. The spectra were recorded on the DEF film by time integration over 5 shots. The field of view of both the spectrometers covered the same volume.

An example of the results is shown in Fig. 8.5; this is for the case of the dispersion planes directed perpendicular to each other. The apparent intensity of the resonance line ($w$) is different in the two spectra. The polarization degree of the $w$-line was deduced to be 0.12, while the intercombination line ($y$) was unpolarized. The authors performed a theoretical interpretation by using a steady-state CR model. Their model was based on a Maxwellian bulk plasma with admixture of 1% of anisotropic non-Maxwellian electrons. The ionization

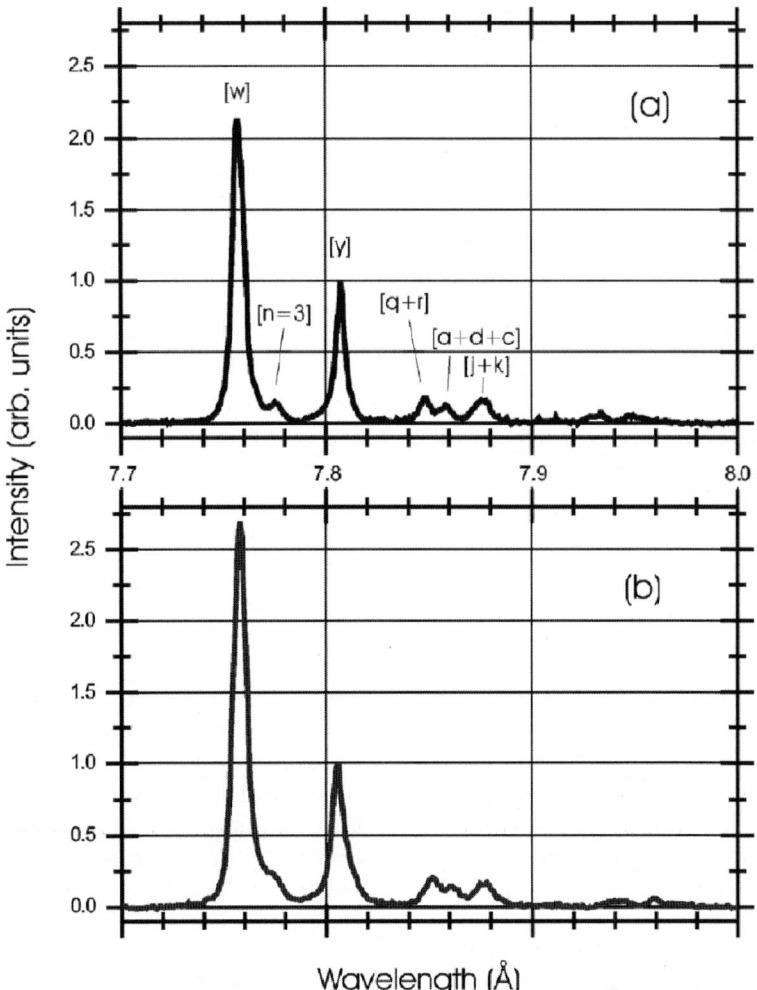

**Fig. 8.5.** Spectra of helium-like Al lines taken by two mutually perpendicular spectrometers in the first order (Quoted from [39], with permission from The American Physical Society.)

temperature for the ions was found to be 250 eV for the Maxwellian electron temperature of 350 eV. They considered the optical thickness of the $w$-line and found that, for $n_e = 10^{26} \, \mathrm{m}^{-3}$, the optical thickness was $\tau \approx 1$ at the line center. They ignored the effect of this opacity on the observed polarization. They succeeded in reproducing theoretically the observed polarization degree from the anisotropic EVDF as assumed above.

Two spectrometers with the dispersion planes perpendicular to each other were used by Shlyaptseva et al. [40] for polarization measurements of helium-like Ti lines, emitted from an X-pinch machine. The relative intensities of the

$w$-, $y$- and $q$-lines were found different for the two spectrometers, where the $q$-line is a lithium-like satellite line. The authors interpreted this as due to polarization of these lines, and they ascribed the polarization to the presence of a strong electron beam with energy above 5 keV. The X-ray spectral lines of Be-, B-, C-, N-, and O-like Ti ions in the spectral region above 0.26 nm were also found polarized. No polarization was found for Fe lines.

In conventional plasma diagnostics, the relative intensity of the $w$- and $y$-lines ($I_R/I_I$ ratio, where R means "resonance" and I "intercombination") is used to estimate the plasma density. As suggested by (4.5) and discussed in Appendices A and E, an observed intensity is affected by the presence of polarization from two aspects: different sensitivities of the spectrometers for different polarized components and anisotropic intensity distribution of a polarized light. We have seen several examples above. It should therefore be noted that this simple procedure may lead to an incorrect result, and that the polarization of the lines should be properly taken into account for a correct interpretation of the observed line intensities.

### 8.2.2 Plasma Focus and Gas $Z$-Pinch

The Russian word for "focus" has meanings "joke" or "trick" besides the usual meaning. The naming of the "plasma focus" appears to intend to connote even the former meanings. A plasma focus machine focuses a plasma near the geometrical discharge axis and generates hot spots, the number of which is random in each shot and their appearance is irreproducible. These hot spots are very small, i.e., $r \approx 0.1$–$0.5$ mm, and are distributed around the discharge axis.

The systematic polarization study of the helium-like lines, radiated from the Mather-type plasma focus, was carried out in the last decades at the Andrzej Soltan Institute for Nuclear Studies, Poland. Argon was used as the working gas and the machine was operated at the current of 500 kA. Jakubowski et al. [34] found that hot spots were formed independent of each other and their local plasma parameters seemed to be of a stochastic nature. Even so, space-resolved spectra demonstrated that they had a certain correlation with the time-integrated pinhole images of the individual hot spots. It was thus possible to determine the hot spot position with respect to the anode end, and at the same time, to estimate plasma parameters of the individual hot spots.

In their succeeding paper [28], they concluded that the intensity pattern of the helium-like lines is correlated with the lifetime of the hot spot. They found that strong pulsed electron beams are emitted in the directions perpendicular to the discharge axis besides the beam along the axis. An interesting observation was that the observed spectra of the helium-like lines appeared to be correlated also with the presence of these electron beams.

The study of polarized helium-like argon lines is described in [32]. Two focusing Johann spectrometers were used; they were equipped with quartz

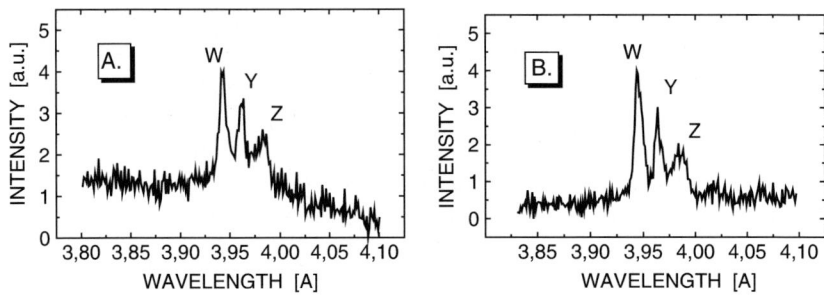

**Fig. 8.6.** Spectra of helium-like ArXVII lines taken by two spectrometers. The dispersion plane is parallel to the discharge axis for both the spectrometers. $I = 500\,\text{kA}$ (Quoted from [32].)

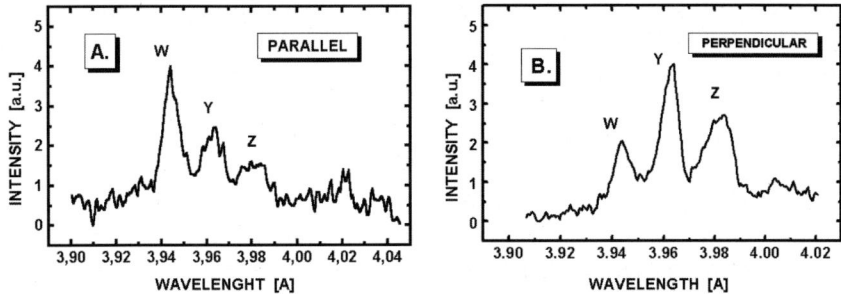

**Fig. 8.7.** Spectra of helium-like ArXVII lines taken by two spectrometers. (**a**) The same as Fig. 8.6. (**b**) Dispersion plane perpendicular to the discharge axis (Quoted from [32].)

crystals having $2d = 0.851\,\text{nm}$ (spectrometer A) and $0.667\,\text{nm}$ (spectrometer B), respectively, so that the Bragg angles were $\theta_B = 27.7°$ and $36.4°$, respectively, for the $w$-line. In the first experiment, the dispersion planes of both the spectrometers were oriented parallel to the discharge axis. Figure 8.6 shows the spectra of the ArXVII lines: the $w$-line is well resolved and the $y$-line is blended with the lithium-like satellite lines.

In the second experiment, spectrometer B was rotated by 90°, so that its dispersion plane is perpendicular to the discharge axis. The two spectrometers were carefully aligned to observe the same plasma region that contained only a single hot spot. The intensity of radiation was high enough to record the spectrum in a single shot. Figure 8.7 shows the result. The relative intensity of the $w$-line in Fig. 8.7b is substantially different from that in Fig. 8.7a as well as from Fig. 8.6. From the same reasoning as "Vacuum sparks," we may conclude that this difference is interpreted as due to strong polarization of the $w$-line. Unfortunately, however, since the crystals were not calibrated for efficiencies for different polarizations, it was not possible to estimate quantitatively the polarization degree. A further experiment was performed [35], and the above conclusion was confirmed.

Space-resolved and time-resolved studies of polarization of the helium-like lines were carried out on the same plasma focus machine and with the same spectrometers. In [41], a space-resolved spectrum was obtained with spectrometer B, while a space-integrated spectrum was obtained with A. The $I_R/I_I$ ratio from B was 1.5, which was the average over three hot-spots. This ratio is substantially different from $I_R/I_I = 2.01$ of the space-integrated spectrum from A. Possibilities of experimental errors due to photon statistics were investigated in numerous plasma focus experiments and denied. This difference was therefore explained as due to polarization of the line ($w$).

In [42], the X-ray films were replaced by two sets of miniature scintillators of the NE102A type. An individual detector recorded a single spectral line, i.e., $w$, $y$, or some other lines. Two spectrometers were arranged so that the dispersion planes were parallel and perpendicular to the discharge axis, respectively. Output signals were recorded with a Tektronix 4-channel oscilloscope simultaneously. The traces of the signals had FWHM of the order of 10 ns. The statistical analysis was performed on the results over 300 shots. The $I_R/I_I$ ratio was 1.17 from the perpendicular spectrometer and it was 2.0 from the parallel one. The difference may again be explained from different polarizations of these two lines.

Besides the presence of polarization as discussed above, other important features of the observed spectra were: (1) the width of the $w$-line is larger than that of the $y$-line for ArXVII lines at the current 500 kA and 3 MA (at the current 150 kA, the $y$-line was too weak to be resolved), and (2) the relative intensity of the $y$-line was higher than expected.

The Zeeman and Doppler widths were estimated to be smaller than the measured width of the $w$-line. It was estimated that the Stark broadening (shifts) could become consistent with the measured width (10 mÅ) of the $w$-line only if we have an extremely strong field of $E > 10^{12}\,\mathrm{V\,m^{-1}}$. No experimental evidence exists that confirms the presence of such a strong field. A theoretical estimate [43], however, predicts the electric field strength of $E \approx 10^{12}\ \mathrm{V\,m^{-1}}$. Thus, the broadening of the $w$-line may be ascribed, at least partly, to the plasma electric field. Another factor to be considered is the opacity. The dimension of the hot spot is much larger than the estimated absorption length of the $w$-line, $L_w \approx 2\,\mu\mathrm{m}$, at the line center. The apparent broadening of the $w$-line may be due to the opacity effect. It is noted, however, that the observed intensity is not affected by this opacity effect. This is because the upper level of the $w$-line is in the density region of corona equilibrium, and the decrease in the effective transition probability is compensated exactly by the increase in the upper-level population.

Under the experimental conditions, the apparent $y$-line intensity may have the contribution from the forbidden line $1^1S_0$–$2^1S_0$; the strong electric field mixes the 2P wavefunction into the 2S wavefunction, and the energy of $2^1S_0$ is close to the upper level of the $y$-line, $2^3P_1$. It was also calculated how the intensity of the $w$-line is affected by a strong electric field. How the forbidden line affects the polarization characteristics of the $y$- and $w$-lines is under study.

Summarizing the polarization measurements on $z$-pinches, polarization was observed at the current 150 kA (vacuum spark) and 500 kA (plasma focus), but no sign of polarization was found at the current 3 MA (gas pinch).

As noted near the close of the introduction before Sect. 8.2.1, the observed spectra including polarization characteristics would be interpreted by an adequate theoretical framework, which consists of an MHD model and a CR model. An attempt in this direction was presented by Baronova et al. in their review of the polarization measurements on $z$-pinches [35]. We discuss briefly their results. Polarization of lines is a consequence of directional electron beams and/or the strong electric/magnetic fields. As noted already, in time-integrated measurements, the important point is how long the line emission coexists with these anisotropy factors. Figures 8.8 and 8.9 show examples of the results of an MHD calculation incorporated with a CR model, although they are still on a preliminary stage. These figures show the temporal developments of various plasma parameters, including the electron beam, and the intensities of the helium-like and hydrogen-like lines. These figures suggest the "overlap" of these anisotropy factors with the line emissions, so that polarization of the lines, is larger for smaller discharge currents. However, even for the discharge current of 500 kA, the space and time averaged degree of plasma anisotropy would be small. A time- and/or space-resolved data could reveal these changing polarization degrees.

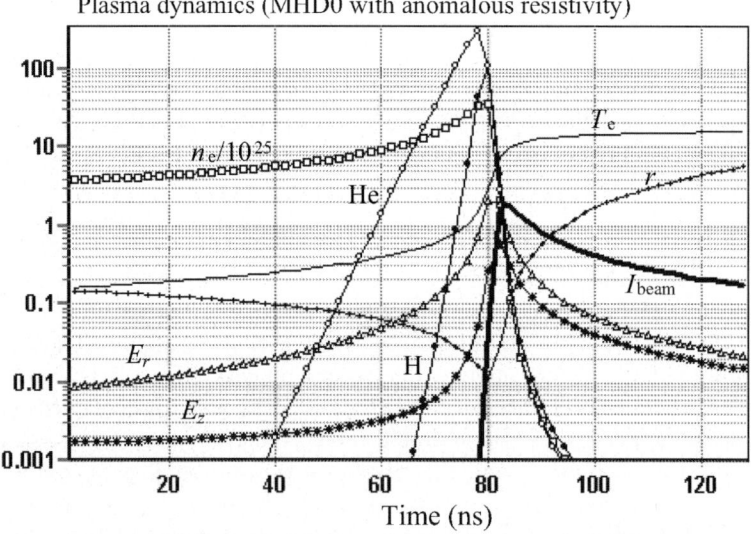

**Fig. 8.8.** Results of MHD calculation for $I = 500$ kA. $n_e$: electron density in m$^{-3}$; $T_e$: electron temperature in keV; $E_z$, $E_r$: axial and radial electric fields in MV cm$^{-1}$; $r$: plasma neck radius in cm; $I_{beam}$: electron beam in kA generated at the neck. Results of the time-dependent CR model. H, He: intensity of resonance lines of H- and He-like ions of argon (arbitrary units)

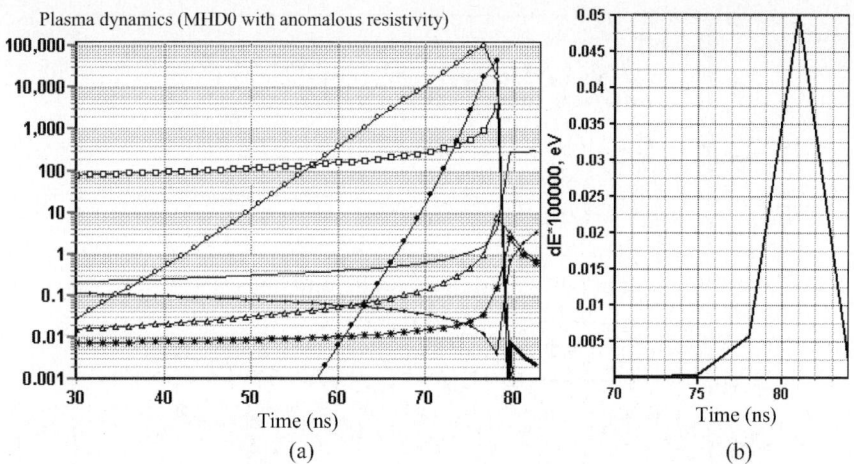

**Fig. 8.9.** (a) Similar result to Fig. 8.8 except that $I = 3$ MA. (b) Stark shift of the $^1P_1$ level of ArXVIII

We also have to note that the two-crystal scheme, i.e., two spectrometers are mounted in an azimuthal plane, is valid under the condition of axial symmetry of the plasma. Otherwise, the only way to investigate polarization without any ambiguity is to use the X-ray polarimeter which is described in Sect. 15.2.

As stated at the beginning of this section, we are still at the developing stage in both the experimental and theoretical techniques of polarization spectroscopy. When these techniques reach a satisfactory level, polarization measurements will become a powerful tool to study electric fields and electron beams generated in the plasma.

(E.O. Baronova)

## 8.3 Laser-Produced Plasmas

Kieffer et al. [44, 45] illuminated an aluminum target with a $1.053\,\mu$m laser pulse of 1 ps duration ($8 \times 10^{14}$ Wcm$^{-2}$: only in this section and Chap. 9 we use different units from the SI units, because it is a common practice to use these units in this research field) preceded by a prepulse of 60 ps and $8 \times 10^{11}$ Wcm$^{-2}$; the preplasma of $kT_e \simeq 100$ eV produced by the prepulse was heated by the main pulse to 500 eV. They observed helium-like lines and lithium-like satellite lines of aluminum with a Johann-type spectrometer based on an ADP crystal ($2d = 1.064$ nm) with 300-mm radius of curvature. The Bragg angle was 46.8°, so that the polarization separation efficiency was quite high (see Chap. 15). The line of sight was 85° with respect to the laser beam which was normal to the target surface. The spectrometer was rotated around its line of sight by 90°, so that intensities of the polarized component in the direction of the incident laser beam and those in the perpendicular direction

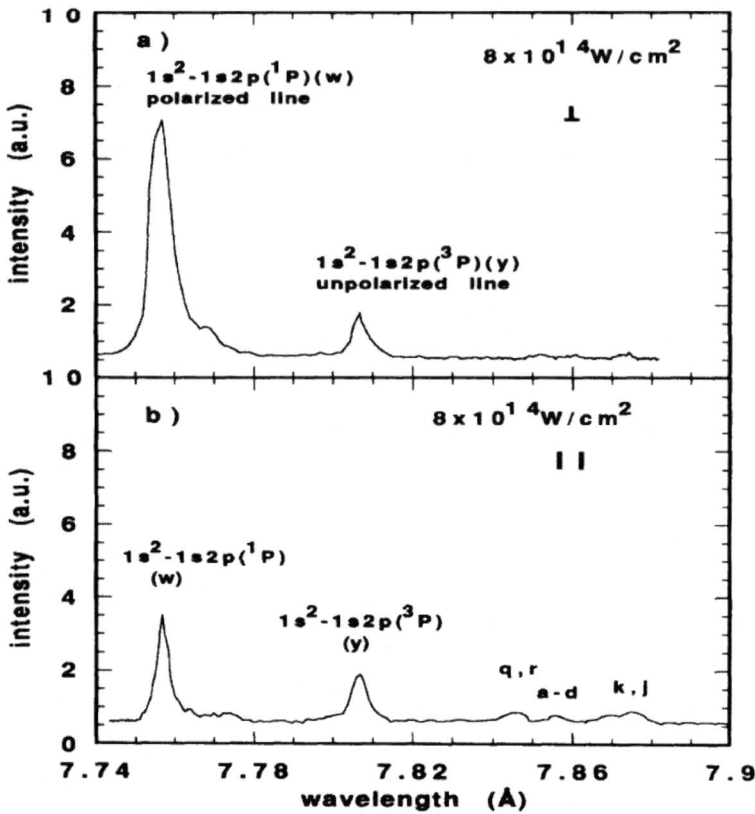

**Fig. 8.10.** Helium-like and lithium-like aluminum lines from a laser-produced plasma. The polarized components are resolved (Quoted from [44], with permission from The American Physical Society.)

were recorded separately. The plasma was located off the Rowland circle and a diaphragm was used so that, in the both configurations, exactly the same area of the crystal was illuminated. Figure 8.10 shows the recorded spectra accumulated over 30 shots: (a) the polarized component in the perpendicular direction and (b) in the parallel direction.

These spectra show that the $y$-line ($1^1S_0$–$2^3P_1$) is almost unpolarized, while the $w$-line ($1^1S_0$–$2^1P_1$) is strongly polarized in the direction perpendicular to the laser beam with the polarization degree $(-25 \pm 7)\%$. The lithium-like satellite lines in the longer wavelength region are weak, but seem to show positive polarization. From the line intensity ratio $y/w$, they estimate the emission zone to have electron densities of $(0.3$–$0.5)n_c$, where $n_c = 10^{27}\,\mathrm{m}^{-3}$ is the critical density for the laser wavelength. As is shown in Fig. 6.4, the polarization of the $y$-line excited by beam electrons is quite small near the excitation threshold. Note that this figure is for helium-like iron, but the overall feature is quite similar for aluminum. Aluminum has a nuclear spin of 5/2, so that the hyperfine interaction further reduces drastically the polarization of the $y$-line. Thus, the present observation of the unpolarized $y$-line is quite reasonable. The negative polarization of the $w$-line indicates that EVDF is oblate (see Fig. 6.4).

The polarization of emission lines was independent of the polarization direction of the laser light. So, the observed polarization was of a plasma origin. They calculated EVDF with the 1-D Fokker-Planck code. Negative $f_2/f_0$, i.e., oblate distributions were obtained in the underdense plasma region which was the emission zone, while positive and small $f_2/f_0$ beam-like distributions were observed in the overdense regions close to the target surface. This was interpreted as due to the nonlocal transport of high-energy electrons from the underdense region into the low-temperature overdense plasma, resulting in a depletion of the axial component of hot electrons in the emission zone. The estimated polarization degree of the $w$-line was $-(10-15)\%$. The measured polarization degree is higher than the calculation, indicating that the actual EVDF is more oblate than calculation. The positive polarization of the lithium-like satellites, which are produced in the overdense region, is consistent with this picture.

Another experiment [45] was reported in which a nanosecond prepulse of $2 \times 10^{10}$ Wcm$^{-2}$ was followed by the main pulse of 400 fs with $10^{16}$ Wcm$^{-2}$. The polarization degree of the $w$ line was observed to be $(-5 \pm 5)\%$. The corresponding calculation gave the value $-5\%$, being consistent with the experiment.

In the context of the fast-ignitor plasma research, Inubushi et al. [46] investigated EVDF in a plasma produced by ultrashort laser irradiation. A $C_8H_7Cl$ tracer layer with thickness of 0.11 μm is sandwiched by a CH overcoat (thickness 0–0.16 μm) and the $C_3H_4O_3$ substrate (thickness 1.2 mm). An 800-nm, 130-fs laser pulse irradiated the target with irradiance of $(1-1.3) \times 10^{17}$ Wcm$^{-2}$. The irradiation was almost normal to the target surface. The main pulse was preceded by a prepulse of 130 fs and $10^{11}$ Wcm$^{-2}$ by 13 ns: the prepulse produced a preplasma of the scale length of 3 μm. A toroidally curved quartz crystal spectrometer observed the helium-like Cl $1^1S$–$2^1P$ resonance line from the direction 85° with respect to the target normal. The detector was a cooled CCD and 6,000 shots were accumulated to obtain enough signals. The Bragg angle was 41.7°, so that the s-light was predominantly reflected. The spectrometer was rotated by 90° to observe the perpendicularly polarized components. Figure 8.11 shows examples of the spectra, where the quantization axis is taken as target normal. The relative intensities of the resonance line depend on the overcoat thickness. Figure 8.12 shows the polarization degree against the depth of the tracer layer, where the $\pi$ and $\sigma$ component intensities were calculated from the observed intensities on the assumption of perfect crystal (see Chap. 15). According to the simulation, $n_e$ of the tracer plasma with 0.11-μm depth is about $1.5 \times 10^{21}$ cm$^{-3}$, and higher for larger depths. Figure 8.12 suggests that EVDF in the plasma with 0.15 and 0.21-μm tracer depths is cigar-like along the surface normal, and that with

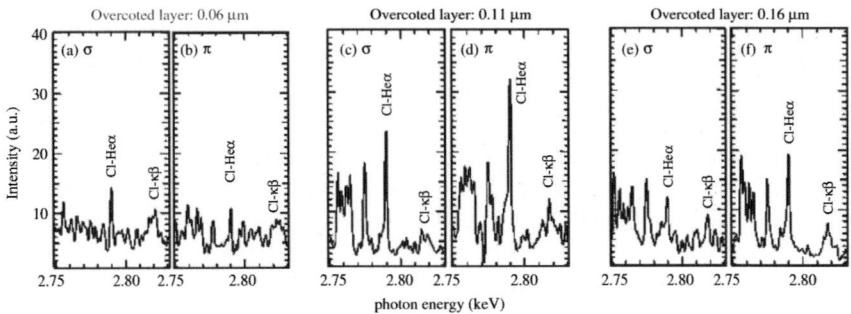

**Fig. 8.11.** Polarized components of spectra including resonance line of hydrogen-like chlorine for various depths (Quoted from [46], with permission from Elsevier.)

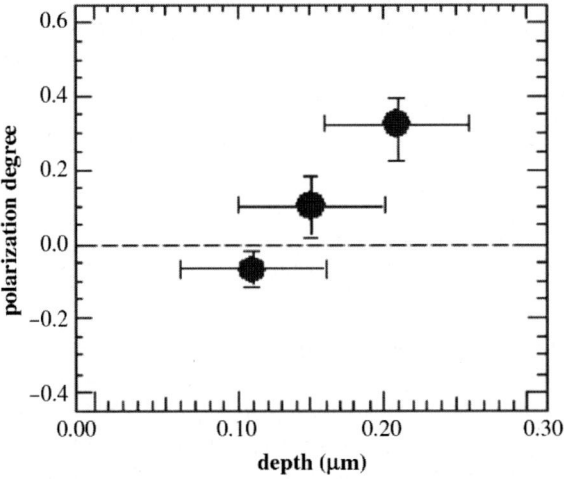

**Fig. 8.12.** Polarization degree against the depth of chlorine (Quoted from [46], with permission from Elsevier.)

0.11-μm is pancake-like. This result is consistent with the particle simulation result which shows that hot electrons created by absorption of the laser field are transported into the target.

(T. Fujimoto)

## 8.4 Magnetically Confined Plasmas

### 8.4.1 Tokamak Plasmas

In tokamak plasma diagnostics, the possibility of polarization of X-ray emission lines from the levels excited by run-away electrons is discussed in [47]. The magnetic sublevels of singlet and triplet states of helium-like ions $^1P_1$, $^3P_{1,2}$, and $^3S_1$ may not be populated statistically with the preferred direction for the electron impact excitation, and the alignment in these levels results in polarization of the emission lines.

The line intensity ratio between the fine structure levels has been expected to be proportional to the statistical weights in tokamaks and solar flares where the effect of ion collisions can be neglected. However, the line intensity ratio is sometimes found to deviate from the statistical weights. The intensity ratio of the Lyman $\alpha$ fine structure splitting doublet lines $\beta = I(1^2S_{1/2}-2^2P_{1/2})/I(1^2S_{1/2}-2^2P_{3/2})$ of hydrogen-like MgXII observed in a solar flare, which is expected to be one half, is reported to be $\beta = 0.67$ [48]. In the Alcator C tokamak, the observed intensity ratio for hydrogen-like SXVI varies from $\beta = 0.5$ to $0.8$ for densities $1-4 \times 10^{20}\,\text{m}^{-3}$ [49]. The line ratio $\beta$ for hydrogen-like TiXXII emitted from the JT-60 tokamak is measured with a crystal spectrometer, and $\beta$ varies from 0.494 to 0.572 [50]. One possible origin of this intensity ratio anomaly is considered to be the presence of polarization in the second component line. The Bragg angle of the crystal spectrometer for the TiXXII experiment is $40.6°$, which is close to $45°$, and the light polarized perpendicularly to the incidence plane is selectively reflected. On the assumption that all the electrons are anisotropic fast electrons with their energy of $100\,\text{keV}$, the emission line is found polarized and the apparent intensity ratio is estimated to be 0.53 [50].

The first application of PPS on a magnetically confined plasma is performed on the WT-3 tokamak [51]. The major and minor radii are $1.3\,\text{m}$ and $0.4\,\text{m}$, respectively, and the toroidal magnetic field is $1.5\,\text{T}$. The electron density is typically $3 \times 10^{18}\,\text{m}^{-3}$ at the center, and the electron temperature is in the range of 100–400 eV. The central region of the plasma is viewed with a uv-visible spectrometer with steering mirrors. A 5.4-mm thick calcite plate is placed behind the entrance slit for polarization separation. The toroidal direction is taken as the quantization axis. The $\pi$ light is the e-ray and displaced by $1\,\text{mm}$ from the $\sigma$ light, the o-ray in the direction of dispersion. The dispersed light is detected by a microchannel plate (MCP) equipped with a 512-channel photodiode array detector. Examples of the observed spectra in the ultraviolet wavelength region of helium-like CV and boron-like OV are shown in Figs. 8.13 and 8.14, respectively. The relative intensities of $\pi$ and $\sigma$ components apparently change during the course of discharge, indicating the presence of and a change in polarization. In this experiment, however, an uncertainty analysis as will be discussed in Sect. 14.2 is not done, and the presence of polarization may not be conclusive.

### 8.4.2 Cusp Plasma

Figure 1.1 of this book introduced the readers to the world of PPS. That picture was of a cusp plasma. The procedure to obtain the polarization map will be given in Chap. 14. In the present chapter we review other aspects of the experiment [52]. The experimental setup for PPS is shown in Fig. 8.15. A pair of coils produce a magnetic field with a cusp configuration, and a microwave of

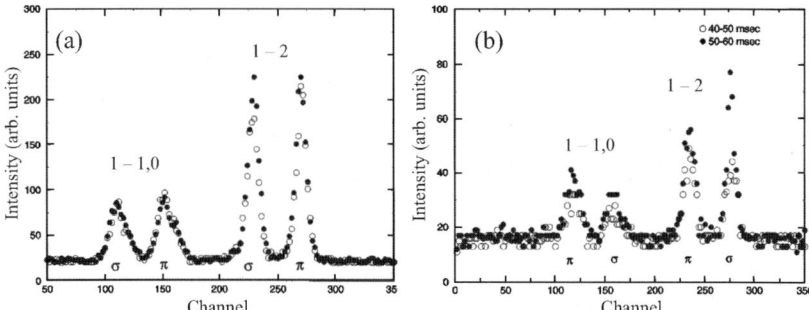

**Fig. 8.13.** Polarization resolved spectra of the helium-like carbon lines, CV (2s $^3$S$_1$–2p $^3$P$_{1,0,2}$) observed from plasmas. (**a**) In the joule heating mode for different shots. The two peaks at right are the $\pi$ (e-ray) and $\sigma$ (o-ray) components of the $J = 1$–2 line. The weaker lines ($J = 1$–1, 0) are unresolved and the $\pi$ and $\sigma$ components each appears as aggregates. (**b**) In the lower-hybrid current drive mode (*open circle*) and in the (lower-hybrid current drive + electron cyclotron heating) mode (*filled circle*). The relative positions of the $\pi$ components have been interchanged from (**a**) (Quoted from [51], with permission from The American Physical Society.)

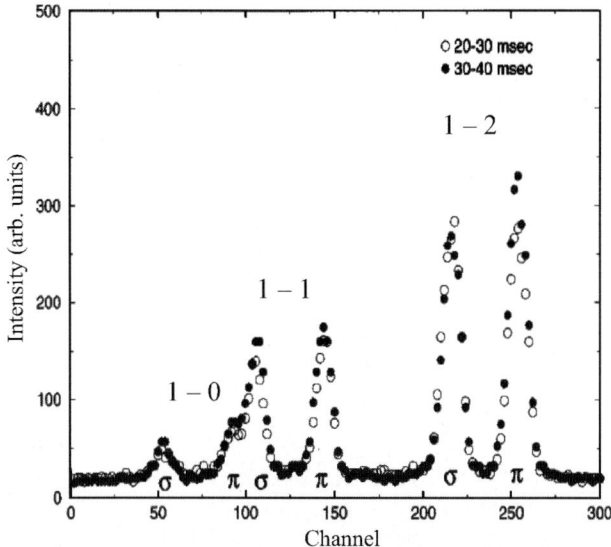

**Fig. 8.14.** Polarization resolved spectra of the beryllium-like oxygen lines OV (3s $^3$S$_1$–3p $^3$P$_{0,1,2}$) from tokamak plasma in the joule heating mode (*open circle*) and in the lower-hybrid current drive mode (*filled circle*). The two peaks at right are the $\pi$ and $\sigma$ polarized components of the $J = 1$–2 line (Quoted from [51], with permission from The American Physical Society.)

2.45 GHz is fed into the vessel containing helium with pressure $2.3 \times 10^{-2}$ Pa. The electron cyclotron resonance (ECR) field strength is 87.5 mT, and the ECR surface is a spheroid with short diameter of 92 mm along the central axis

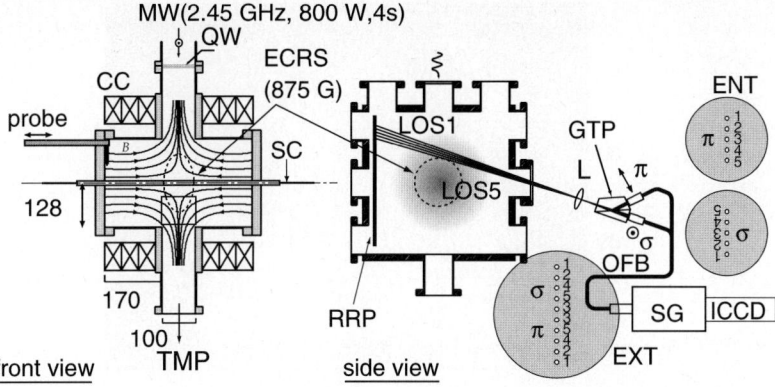

**Fig. 8.15.** The experimental setup for PPS on ECR helium cusp plasma. Dimensions are in millimeters. MW: microwave, QW: quartz window, CC: cusp coils, ECRS: electron cyclotron resonance surface of 87.5 mT, SC: Stuffer conductor, no currents in the present experiment, RRP: reflection reducing plate made of knifes edge, LP: linear polarizer, BPF: bandpass filter, DC: digital camera, LOS: line of sight, L: lens, GTP: beam splitting Glan-Thompson prism, OFB: optical fiber bundle, SG: spectrograph, ICCD: image-intensified charge-coupled device (Quoted from [52], with permission from IOP Publishing.)

and long diameter of 168 mm on the central plane. The plasma is observed through the side viewing port. The plasma image is recorded with a digital camera to produce the polarization map, as discussed in Chap. 14.

Figure 8.16 shows the image of the plasma which is at the position of line-of-sight (LOS) 1–5 as discussed below. Figure 8.16a, b is the plasma image and the polarization map by the 501.6-nm ($2^1S – 3^1P$) line, respectively; this figure is the same as Fig. 1.1. The polarization directions are perpendicular to the magnetic field. The polarization degree $P$ increases toward the outer region. $P$ exceeds 10% in the line cusp, or the extreme edge of the cusp magnetic field, and is 5% around the ECR surface.

As shown in Fig. 8.15, the polarization separation optics (PSO) consists of a lens, a beam splitting Glan-Thompson polarizer and an optical fiber bundle. The PSO is tilted by 15° to resolve the polarized components which are parallel ($\pi$ light) and perpendicular ($\sigma$ light) to the magnetic field line (see Fig. 8.16a). The light emanating from the fiber bundle goes into the spectrometer to be dispersed and recorded.

Figure 8.17a shows an example of the polarization-resolved spectra taken with 30 ms exposure on LOS 1. Three emission lines are He I 492.2 nm ($2^1P$–$4^1D$), 501.6 nm ($2^1S$–$3^1P$), and 504.8 nm ($2^1P$–$4^1S$). Since the upper level of the 504.8-nm line is a $^1S$ state, this line emission is never polarized and is used as the reference for polarization determination of other lines. The $\sigma$-component intensities $I_\sigma$ of the 501.6 and 492.2-nm lines are found to be higher than the corresponding $\pi$-component intensity $I_\pi$. The total intensity and the

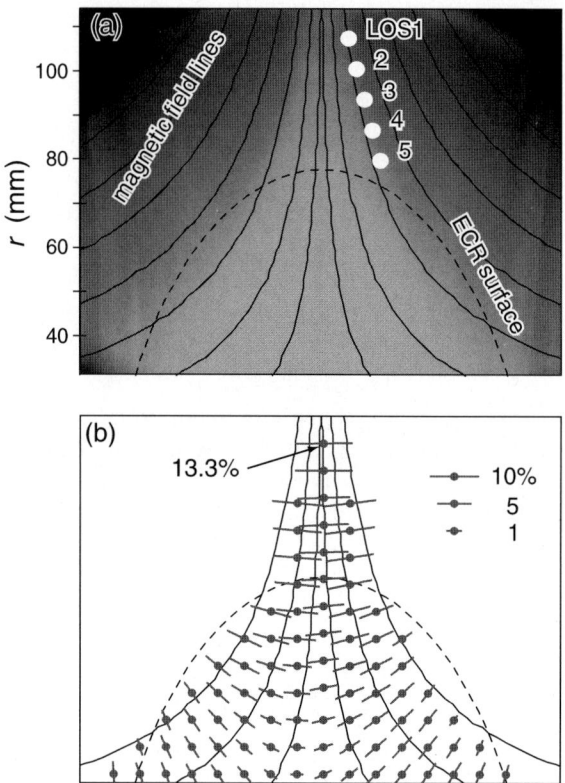

**Fig. 8.16.** (a) Plasma image with a bandpass filter for the $\lambda$501.6-nm line. (b) Polarization map of the He I 501.6-nm ($2^1$S–$3^1$P) line. *Solid* and *dashed curves* are magnetic field lines and the ECR surface on the perpendicular plane. The direction of the line indicates the polarization direction and the length is proportional to the square root of the polarization degree. The positions of LOS 1–5 are also given (Quoted from [52], with permission from IOP Publishing.)

longitudinal alignment of an emission line are defined as $I = \frac{2}{3}(I_\pi + 2I_\sigma)$, $A_L = (I_\pi - I_\sigma)/(I_\pi + 2I_\sigma)$, respectively according to (4.6) and (4.7). A total 224 spectra in the four discharges are processed to construct a histogram of the observed $A_L$. Figure 8.17b shows an example of the histograms of $A_L$ for the three lines. The intensities of the 492.2-nm line and the 504.8-nm line are nearly equal and the dispersions of their distributions are almost equal, too. The intensity of the 501.6-nm line is higher than those of the 492.2- and 504.8-nm lines. The dispersion of 501.6 nm is narrower. This fact indicates that the longitudinal alignments or polarization degrees of the 501.6- and the 492.2-nm lines are quite reproducible. The values of $A_L$ for both the 492.2- and 501.6-nm lines are negative, all from LOS 1 to LOS 5. The absolute value of $A_L$ decreases with an increase in helium pressure to almost diminish at $p_{He} = 1.1$ Pa.

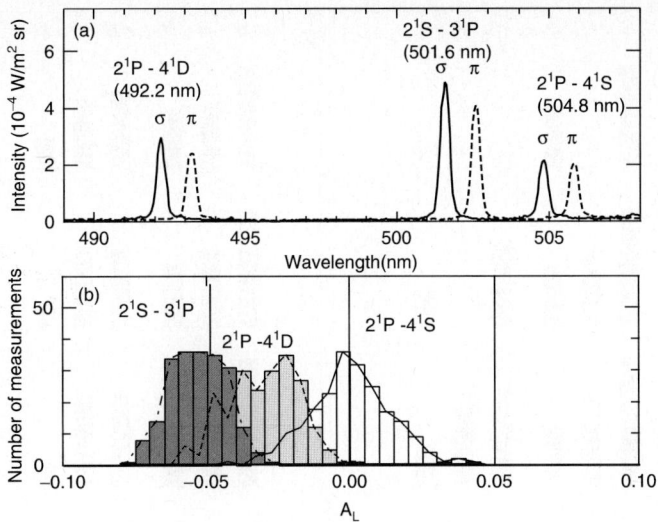

**Fig. 8.17.** (a) Polarization-resolved spectra observed on LOS 1. The $\pi$ and $\sigma$ components are shown by the *dashed* and *solid curves,* respectively. The $\pi$ component is displaced by 1 nm. (b) Histograms of the longitudinal alignment $A_L = (I_\pi - I_\sigma)/(I_\pi + 2I_\sigma)$. The $2^1P$–$4^1S$ line is used to calibrate relative sensitivities of the observation system (Quoted from [52], with permission from IOP Publishing.)

The total intensity $I_0$ of emission lines $s \leftarrow p$ is proportional to the population $n(p)$ of the upper level $p$ (4.6):

$$I_0(p, s) = n(p)A(p, s)\hbar\omega\frac{\mathrm{d}V\mathrm{d}\Omega}{4\pi}, \qquad (8.4)$$

where $A(p, s)$ is Einstein $A$ coefficient, $\mathrm{d}V$ is the volume of the observed region, and $\mathrm{d}\Omega$ is the solid angle subtended by our optics. Figure 8.18a shows $n(p)/g(p)$ in a logarithmic scale, where $g(p)$ is the statistical weight. The longitudinal alignment $A_L$ is proportional to the relative alignment, i.e., alignment $a(p)$ divided by $n(p)$ (4.8),

$$A_L(p, s) = (2L_p + 1)\left(\frac{3}{2}\right)^{1/2}(-1)^{L_p+L_s}\left\{\begin{array}{ccc} L_p & L_p & 2 \\ 1 & 1 & L_s \end{array}\right\}\frac{a(p)}{n(p)}, \qquad (8.5)$$

where {} is the 6-$j$ symbol and $L_i$ is the orbital angular momentum quantum number of level $i$. Note that we assume singlet levels here. We convert the observed $A_L$ into $a(p)/n(p)$. For triplet levels, depolarization caused by fine structure is corrected for. Figure 8.18b shows the result of the measurement on thirteen emission lines observed.

The population-alignment collisional-radiative (PACR) model for He is developed. Alignment is considered on the levels $n^1P$, $n^1D$, $n^3S$, $n^3P$, and $n^3D$ with $n \leqslant 7$. Two sets of alignment production cross sections $Q_0^{0,2}$, (4.40b)

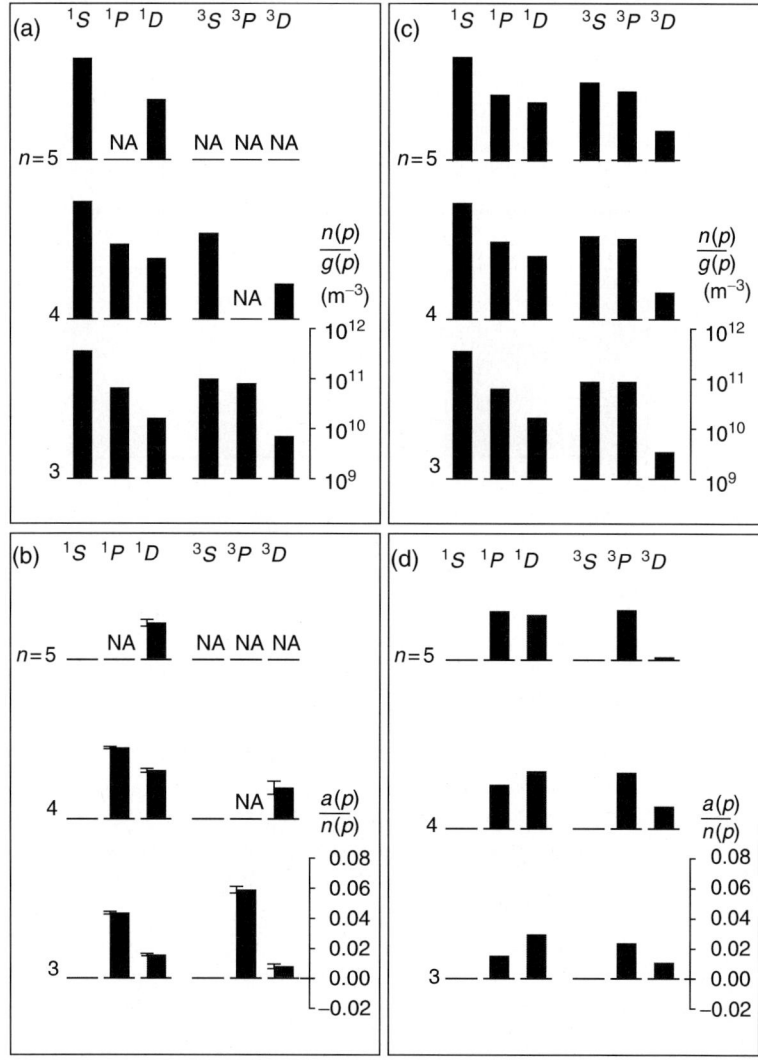

**Fig. 8.18.** The experimentally obtained intensity and longitudinal alignment are converted, respectively, into (**a**) the population per unit statistical weight, $n(p)/g(p)$, and (**b**) the relative alignment, $a(p)/n(p)$. The synthesized (**c**) $n(p)/g(p)$ and (**d**) $a(p)/n(p)$ by the population-alignment collisional-radiative model for helium. NA indicates that $n(p)/g(p)$ and $a(p)/n(p)$ are not available because of low signal intensity (Quoted from [52], with permission from IOP Publishing.)

or (5.41), are based on two theoretical calculations; one is calculated by the distorted wave approximation (DWA) [53] and the other by the convergent close-coupling (CCC) method [54].

The results of the PACR model with an assumption of isotropic thermal EVDFs are confirmed to coincide with those from the conventional CR model. We try to interpret the experimental population and alignment distributions shown in Fig. 8.18a, b in terms of a single set of plasma parameters. As a model EVDF of gyrating electrons accelerated by the ECR microwave field, we adopt a Saturn-type EVDF: an isotropic Maxwell velocity distribution $f_{th}(v)$ with temperature parameter $T_{eth}$ superimposed by a ring component of a shifted Maxwell distribution $f_r(v, \theta)$ with temperature parameter $T_{er}$. The direction of the displacement $V_r$ is perpendicular to the quantization axis in the velocity space. We assume $f(v, \theta) = f_{th}(v) + f_r(v, \theta)$, where the central-thermal and ring components are given by

$$f_{th}(v, \theta) = (1 - \alpha)2\pi \left( \frac{m}{2\pi k_B T_{eth}} \right)^{3/2} \exp \left( \frac{-mv^2}{2k_B T_{eth}} \right) \qquad (8.6)$$

and

$$f_r(v, \theta) = \alpha 2\pi A_r \exp \left( \frac{-m(v^2 - 2vV_r \sin\theta + V_r^2)}{2k_B T_{er}} \right), \qquad (8.7)$$

respectively. Here $m$ is the electron mass. $A_r$ is the normalization factor defined by the normalization $\iint f_r(v, \theta)v^2 \sin\theta dv d\theta = \alpha$, where $\alpha$ represents the fraction of the electron densities of the ring component.

An ionizing plasma is assumed. The effect of radiation trapping on the resonance series lines $1^1S$–$n^1P$ is taken into account by means of the escape factor $g_0$ [55]. Alignment relaxation, or disalignment, by radiation reabsorption (see Chap. 7) is also included in the model. It is remembered, however, that these treatments apply only to a plasma with a simple structure like a slab or a cylinder. For the present cusp plasma, a more realistic picture for the edge region would be that this plasma is illuminated by the central luminous plasma (see Fig. 8.16a). It is extremely difficult, however, to treat this problem properly without solving the radiation transport equation. For this reason, we exclude these $n^1P$ levels in the following fitting procedure.

The diffusion loss rates of $2^1S$ and $2^3S$ metastable atoms are estimated. Since our PACR model does not include the diffusion loss, we effectively take this effect into account by reducing the population of these two levels. As a result of least squares analysis between the synthesis and the experiment for $n(p)$ and $a(p)$, it is found that the reduction to one-twentieth of the original $2^3S$ population brings the calculated $n(p)$ of the triplet system to overall agreement with the experiment. We fit our calculated $n(p)/g(p)$ and $a(p)/n(p)$ to experiment by adjusting the parameters of the Saturn-type EVDF. The best-fit result based on the CCC cross sections is shown in Fig. 8.18c and d; this is for $T_{eth} = T_{er} = 14\,eV$, $V_r = 1.8 \times 10^6\,ms^{-1}$ (9.2 eV) and $\alpha = 0.40$ with $n_e = 2.0 \times 10^{16}\,m^{-3}$. The agreement of the population is satisfactory. It is noted that the population distribution is almost equal to that for thermal EVDF with $T_e = 20$ eV; this is understood from the fact that as shown in Fig. 8.19a, the EVDF for this thermal plasma is almost equal to the fundamental moment $f_0$ of the present plasma.

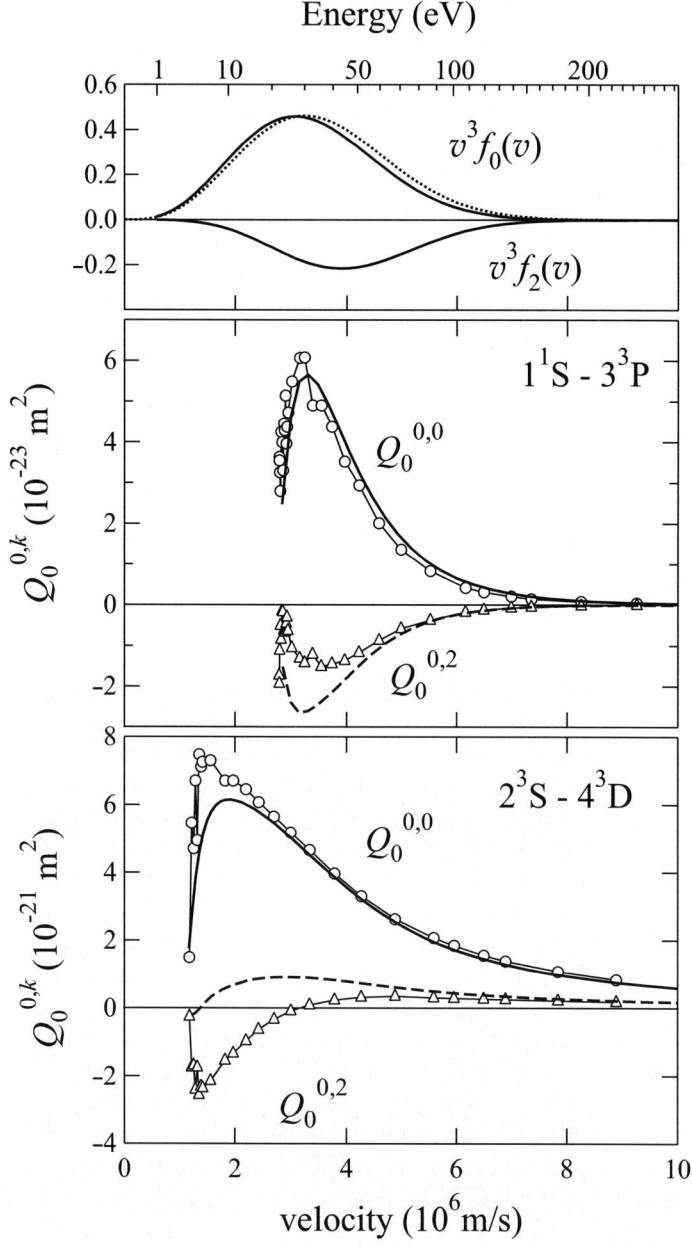

**Fig. 8.19.** (a) The *solid curves*: $K = 0$ (*upper*) and $K = 2$ (*lower*) terms of Legendre expansion of the Saturn-type EVDF, $v^3 f_K(v, \theta)$, with $T_{\text{eth}} = T_{\text{er}} = 14\,\text{eV}$, $V_r = 1.8 \times 10^6\ \text{m s}^{-1}$, and $\beta = 0.40$. The *dotted curve*: Maxwell of $T_e = 20\,\text{eV}$. (**b**) Excitation and alignment creation cross sections from $1^1\text{S}$ ground state to $3^3\text{P}$ level, and (**c**) from $2^3\text{S}$ metastable to $4^3\text{D}$ level. *Solid curve*: $Q_0^{0,0}$. *Dashed curve*: $Q_0^{0,2}$ by DWA [53]. *Open circles* and *triangles* are $Q_0^{0,0}$ and $Q_0^{0,2}$, respectively, by CCC [54] (Quoted from [52], with permission from IOP Publishing.)

Substantial disagreement still remains with $a(p)/n(p)$. Here we disregard $n^1P$ levels owing to the difficulty stated above. For example, the calculated $a(p)/n(p)$ of $3^3P$ is smaller than that of the observation by a factor three. The alignment creation rate coefficient $C^{0,2}(r,p)$ is given from the integration of $Q_0^{0,2}$ over $v^3 f_2(v)$ (4.42b), (see also Fig. 8.19). The alignment creation flux $C_0^{0,2}(r,p)n_e n(r)$ from the ground state predominates over that from the metastable level by a factor 70. As Fig. 8.19 shows, DWA gives larger $Q_0^{0,2}$ values (negative) than CCC and thus a larger $C_0^{0,2}(r,p)$ by a factor 1.5.

If we employ the DWA cross sections in place of CCCs, $a(3^3P)/n(3^3P)$ is larger by the same factor. Instead, the overall agreement of $a(p)/n(p)$ deteriorates, e.g., $a(n^3D)$ becomes negative. We find a large difference in the alignment creation cross section $Q_0^{0,2}$ from $2^3S$ metastable to $n^3D$ state between CCC and DWA as shown in Figs. 6.3 and 8.19c. In addition to the Saturn-type, we try a double-temperature EVDF described by two different temperatures in the parallel and perpendicular directions to the magnetic field. The present EVDF is well reproduced by the $T_\parallel = 14\,\text{eV}$ and $T_\perp = 21\,\text{eV}$; in fact, both the Saturn-type and double-temperature EVDFs are almost indistinguishable. Figure 8.20 illustrates the best-fitted EVDF in a contour plot in the velocity space. The best-fitted EVDF is oblate.

(A. Iwamae)

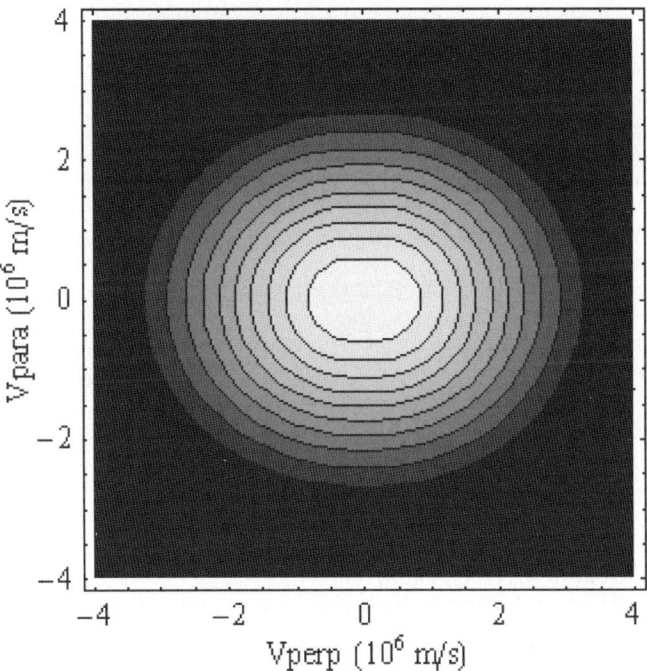

**Fig. 8.20.** The best-fit EVDF shown in a contour plot in the velocity space. The ordinate and abscissa are, respectively, the parallel and perpendicular components of the electron velocity to the magnetic field. The EVDF is oblate

# References

1. S.A. Kazantsev, Sov. Phys. - Usp. **26**, 328 (1983)
2. S.A. Kazantsev, N.Ya. Polynovskaya, L.N. Pyatnitskii, S.A. Edel'man, Sov. Phys. - Usp. **31**, 785 (1988)
3. C.G. Carrington, A. Corney, Opt. Commun. **1**, 115 (1969)
4. Kh. Kallas, M. Chaika, Opt. Spectrosc. **27**, 376 (1969)
5. S.A. Kazantsev, V.I. Eiduk, Opt. Spectrosc. **45**, 735 (1978)
6. S.A. Kazantsev, A.G. Rys, M.P. Chaika, Opt. Spectrosc. **54**, 124 (1983)
7. L. Tonks, I. Langmuir, Phys. Rev. **34**, 876 (1929)
8. S.A. Kazantsev, Sov. JETP Lett. **37**, 159 (1983)
9. A.I. Drachev, S.A. Kazantsev, A.G. Rys, A.V. Subbotenko, Opt. Spectrosc. **70**, 159 (1991)
10. D.Z. Zhechev, M.P. Cahika, Opt. Spectrosc. **43**, 352 (1977)
11. D.Z. Zhechev, M.P. Cahika, Opt. Spectrosc. **45**, 735 (1978)
12. A.G. Petrashen, V.N. Rebane, T.K. Rebane, Sov. Phys. - JETP **60**, 84 (1984)
13. A.G. Petrashen, V.N. Rebane, T.K. Rebane, Opt. Spectrosc. **58**, 22 (1985)
14. A.G. Petrashen, V.N. Rebane, T.K. Rebane, Opt. Spectrosc. **62**, 125 (1987)
15. S.A. Kazantsev, A.G. Petrashen, N.T. Polezhaeva, V.N. Rebane, T.K. Rebane, JETP Lett. **45**, 17 (1987)
16. S.A. Kazantsev, N.T. Polezhaeva, V.N. Rebane, Opt. Spectrosc. **63**, 15 (1987)
17. S.A. Kazantsev, A.G. Petrashen, N.T. Polezhaeva, V.N. Rebane, Opt. Spectrosc. **68**, 740 (1990)
18. L.Ya. Margolin, N.Ya. Polynovskaya, L.N. Pyatnitskii, R.Sh. Timergaliev, S.A. Édel'man, High Temp. **22**, 149 (1984)
19. M. Lombardi, J.-C. Pebay-Peyroula, C. R. Acad. Sci., Paris **261**, 1485 (1965)
20. S.A. Kazantsev, N.Ya. Polynovskaya, L.N. Pyatnitskii, S.A. Edelman, Opt. Spectrosc. **58**, 28 (1985)
21. S.A. Kazantsev, A.V. Subbotenko, J. Phys. D **20**, 741 (1987)
22. L. Danielsson, N. Brenning, Phys. Fluids **18**, 661 (1975)
23. W.H. Bennet, Phys. Rev. **45**, 890 (1934)
24. V.V. Vikhrev, S.I. Braginski, Rev. Plasma Phys. **10**, 425 (1986)
25. M.A. Liberman, J.S. De Groot, A.Toor, R.B. Spielman, *Physics of High Density Z-Pinch Plasmas* (Springer, New York, 1998)
26. V.V. Vikhrev, E.O. Baronova, Proc. 6th International Conference on Dense Z-Pinches, AIP Conf. Proc. **808**, 354 (2006)
27. E.O. Baronova, V.A. Rantsev-Kartinov, M.M. Stepanenko et al., Plasma Phys. Rep. **1**, 86 (1994)
28. L. Jakubowski, M. Sadowski, E. Baronova, Czech. J. Plasma Phys. **54**, (Suppl.C) SPPT271, 1,C1 (2004)
29. T. Fujimoto, T. Kato, Phys. Rev. A **30**, 379 (1984)
30. E.V. Aglitskiy et al., Kvantovaia Elektronika **1**, 579 (1974)
31. M.K. Inal, J. Dubau, J. Phys. B **20**, 4221 (1987)
32. E.O. Baronova, G.V. Sholin, L. Jakubowski, JETP Lett. **69**, 921 (1999)
33. V.V. Vikhrev, V.V. Ivanov, K.N. Koshelev, Plasma Phys. Rep. **8**, 1211 (1982)
34. L. Jakubowski, M. Sadowski, E.O. Baronova, Czech. J. Phys. **50** (Suppl. S3), 173 (2000)
35. E.O. Baronova, M.M. Stepanenko, G.V. Sholin, L. Jakubowski, T. Fujimoto, in *Proceedings of the Japan–US Workshop on Plasma Polarization Spectroscopy, NIFS-PROC-57,* ed. by T. Fujimoto, P. Beiersdorfer, 2004, p. 11. http://www.nifs.ac.jp/report/nifsproc.html

36. E.O. Baronova, M.M. Stepanenko, L. Jakubowski, H. Tsunemi, J. Plasma Fusion Res. **78**, 759 (2002) (in Japanese)
37. L. Jakubowski, M. Sadowski, E. Baronova, Nucl. Fusion **44**, 395 (2004)
38. V.V. Veretennikov, A.N. Gurei et al., Plasma Phys. **7**, 1199 (1989).
39. F. Walden, H. -J. Kunze, A. Petoyan et al., Phys. Rev. E **59** 3562 (1999).
40. A.S. Shlyaptseva, S.B. Hansen, V.L. Kantsyrev et al., Rev. Sci. Instrum. **72**, 1241 (2001)
41. E.O. Baronova, G.V. Sholin, L. Jakubowski, Plasma Phys. Contr. Fusion **45**, 1071 (2003)
42. L. Jakubowski, M. Sadowski, E. Baronova, in *Proceedings of the Japan–US Workshop on Plasma Polarization Spectroscopy, NIFS-PROC-57*, ed. by T. Fujimoto, P. Beiersdorfer, 2004, p. 30. http://www.nifs.ac.jp/report/ nifsproc.html
43. A.V. Gordeev, T.V. Loseva, Plasma Phys. Rep. **29**, 809 (2003)
44. J.C. Kieffer, J.P. Matte, H. Pépin, M. Chaker, Y. Beaudoin, T.W. Johnston, C.Y. Chien, S. Coe, G. Mourou, J. Dubau, Phys. Rev. Lett. **68**, 480 (1992)
45. J.C. Kieffer, J.P. Matte, M. Chaker, Y. Beaudoin, C.Y. Chien, S. Coe, G. Mourou, J. Dubau, M.K. Inal, Phys. Rev. E **48**, 4648 (1993)
46. Y. Inubushi, H. Nishimura, M. Ochiai, S. Fujioka, T. Johzaki, K. Mima, T. Kawamura, S. Nakazaki, T. Kai., S. Sakabe, Y. Izawa, J. Quant. Spectrosc. Radiat. Transf. **99**, 305 (2006)
47. E. Källne, J. Källne, Phys. Scripta T **17**, 152 (1987)
48. K.J.H. Phillips, J.W. Leibacher, C.J. Wolfson et al., Astrophys. J. **256**, 774 (1982)
49. E. Källne, J. Källne, J.E. Rice, Phys. Rev. Lett. **49**, 330 (1982)
50. H. Kubo, A. Sakasai, Y. Koide, T. Sugie, Phys. Rev. A **46**, 7877 (1992)
51. T. Fujimoto, H. Sahara, T. Kawachi, T. Kallstenius, M. Goto, H. Kawase, T. Furukubo, T. Maekawa, Y. Terumichi, Phys. Rev. E **54**, R2240 (1996)
52. A. Iwamae, T. Sato, Y. Horimoto, K. Inoue, T. Fujimoto, M. Uchida, T. Maekawa, Plasma Phys. Contr. Fusion **47**, L41 (2005)
53. G. Csanak, D.C. Cartwright, J. Phys. B: Atom. Mol. Phys. **22**, 2769 (1989); G. Csanak, D.C. Cartwright, S.A. Kazantsev, I. Bray, Phys. Scripta. T **78**, 47 (1998)
54. I. Bray, Phys. Rev. A **49**, 1066 (1994); D.V. Fursa, I. Bray, Phys. Rev. A **52**, 1279 (1995); CCC Data Base http://atom.murdoch.edu.au/CCC-WWW/index. html
55. T. Fujimoto, *Plasma Spectroscopy* (Oxford University Press, Oxford, 2004), p. 250

# Experiments: Recombining Plasma

A. Iwamae

In Chap. 8, we reviewed various PPS experiments on ionizing plasmas. In this chapter, we review a couple of experiments performed on recombining plasmas.

## 9.1 Introduction

In the ionizing plasma, excitation and ionization by electron impact are the principal mechanisms to determine the excited level populations. The alignment creation by the anisotropic excitation is well understood, at least qualitatively, and the formulation is well established which describes the relation between the polarization of emission lines and the anisotropic EVDF. These points were discussed in Chaps. 4, 6, and 8. In recombining plasma, radiative recombination and three-body recombination, and dielectric recombination as well, populate the excited levels. When the EVDF of the recombining electrons is anisotropic, the recombination continuum is polarized, and the excited atoms (ions) produced by the recombination would be aligned. As suggested in Sect. 6.1.3, alignment creation and destruction by elastic collisions tend to be important in this situation and complicate the problem further. The implementation of the PACR model, the formulation of which is given in Sect. 4.4.2, is still to be developed. On the other hand, a few experimental observations have already been reported on the polarization of emission lines from recombining plasmas.

## 9.2 Laser-Produced Plasmas

Yoneda et al. report the polarization of the emission lines of the resonance transitions in helium-like and hydrogen-like fluorine from a high-density plasma which is created by a pulsed KrF ($\lambda$248 nm) laser radiation [1, 2]. Figure 9.1 shows the experimental setup. The laser radiation with 2-ps pulse

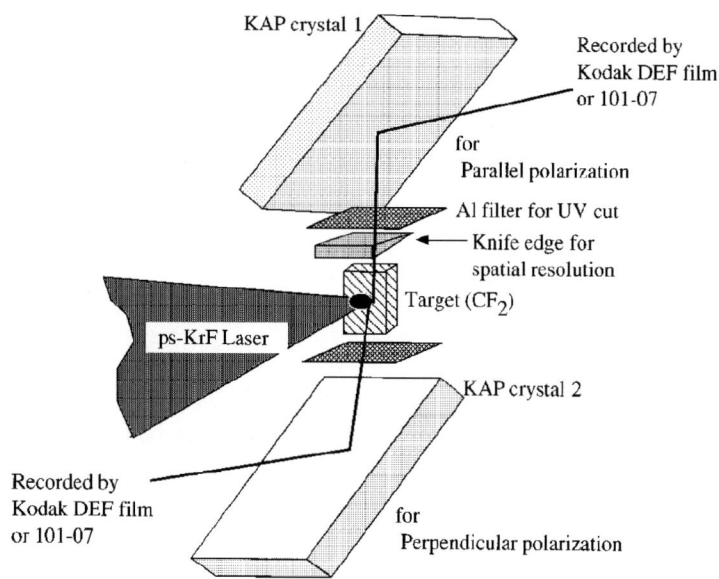

**Fig. 9.1.** Experimental setup for measurement of the polarization of the line emission. A pair of crystal spectrometers are located at almost the same distance on the same line of sight, and consist of the same Bragg crystals and uv-cut filters. The surfaces of the crystals are aligned orthogonally with each other. A knife edge is set near the plasma for measuring the spatial profile of the line emission (Quoted from [1], with permission from The American Physical Society.)

is focused on the $CF_2$ foil target with an $F/3$ aspherical lens. The incident angle of the laser beam on the target surface is $40°$; the effect of resonance absorption is not dominant under this condition. The intensity on the target is varied from $10^{14}$ to $10^{16}\,\mathrm{W\,cm^{-2}}$. The power contrast ratio of the prepulse and the main pulse is well controlled with a saturable absorber and is $10^{-8}$ on the target. A pair of spectrometers with flat cleaved KAP crystals are used to select each of the orthogonal linearly polarized components. The crystals are located at the same distance, on almost the same line of sight. As shown in Fig. 9.1, the plane of incidence of the upper crystal is aligned to be parallel to the target surface and that of the lower crystal is aligned parallel to the target normal, which is taken as the quantization axis. At the incidence angle of $40°$–$50°$ to the crystal, the reflectivity ratio of the s- and p-polarized light is greater than 10 (see Sect. 15.2). Aluminum filters are used for uv-cut. Each of the orthogonal polarized components, parallel and perpendicular to the target normal, is recorded with Kodak Direct Exposure Film or the 101-07 X-ray film.

Figure 9.2 shows an example of typical spectra of the perpendicular and parallel components. The resonance-series lines of helium-like and hydrogen-like fluorine and the recombination continuum are observed. The intensity

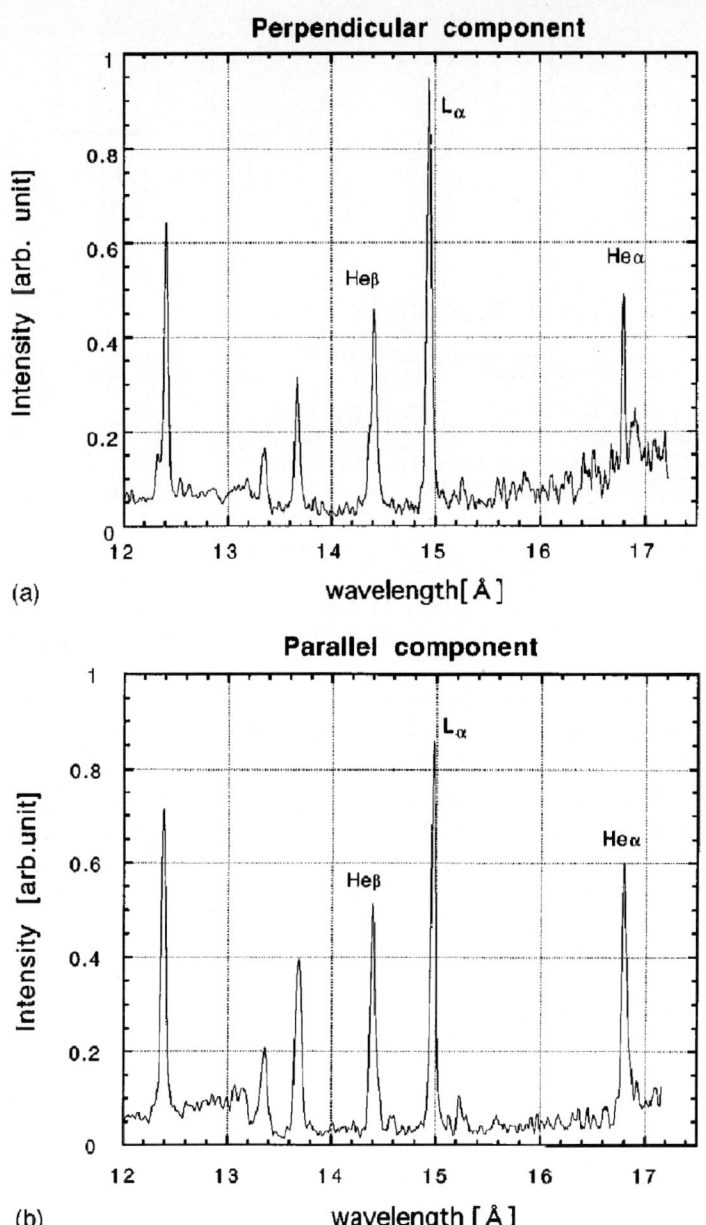

**Fig. 9.2.** Typical example of the polarization-resolved spectra. Helium-like FVIII and hydrogen-like FIX resonant lines are observed. The ratio of the Heα and Lyα line intensities clearly shows a large difference between (**a**) the perpendicular and (**b**) the parallel component intensities (Quoted from [1], with permission from The American Physical Society.)

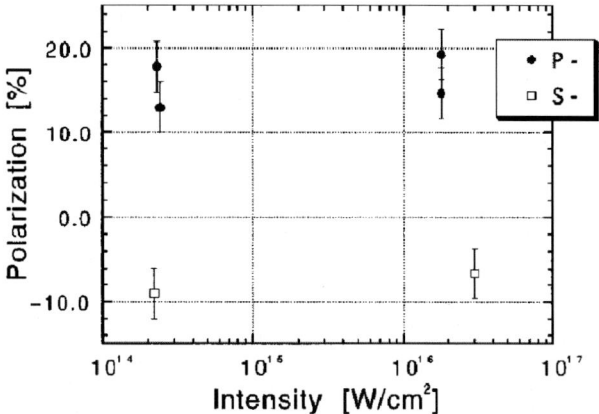

**Fig. 9.3.** The dependence of the polarization of Heα line on the irradiation intensity and the laser polarization (Quoted from [1], with permission from The American Physical Society.)

distribution pattern of the resonance-series lines and the presence of the recombination continuum indicate that the observed plasma is in the recombining phase (see Chap. 3). The electron density of the emitting plasma is estimated from the Stark broadening of the Heβ line to be $0.7 - 1.5 \times 10^{22}$ cm$^{-3}$. The authors assume that the Lyα line is unpolarized. The polarization degree of the Heα ($1\,^1S_0 - 2\,^1P_1$) line is determined from $P = (I_\parallel - I_\perp)/(I_\parallel + I_\perp)$ to be $+0.25$ for the spectra in Fig. 9.2. As has been mentioned in Sect. 1.2, all the resonance-series lines ($1\,^1S_0 - n\,^1P_1$) are polarized and the recombination continuum is also polarized. The polarization of the continuum indicates that the velocity distribution of low-energy electrons is anisotropic: more directional to the parallel direction, i.e., the target surface normal.

The dependence of the polarization degree of the Heα line on the polarization and intensity of the incident laser irradiation is shown in Fig. 9.3. The incident angle of the laser beam on the target is 40°. Positive polarization degree is observed in the p-polarized laser irradiation and negative polarization degree for the s-polarized irradiation. This is a remarkable evidence of the dependence of polarization of emission line on the laser light polarization. In experiments with much longer pulse duration, almost no difference is found between the p- and the s-polarized laser irradiation. It is apparent that the anisotropy is directly driven by the applied laser field. The detailed mechanism of the anisotropy of EVDF in this experiment is not known yet.

J. Kim and D.E. Kim examine the polarization of a spectral line from low-temperature aluminum plasmas produced by low-power laser pulses with a polarization resolved uv–visible spectrometer [3, 4]. A pulse of duration 3 ns and energy per pulse 6 mJ from a Q-switched Nd/glass laser is focused normal on an aluminum target surface. The power density on the target surface is about $4 \times 10^9$ W cm$^{-2}$, which is much low compared with

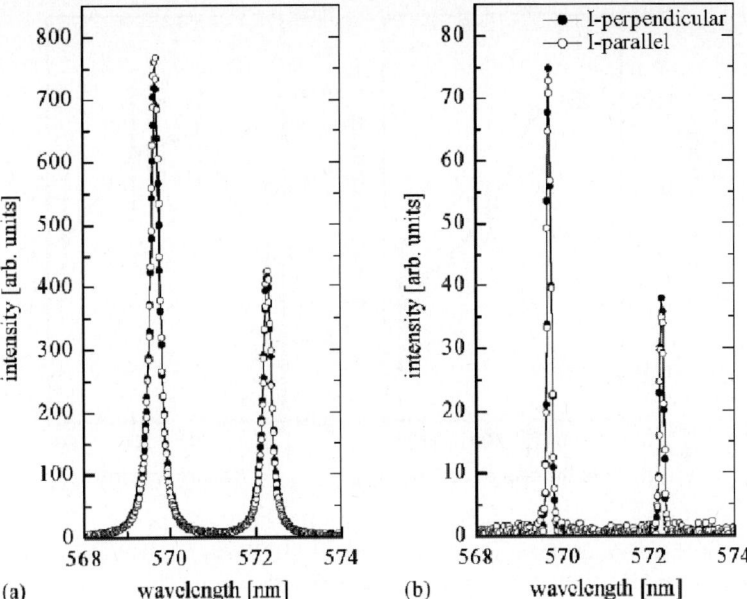

**Fig. 9.4.** Polarization-resolved spectra of AlIII $2s\,^2S_{1/2}-4p\,^2P^o_{1/2,3/2}$ transition based on a dichroic polarizer. Laser energy is 6 mJ. (**a**) Taken at 220 μm from the target surface and (**b**) Taken at 1.3 mm from the target surface. The *open circles* are the polarized component parallel to the laser incidence axis and *closed circles* are perpendicular (Quoted from [3], with permission from The American Physical Society.)

the previous X-ray emission experiment. The emission from the plasma is observed from the direction perpendicular to the laser incident axis. Figure 9.4a, b shows polarization-resolved spectra of AlIII 572.27 and 596.6 nm $(4s\,^2S_{1/2}-4p\,^2P^o_{1/2,3/2})$ transition line taken at different distances from the target surface, 220 μm and 1.3 mm, respectively. The polarization was resolved with a dichroic polarizer. The intensities are obtained by fitting the emission lines with a Voigt profile. The relative sensitivity of the observation system for the polarized components is calibrated by the unpolarized 572.27-nm line $(4s\,^2S_{1/2}-4p\,^2P^o_{1/2})$. The laser incident axis is taken as the quantization axis. The polarization degree is defined by $P = (I_\pi - I_\sigma)/(I_\pi + I_\sigma)$, and its dependence on the distance from the target surface is plotted in Fig. 9.5a. At the distance 220 μm from the target surface $P$ is 2.1% and it gradually decreases with an increase in the distance to almost diminish at 1.3 mm. The second measurement is performed with a calcite plate to resolve the two polarized components and it is confirmed that the observed polarization characteristics are the same and unaffected by the difference in the instrumentation and by the shot-to-shot variation. Figure 9.5b shows the time-resolved measurement on the polarization degree for the same lines. The higher degree of polarization is observed.

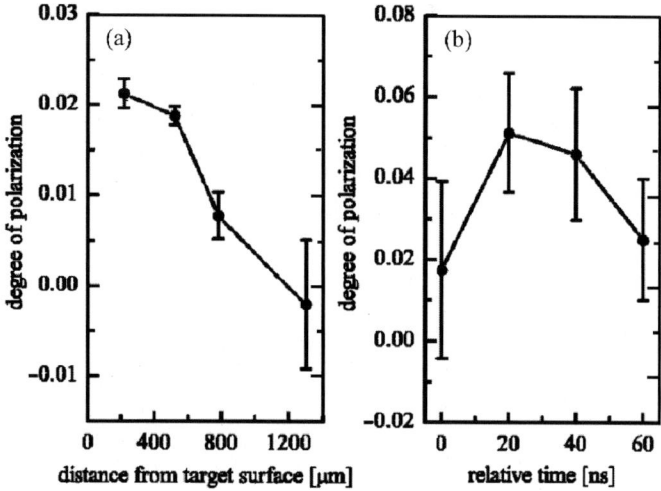

**Fig. 9.5. (a)** The degree of polarization of AlIII $2s\,^2S_{1/2}-4p\,^2P^o_{1/2,3/2}$ lines as a function of the distance from the target surface. **(b)** Time resolved degree of polarization measurement (Quoted from [3], with permission from The American Physical Society.)

The authors assume LTE populations for high-lying levels and interpret the intensities of AlIII 361.235 nm $(3d\,^2S - 4p\,^2P^o)$, 447.993 nm $(4f\,^2F^o-5g\,^2G)$, and 451.253 nm $(4p\,^2P^o-4d^2D)$ lines observed at the distance 220 μm as emitted from the plasma of effective electron temperature 3 eV. The electron density is estimated from the Stark broadening to be $2.2 \times 10^{17}\,cm^{-3}$. The plasma parameters give the electron–electron relaxation time of a few ps, which is much shorter than the laser pulse duration and the observation time. The EVDF is thus unlikely to be anisotropic. The origin of the polarization may not be an anisotropic EVDF. Quantitative evaluation including the recombination process is needed.

# References

1. H. Yoneda, N. Hasegawa, S. Kawana, K. Ueda, Phys. Rev. E **56**, 988 (1997)
2. H. Yoneda, N. Hasegawa, S. Kawana, K. Ueda, Fusion Eng. Design **44**, 141 (1999)
3. J. Kim D.-E Kim, Phys. Rev. E **66**, 017401 (2002)
4. J. Kim D.-E. Kim, Appl. Surface Sci. **197–198**, 188 (2002)

# Various Plasmas

Y.W. Kim, T. Kawachi, and P. Hakel

In the preceding chapters, we have exhausted the theoretical framework of PPS, i.e., the PACR model, and its ingredient, i.e., the cross sections and the PPS experiments on ionizing plasmas and recombining plasmas performed so far. In this chapter, we review several other facets which are quite interesting from the standpoint of PPS. The first is a PPS observation on a laser-produced-plasma plume expanding in a gas. Remarkable polarization of emission lines is observed. The second topic is the recent experimental and theoretical study on the polarization of the laser-driven X-ray lasers, which implies that the anisotropic radiation trapping may cause substantial alignment in the lasing levels. The third presents another approach to calculate the polarized line emission on the basis of the multipole radiation fields; this is an alternative method to the photon density-matrix formalism presented in Chap. 4.

## 10.1 Charge Separation in Neutral Gas-Confined Laser-Produced Plasmas

In high-density plasmas, e.g., with plasma temperature of $10\,\mathrm{eV}$ and electron number density of $10^{24}\,\mathrm{m}^{-3}$, the Debye length, say in aluminum plasma, becomes comparable to the mean distance between electrons. This is in fundamental conflict with the notion that Debye shielding results from the electron charge cloud surrounding the nucleus. In this regime of plasma the Coulomb potential energy for a charged particle pair becomes no longer negligible compared with the thermal kinetic energy of a single particle. Thus, the plasma becomes nonideal. The equation of state must include the nonideality contributions and the transport properties must be similarly corrected for.

Nonuniform, nonideal plasmas thus present interesting open problems in plasma physics. Well-characterized dense plasmas are needed for critical examination of the problems of self-absorption, nonideality, and charge separation. To this end, we investigate the structure and evolution of such plasma that is produced by a single laser pulse incident on a solid aluminum target, i.e., the laser-produced plasma (LPP) plume. A neutral gas of argon confines the LPP plume to attain higher plasma density [1–4]. The peak temperature and number density of the plasma, which is comprised of multiple ionized species, range up to $60\,\mathrm{eV}$ and $10^{27}\,\mathrm{m}^{-3}$, respectively.

Our LPP research program has been focused on developing experimental tests of new theories of plasma transport properties and atomic structures appropriate for the plasma density regime where the nonideality of the equation of state becomes significant. New estimates for the screening length [2, 5–7] need to be examined to facilitate accurate calculation of the lowering of ionization potentials and the equation of state, both necessary for equilibrium plasma calculations. Plasma absorption is substantially affected, as we have discovered [2], and this plays out in the evolution of LPP plume.

### 10.1.1 Nonideal Plasmas and Their 3D Plasma Structure Reconstruction

#### Weakly Nonideal LPP Plumes in Vacuum

The early-time spectral emissions from weakly nonideal LPP plumes are essentially a continuum, consisting of bremsstrahlung radiation and severely Stark-broadened line emissions [8–10]. Our plasma diagnostic method is novel in that the LPP plume is imaged entirely with the continuum plasma emissions. Due to the presence of significant self-absorption, the analysis leading to the plasma's density and temperature as function of space and time is carried out in close coupling to plasma equilibrium calculations. We start from the scaling relations relating the specific continuum emission intensity $I$ at a point within the LPP plume to the local temperature $T$ and pressure $p$: $T = C_\mathrm{T} I^\alpha$ and $p = C_\mathrm{p} I^{(\alpha+\beta)}$.

The program of continuum-based plasma diagnostics has consisted of measuring the side-view plasma luminosity profiles as function of time; these are in the form of streak photographs taken at different distances from the target surface (see Fig. 10.1). The attenuation by the plasma plume of laser beams at its fundamental and second harmonic frequencies, the total mass contained in each LPP plume, and the total energy deposited into the plume are determined from various information obtained in the experiment.

For an LPP plume produced in a vacuum, it is axially symmetric but the plasma medium is optically thick and the measured luminosity is complicated by self-absorption of the plasma emissions on the way out of the plasma. The modified Abel inversion algorithm, as described below, makes solutions of $I(r, z, t)$ corrected for the effect of self-absorption. Here $I(r, z, t)$ is the specific continuum emission intensity, emanating from a differential plasma volume element at radial distance $r$ and axial distance $z$ from the target surface at time $t$. The basic steps of the process are outlined in Fig. 10.2. The two scaling constants and two scaling exponents are first selected. The

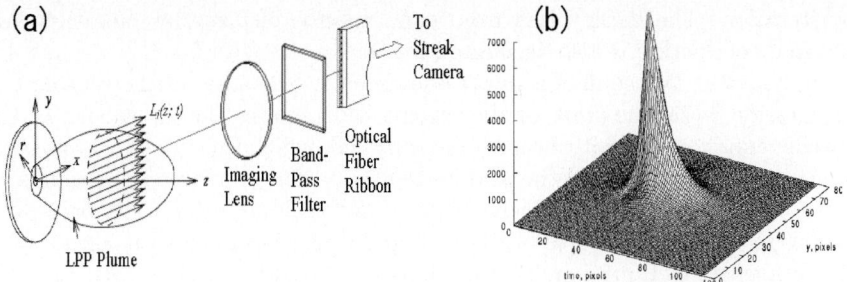

**Fig. 10.1.** (a) The arrangement for streak measurement of a side-view luminosity profile at distance $z$ from the target surface. (b) Luminosity profile at 0.5 mm from the aluminum target surface as function of time when the laser pulse energy is 2.3 J

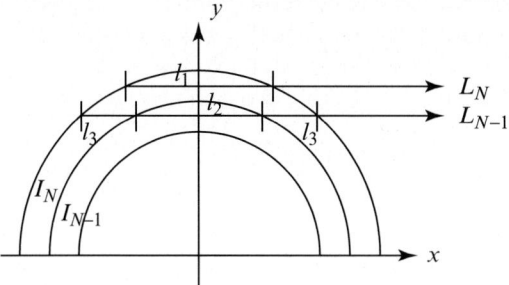

**Fig. 10.2.** Modified Abel inversion algorithm that includes the effect of self-absorption on the line-of-sight integration of the specific plasma continuum intensity. The cross-sectional slice of the LPP plume is divided into $N$ equal-width shells, and the specific continuum intensity $I_i$ is integrated along the line of sight for $i = N$ to 1. Self-absorption is included, assuming that the plasma absorption coefficient is constant within a given shell

inversion of the side-view luminosity $L(y, z)$ into $I(r, z, t)$ is performed: first the slice is divided into $N$ shells; here $i$ goes from 1 to $N$:

$$L_N = \frac{I_N}{a_N}(1 - e^{-a_N l_1}),\tag{10.1}$$

$$L_{N-1} = \frac{I_N}{a_N}(1 - e^{-a_N l_3})e^{-a_{N-1} l_2}\,e^{-a_N l_3} + \frac{I_{N-1}}{a_{N-1}}(1 - e^{-a_{N-1} l_2})e^{-a_N l_3}$$
$$+ \frac{I_N}{a_N}(1 - e^{-a_N l_3}).\tag{10.2}$$

Here, $a_i$ denotes the plasma absorption coefficient in the $ith$ plasma shell. One then proceeds to solve for $I_N$, $I_{N-1}, \ldots$, and $I_1$ for the selected cross-sectional slice. The plasma temperature and density and the specific continuum intensity from each differential plasma volume element are found self-consistently by iteration. The energy loss from each volume element by radiation and thermal conduction is also accounted for.

Solving for $I_N$ requires a priori knowledge of the state of the plasma cell in question while the state of the plasma cell requires the knowledge of the specific emission intensity from the plasma cell, which then gives the pressure and temperature through the scaling relations. The workable approach is to solve (10.1) or (10.2) iteratively. That is, we evaluate $a_N$ on the basis of a proposed value of $I_N$, solve (10.1) to find $I_N$ and repeat until it agrees with the proposed $I_N$. Equation (10.2) is similarly solved for each succeeding value of $i$ until the innermost shell has been reached, i.e., $i = 1$. The typical value of $N$ is 96 in this study.

The equilibrium ionization is given in terms of the set of Saha equations, which are incorporated into the inversion algorithm. Each small element of the plasma can be asserted to be in local thermodynamic equilibrium (LTE), given that the collision times between particles in the plasmas of our interest are of the order of ten fs whereas the plasma evolves in the nanosecond timescale. For aluminum plasma, we have thirteen Saha equations:

$$\frac{\alpha_i(1 - \alpha_{i+1})}{(1 - \alpha_i)} = \frac{2k_{\mathrm{B}}T(2\pi mk_{\mathrm{B}}T)^{3/2}}{p_{\mathrm{e}}h^3} \frac{(q_{\mathrm{e}})_{\mathrm{Al}^{i+}}}{(q_{\mathrm{e}})_{\mathrm{Al}^{(i-1)+}}} \, , \quad i = 1, 2, \ldots, 12, \quad (10.3)$$

$$\frac{\alpha_{13}}{(1 - \alpha_{13})} = \frac{2k_{\mathrm{B}}T(2\pi mk_{\mathrm{B}}T)^{3/2}}{p_{\mathrm{e}}h^3} \frac{(q_{\mathrm{e}})_{\mathrm{Al}^{13+}}}{(q_{\mathrm{e}})_{\mathrm{Al}^{12+}}} \quad (10.4)$$

Here, $\alpha_i$s denote the degree of ionization for the atom in the $i$th stage of ionization. $(q_{\mathrm{e}})_{\mathrm{Al}^{i+}}$ is the electronic partition function for aluminum in the $i$th stage of ionization. $m, p_{\mathrm{e}}$, and $k_{\mathrm{B}}$ denote the electron mass, partial pressure of electrons in the plasma, and the Boltzmann constant, respectively. These coupled Saha equations have to be solved iteratively because the right-hand side of each equation requires the full knowledge of the degrees of ionization. The equation of state gives the electron pressure as

$$p_{\mathrm{e}} = p\frac{\bar{Z}}{1 + \bar{Z}}, \quad (10.5)$$

where the mean ion charge is given by

$$\bar{Z} = \alpha_1 + \alpha_1\alpha_2 + \alpha_1\alpha_2\alpha_3 + \cdots + \alpha_1\alpha_2 \cdots \alpha_{13} \, . \quad (10.6)$$

The partition function requires the electronic level at which the sum may be cutoff. The cutoff of the sum over states is found by calculating the lowering of ionization potential according to the Debye model [11], while any impending breakdown of Debye length $\rho_{\mathrm{D}}$ due to plasma nonideality is checked for against the screening radius $r_s$. Depending on the plasma condition, we make use of either

$$\Delta\Phi_{\mathrm{I}}(\mathrm{Debye}) = \frac{(z + 1)e^2}{4\pi\varepsilon_0\rho_{\mathrm{D}}}, \quad (10.7)$$

or

$$\Delta\Phi_{\mathrm{I}}(r_{\mathrm{s}}) = n_{\mathrm{e}}k_{\mathrm{B}}T \int_0^{r_{\mathrm{s}}} (e^{-\phi_{ei}/k_{\mathrm{B}}T} - 1)\mathrm{d}^3r \,, \tag{10.8}$$

where $z$ is the nuclear charge of the ion under consideration. Here the Debye length and its replacement screening length, $r_{\mathrm{s}}$, are given, respectively, by

$$\rho_{\mathrm{D}} = \sqrt{\frac{\varepsilon_0 k_{\mathrm{B}}T}{(n_{\mathrm{e}} + \sum_z z^2 n_z)e^2}}, \tag{10.9}$$

and

$$\frac{r_{\mathrm{s}}}{\rho_{\mathrm{D}}} = 1 + \frac{\Gamma^{3/2}\bar{Z}^{1/2}\pi^{3/2}}{\sqrt{2}} \ln\left[1 + \left(\frac{r_{\mathrm{s}}}{\rho_{\mathrm{D}}}\right)^2 \frac{2\bar{Z}}{3\pi\Gamma^3}\right], \tag{10.10}$$

where the nonideality parameter is

$$\Gamma = \frac{\bar{Z}e^2 n_{\mathrm{e}}^{1/3}}{4\pi\varepsilon_0 k_{\mathrm{B}}T}. \tag{10.11}$$

Note that $r_{\mathrm{s}}$ appears on both sides of (10.10) but can be found by solving for it iteratively. The plasma absorption coefficient $a$ is then calculated to obtain the luminosity over the detector's spectral response and to track the energy loss by radiation as function of frequency $\omega$: [12,13]

$$a = \frac{2\omega}{\sqrt{2}c} \left[\sqrt{\left(1 - \frac{\omega_{\mathrm{p}}^2}{\omega^2 + \nu^2}\right)^2 + \left(\frac{\nu}{\omega}\frac{\omega_{\mathrm{p}}^2}{\omega^2 + \nu^2}\right)^2} - \left(1 - \frac{\omega_{\mathrm{p}}^2}{\omega^2 + \nu^2}\right)\right]^{1/2}, \tag{10.12}$$

where the collision frequency is

$$\nu = \frac{n_{\mathrm{e}}}{(k_{\mathrm{B}}T)^{3/2}} \frac{\bar{Z}\pi^{3/2}e^4}{m_{\mathrm{e}}^{1/2}2^{5/2}\gamma(\bar{Z})} \ln(\Lambda) \tag{10.13}$$

with Spitzer's correction factor [11,12]

$$\gamma(\bar{Z}) \approx \frac{\bar{Z}n_{\mathrm{e}}}{k_{\mathrm{B}}T} \ln\left(\frac{k_{\mathrm{B}}T\rho_{\mathrm{D}}}{\bar{Z}e^2}\right). \tag{10.14}$$

Here $\omega_{\mathrm{p}}$ is the plasma frequency, and $\Lambda$ is the Coulomb length.

The inversion calculation is continued to successive time intervals for the entire lifetime of the LPP plume. In the end, we have on hand a full set of calculated plasma mass, plasma energy, and transmitted intensities of the probing laser beams at the two wavelengths, all as a function of time. The results are compared with the corresponding measurements. The search for the optimal values of the scaling constants and scaling exponents takes place in a hierarchical manner – coarse-grained searches followed by fine-grained searches. The final values for the scaling exponents and scaling constants, which are globally applicable to the entire system of data are $\alpha = 0.45 \pm 0.03$, $\beta = 1.0 \pm 0.03$, $C_T = 0.190 \pm 0.003$, and $C_p = 340 \pm 5$. The resulting optimization shows evidence for significant modifications to the plasma absorption coefficient through revisions to the collision frequency, whereas only modest corrections to the Debye model of ionization potential lowering are indicated thus far [2, 13]. The maximum temperature and density of the plume shown in Fig. 10.1 are $59.0 \pm 8.0$ eV and $(3.6 \pm 1.1) \times 10^{26}$ m$^{-3}$, respectively. The attenuation coefficient for a laser beam at 530 nm running through the plasma along its axis of symmetry is $0.775 \pm 0.025$. Debye length and the lowering of ionization potential vary widely as functions of position within the plasma. The maximum value of the nonideality parameter $\Gamma$ is 0.122.

## Interfacial Instability in Neutral Gas Confinement of Dense LPP Plumes

In an attempt to gain even higher plasma density, another study of neutral gas (helium or argon) confinement of the LPP plume had been undertaken. When the pressure was raised beyond about 10 atm at room temperature, the LPP plume became less and less reproducible in the evolution of its shape and luminosity and so was the attenuation of the intensities of the probing laser beams through the plume.

The structure of the LPP plume is no longer symmetric about the laser beam axis. This renders the Abel inversion algorithm unworkable as a tool for inverting the 1D plasma luminosity profile into the 2D cross-sectional profile of the plasma's specific continuum intensity. The new inversion method consists of taking two mutually orthogonal side views of the plasma plume at a fixed distance from the target by means of streak photography. The two side views are insufficient, however, for unique reconstruction of the 2D plasma profile. This is because the 2D plasma contains $N_x \times N_y$ unknown elements of the plasma, and the two steaks provide only $N_x + N_y$ measurements at any given time. To overcome this limitation, a snapshot image of the front view (i.e., a cross-sectional view) of the plume is also taken at the start of the plasma. Figure 10.3 shows four sets of such measurements, one each at the neutral gas pressure of 1, 10, 20, and 50 atm. Our approach is different from the Cormack–Hounsefield algorithm in medical X-ray imaging [14], where many

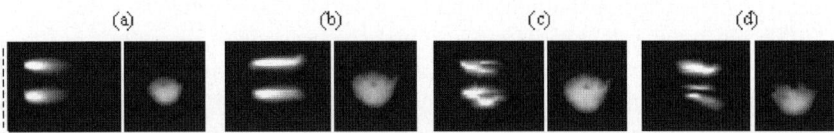

**Fig. 10.3.** A set of two side-view streaks (*left*) and a front-view photographs (*right*) are shown for four different LPP runs, one each from an aluminum target in argon at (**a**) $1.01 \times 10^5$ Pa, (**b**) $1.01 \times 10^6$ Pa, (**c**) $2.03 \times 10^6$ Pa, and (**d**) $5.07 \times 10^6$ Pa. The laser pulse energy is fixed at 2.5 J

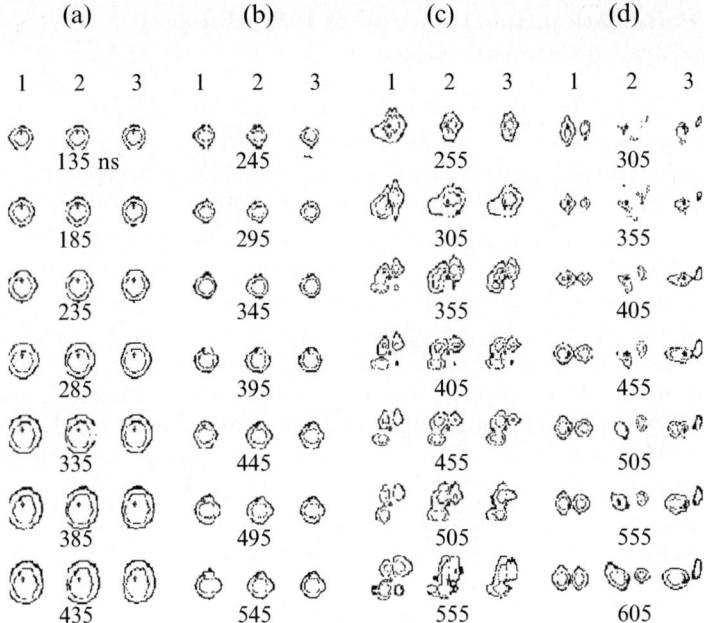

**Fig. 10.4.** Evolution of the constant intensity contours of reconstructed cross section of LPP plume at 50 and 20% of the local maximum is shown, when the target is placed in argon at (**a**) $1.01 \times 10^5$ Pa, (**b**) $1.01 \times 10^6$ Pa, (**c**) $2.03 \times 10^6$ Pa, and (**d**) $5.07 \times 10^6$ Pa. Three runs are shown at each gas pressure; the run number is shown at the top of each column. Each column of contours shows development of the plume at 50 ns intervals. The number above each column of contours corresponds to the repeat run number. The run time is shown just below each set of three contour plots

side views are taken in a scan, but our method is equally effective. The details of the procedure are given in [5].

We first focus on the fact that neutral gas-confined plasma deviates from axial symmetry about the laser beam axis due to intrinsic interfacial instabilities of Rayleigh–Taylor or Richtmyer–Meshkov nature [15–18]. The structure reconstruction of plasma is carried out with the plasma absorption coefficient set to zero in all plasma cells. Figure 10.4 gives a summary of the resulting plasma structures for the LPP plumes from an aluminum target, which is immersed in argon at four different pressures.

For each run, the structure of the plasma is shown as a function of time by two contours of constant specific intensity of the plasma cell at 50 and 20% of the maximum specific intensity, respectively. There are three identical LPP runs shown under identical conditions. With increasing argon density, the run-to-run variation in each set of streaks and image grows larger, indicative of the stochastic nature of the interfacial instability. At the same time, one can show that the intensity of the laser beam transmitted through a pinhole in the target exhibits increasingly larger fluctuations.

## 10.1.2 Polarization Spectroscopy of LPP Plumes Confined by Low-Density Gas

Our interest lies in taking images of the full LPP plumes by means of polarization-resolved emissions from the plasma, be a continuum or line emission, to elucidate further the nature of the interfacial structures such as those shown in Fig. 10.4. The continuum intensity variation in space signifies the existence of large gradients in plasma temperature and pressure, which in turn will drive the plasma flows of complex topology. We anticipate separation of charges throughout the plasma plume, and polarization-resolved imaging will help visualize development of such structures.

The feasibility experiment makes use of an experimental arrangement, shown in Fig. 10.5. The vacuum chamber contains an aluminum target mounted on an electrically floating stage. The emissions from the LPP plume can be detected and analyzed either from the cross section of the LPP plume by looking down on the target through the laser beam focusing lens or from the side through the side wall of fused quartz. We will refer to the first view as an axial view and the latter as a side view. The full uv to near ir spectral range is accessible for both views. The ambient neutral gas density is in the range from high vacuum to 1 atm. Argon is used as the confining gas.

There are basically three measurement configurations. The plasma plume is imaged onto the entrance slit of a spectrograph either for an axial view of the plasma or a side view. In the arrangement of an axial view of the plasma, the center of the plume's image is positioned in the middle of the slit. One half of the slit is covered with an analyzer with its polarization axis parallel to the slit, and the second half is covered with another analyzer with its axis perpendicular to the slit. The central portion of the slit between

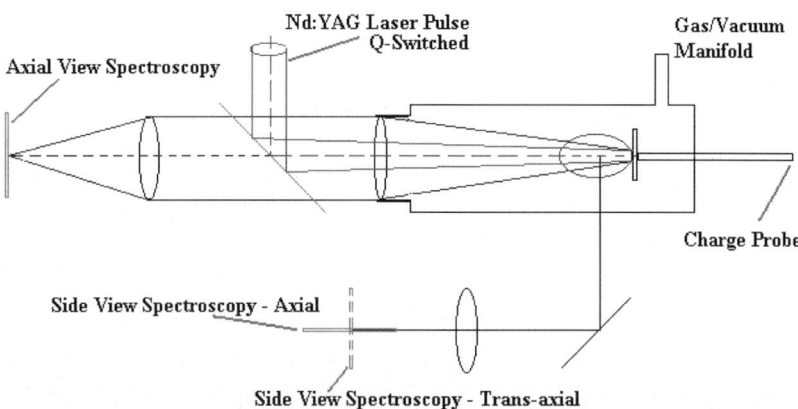

**Fig. 10.5.** The experimental setup used for polarization spectroscopy. Three different orientations of the spectrograph's entrance slit are used to probe the local electric field and particle velocity. The electric potential, to which the target has been driven, is measured by a means of a ×10 probe and a potential divider

the two analyzers is blocked from the LPP plume emissions. Given that at these neutral gas densities the LPP plume retains the axial symmetry, the polarization properties of the plume's emission are measured simultaneously, resolved both in wavelength and radial position.

The side view arrangement entails placing the image of the LPP plume on the slit plane with the plasma axis aligned either parallel or perpendicular to the slit. The first of these alignments is referred to as an axial side view, and the latter as a transaxial side view. In the transaxial side view, the axial symmetry of the plasma plume can be exploited: two analyzers can be placed on the slit in two mutually orthogonal directions. In the transaxial arrangement, the analyzer can be alternated from an alignment with its axis parallel to the slit to another that is orthogonal to it. All three configurations for polarization-resolved imaging are indicated schematically in Fig. 10.5.

Figure 10.6 shows the time-resolved target potentials due to charge separation in the expanding LPP plumes.

Altogether 11 different argon pressures are indicated. Using the common time base the laser intensity is shown as function of time for these runs. They overlap closely, indicating the reproducibility of the laser pulse. The scale of the potential is common to all runs, and the peak value of the potential reaches 14.7 kV. Figure 10.7 shows a set of axial-view measurements for the LPP plume in $5.99 \times 10^2$ Pa argon. The polarization-resolved spectrum is taken using a gated intensified CCD detector placed at the exit plane of the spectrograph. It represents 400 ns integration starting at 300 ns from the onset of the laser

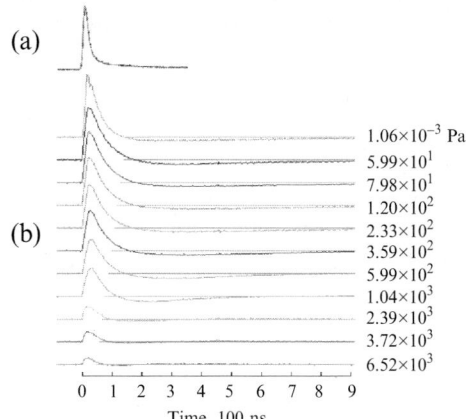

**Fig. 10.6.** Measured time-resolved (**a**) laser pulse and (**b**) target potential as resulting from charge separation within the expanding laser-produced plasma plume from an aluminum target. The target is placed inside the vacuum chamber, which is filled with argon to varying pressures, as shown in units of Pa in (**b**). The conducting target holder is electrically insulated from ground and its potential is measured by using a ×10 probe. The intensity profile of each laser pulse used to generate the plasma is plotted at the *top*, one above the other

(a)

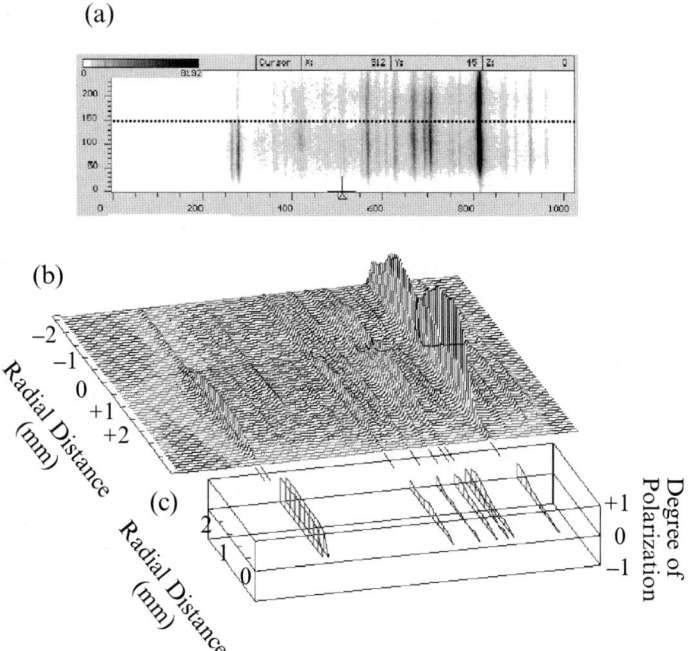

**Fig. 10.7.** Polarization and radial position-resolved axial-view spectrum of the laser-produced plasma plume of aluminum. The target vacuum chamber is filled with argon to $5.99 \times 10^2$ Pa. (**a**) Spectrum displayed as function of radial position, in spatial resolution of 0.025 mm per vertical pixel. Two mutually orthogonal polarization components are obtained simultaneously, the *upper part* being parallel to a radial direction and the *lower part* perpendicular to the direction. The wavelength is shown in pixels, which runs from 0 (389.5 nm) to 1,023 (447.3 nm). The intensity scans (**b**) run as a function of wavelength at each radial position. The *dark line* in the *middle* indicates the plume's center. The line intensities for the two polarization directions are used to evaluate the degree of polarization at each radial position. The results are shown in (**c**) for eight emission lines. The short *solid lines* at the *bottom* of (**b**) identify the line positions in the spectrum of (**c**)

pulse indent on the aluminum target (see Fig. 10.4). The dotted line in the middle of Fig. 10.7a makes the center of the plume and the boundary between the two mutually orthogonal polarization analyzers. The wavelength-resolved emission intensity is shown at different radial distances from the plume's axis in Fig. 10.7b. The dark line running in the middle shows the demarcation for the two polarization directions. Figure 10.7c shows the degree of polarization computed from the spectral intensity data of Fig. 10.7b for a selected group of emission lines. The degree of polarization is shown as function of radial distance from the plume's axis. The positive value means that the line is

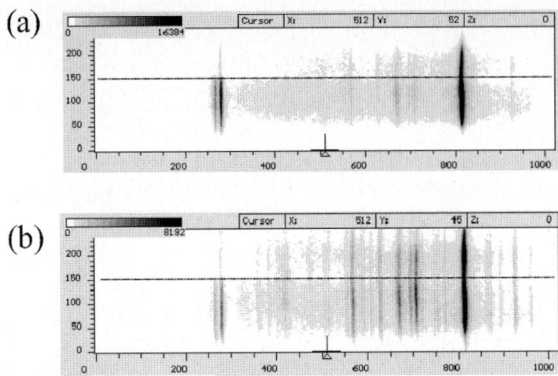

**Fig. 10.8.** Polarization and radial position-resolved axial-view spectrum of the laser-produced plasma plumes from an aluminum target. The target vacuum chamber is filled with argon to two different pressures: (**a**) $6.38 \times 10^3$ Pa and (**b**) $5.99 \times 10^2$ Pa. The *dotted lines* indicate the plume's center, which divides one polarization analyzer parallel to the slit (*bottom*) from another perpendicular to it (*top*). The emissions from the plume's core have been blocked at the slit. The vertical axis indicates the radial distance from the plasma axis, given in detector pixels at resolution of 0.025 mm per pixel. The wavelength axis is displayed horizontally in detector pixel addresses running from 0 (387.4 nm) to 1,023 (477.4 nm)

polarized preferentially in the radial direction. The results show a dependence of the degree of polarization on the wavelength of the emission lines as well as on the radial position from which the plasma emissions emanate.

Figure 10.8 compares the two polarization-resolved spectra from the axial-view imaging of the LPP plume at two different argon pressures, $6.40 \times 10^3$ Pa (a) and $5.99 \times 10^2$ Pa (b), respectively. They show a significant density dependence of the degree of polarization. Also, careful inspection of the individual spectral lines as function of the radial distance from the plume's axis shows that the line centers are blue shifted by amounts that are both wavelength and radial position dependent.

The neutral gas density dependence of the degree of polarization is summarized in Fig. 10.9 for three emission lines, one each from the neutral atoms, singly ionized ions and doubly ionized ions of aluminum, respectively. Figure 10.10 shows a summary of the corresponding dependence of the Doppler shifts on the neutral gas density and the originating species and emission line wavelength. The three emission lines are chosen to probe different aspects of charge separation. The emission lines at 396.152 nm AlII, 466.305 nm AlIII, and 447.997 nm, 447.989 nm AlIII belong to the multiplets, $^2P^o-^2D$ of the neutral aluminum atom, $^1D-^1P^o$ of the singly ionized aluminum ion, and $^2P^o-^2D$ of the doubly ionized aluminum ion, respectively.

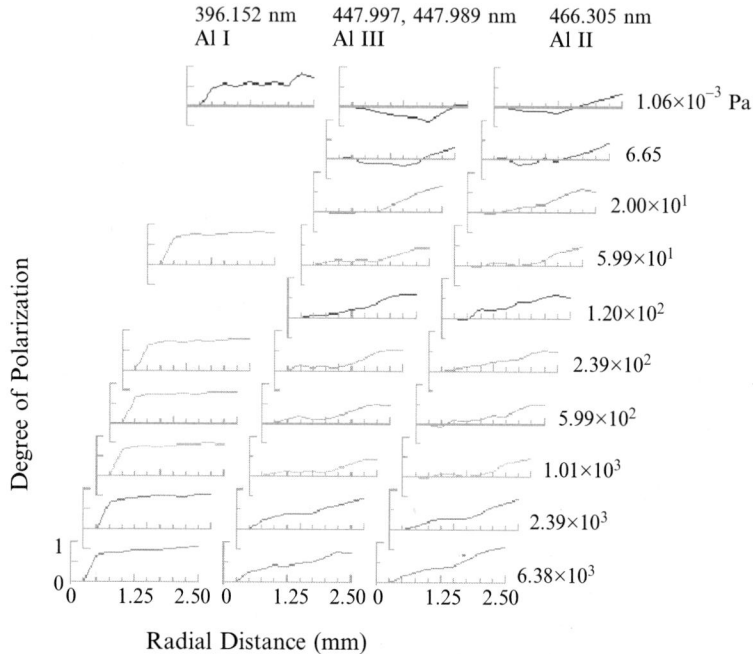

Fig. 10.9. Radial position resolved degree of polarization for three emission lines of laser-produced aluminum plasma. The neutral gas pressure is varied from high vacuum to $6.38 \times 10^3$ Pa, as shown on the right-hand side. The lines are selected from the neutral (396.152 nm AlI), singly ionized (466.305 nm AlII), and doubly ionized aluminum (447.997, 447.989 nm AlIII)

## 10.1.3 Analysis and Discussion

The results of the axial-view spectroscopy indicate that the emission lines are significantly polarized with the plane of polarization predominantly in the radial direction, while the degree of polarization grows larger with the radial distance. The Doppler shifts tend to be greater for doubly ionized ions than for the singly ionized ions. The side-view spectroscopy indicates that the emission lines are polarized preferentially in the transaxial direction but with smaller degrees of polarization. This is consistent with the view that the charge separation takes place in expanding front of the plume where steep density gradients exist. A domed plasma pillar is covered with a sheath of positive charges while a negative charge layer of electrons expands away rapidly. The region of charge separation is lower in plasma density, and the line emissions dominate the plasma spectrum. At the same time the strong electric fields propel charged species and the electric dipole moments become aligned with the field despite considerable thermal fluctuation. In our earlier studies, we have determined that the core temperature of the comparable LPP plumes

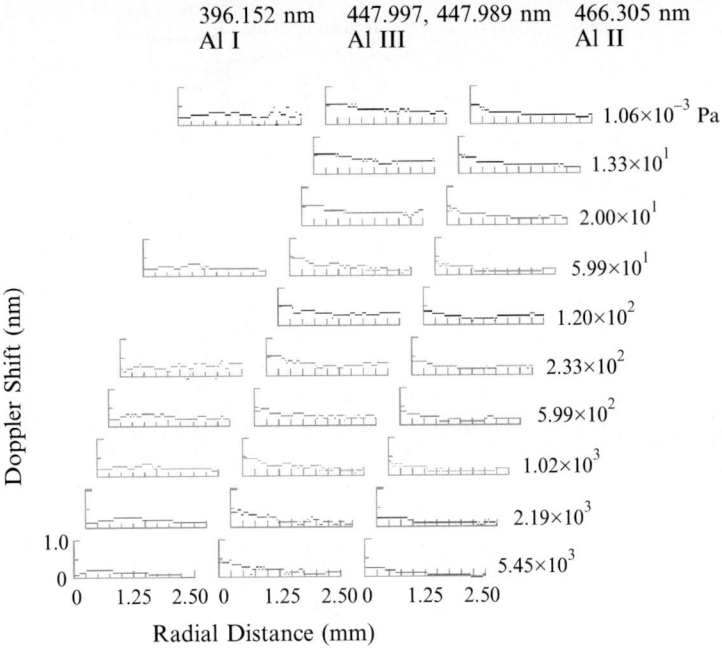

**Fig. 10.10.** Radial position resolved degree of polarization for three emission lines of laser-produced aluminum plasma. The neutral gas pressure is varied from high vacuum to $6.38 \times 10^3$ Pa, as shown on the right-hand side. The lines are selected from the neutral (396.152 nm AlI), singly ionized (466.305 nm AlII), and doubly ionized aluminum (447.997, 447.989 nm AlIII)

reaches $50\,\text{eV}$ at the peak. This translates to a temperature in the range of $10\,\text{eV}$ in the outer layers of the plasma where charge separation takes place.

### 10.1.4 Polarization-Resolved Plasma Structure Imaging

The discussions in Sect. 10.1.3 apply to plasma structure reconstruction based on full lateral-view images, instead of the luminosity streaks. Two mutually orthogonal lateral-view images of an LPP plume may be taken at discrete time intervals. An imaging detector is needed to capture a set of lateral-view images at each selected time interval. If polarization-resolved imaging is desired, the number of detectors may have to be doubled in number. Figure 10.11 shows the optical arrangement suitable for polarization-resolved imaging. The setup for capturing one of the needed two lateral-view images is shown. To round out the completeness of the luminosity data, a front-view snapshot of the LPP plume must be taken at an early moment of the plume's evolution [15, 16].

The dipole moments of the radiating species are by and large aligned radially. This is a direct consequence of the large electrostatic field resulting from the charge separation during the plume's expansion. The preponderance

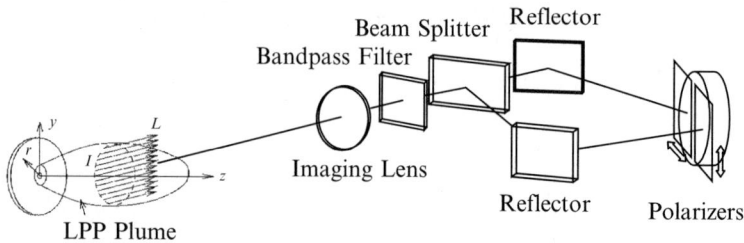

**Fig. 10.11.** The optical setup for capturing polarization- and wavelength-resolved lateral-view images of an LPP plume at a selected time from the start of a laser pulse. A second identical setup is needed to capture another set of lateral-view images in a direction perpendicular to the first at the same time. Here, the image is first split into two images, and the two split images are captured by a single gated, intensified 2D array detector through two separate polarizers of mutually orthogonal polarization

of the large observed degree of polarization suggests that $\boldsymbol{d} \cdot \boldsymbol{E} \geqslant k_\mathrm{B}T$. This can be seen as follows. The local $E$-field is deduced from the measured *degree of polarization*. For an electric dipole in an $E$-field, the mean dipole moment is given by

$$\langle \boldsymbol{M} \rangle = \alpha_a \boldsymbol{E} + \boldsymbol{d}L(y) \,, \tag{10.15}$$

where the Langevin function is defined by

$$L(y) = \left[\coth(y) - y^{-1}\right] \,. \tag{10.16}$$

Here $y = \boldsymbol{d} \cdot \boldsymbol{E}/k_\mathrm{B}T$, and $\alpha_\mathrm{a}$ denotes the atomic polarizability [19].

There are three major considerations for including the distribution of local electric field in the program of plasma structure determination. First, note that the degree of polarization for an emission line must be local, whereas the imaging gives the line-of-sight average degree of polarization, i.e., the degree of polarization for the luminosity. To find the degree of polarization for a local emission line within the plasma, the structure reconstruction needs to be carried out for the two polarization-resolved specific intensity components in such a way that the scaling relations for plasma temperature and pressure are satisfied for the total specific intensity. Second, we remind that the program for electric field determination relies on imaging by line emissions, not continuum emissions. The question is whether the scaling relations are appropriate for the line emissions. We believe that the power-law scaling relations are good approximations, but it is useful to make a critical comparison with an alternative one based on an explicit expression for the specific emission intensity of an emission line. Third, it is necessary to rewrite the expression for the self-absorption coefficient of the plasma with one for an emission line, consisting of the number density in the upper level of the transition, atomic transition probability, emission line profile function, and the photon energy of the emission line [8].

We have examined the feasibility of polarization-resolved imaging of an LPP plume from an aluminum target in low-pressure argon. Two polarization-resolved front-view images of the plume are captured using a single gated intensified CCD array detector, and a 2D distribution of the degree of polarization has been successfully constructed for the plasma luminosity [16]. According to the general approach in our development, the luminosity polarization must now be inverted to the distribution of the polarization of specific line emissions.

The distribution of the electric field in magnitude and direction, however, will have to await a full reconstruction of the 3D plasma structure. By the same token, plasma diagnosis by means of spectral line broadening does not illuminate beyond this point because the apparent emission line profiles are actually the results of line of sight integration through the plasma of strong nonuniformity. Full-scale reconstruction of the 3D plasma structure is needed.

The presence of the neutral background gas confines the plasma plume. The size and shape varies as strong functions of gas pressure at a given moment of the plasma evolution, while strongly influencing the persistence of the electrostatic field due to charge separation. It appears highly likely that the interfacial instability in strongly neutral gas confined LPP plumes imposes complex electric fields near the interface. The robustness of the polarization of emission lines, as we have observed, adds to the confidence that plasma polarization spectroscopy can be further developed as a useful diagnostic tool for analysis of a precipitation of interfacial instability and the ensuing development of 3D plasma structure.

### 10.1.5 Concluding Remarks

The observed Doppler shifts indicate that the energies of neutral aluminum atoms and singly and doubly ionized ions are in the range as large as 15 keV. This far exceeds the peak thermal energy of $k_B T = 50$ eV, suggesting that the species are driven electromagnetically as well as gas-dynamically and thermally. The energy distribution function has an appearance of being bimodal due to the fact that plasma particles are driven spuriously by the electric field. Laser-produced plasmas from atomic clusters are reported to contain MeV-class species through a process known as Coulomb explosion, which results in charge separation [20, 21]. Charge separation is sensitive to other physical processes involved in LPP plume generation, such as the detail of laser-matter interaction in terms of materials structure and transport properties [22–24].

The question is how the dipole moments of the neutral aluminum atoms remain so well aligned radially and their line emissions exhibit such strong Doppler blue shifts. It is quite reasonable to speculate that the line emissions from the neutral atoms result from recombination of the singly ionized species.

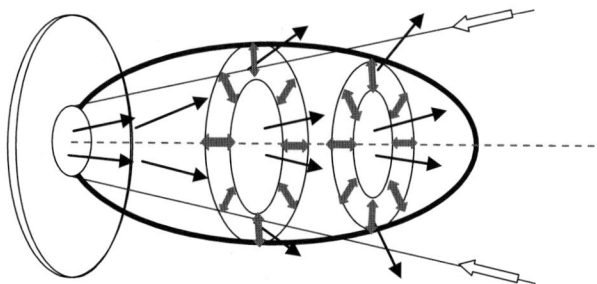

**Fig. 10.12.** The alignment of the electric dipole moments (*double headed arrows*) of radiating neutral and ionized aluminum atoms with the local electric field and the directions of motion (*single-headed arrows*) of super-thermal radiating species. The two *large hollow arrows* on the right-hand side show the rays of the laser beam directed toward the aluminum target

This bodes well with the fact that at the peak of the LPP plume the aluminum ions are three- to tenfold ionized across the core of the plasma plume [1, 2, 5].

The fact that the side-view spectroscopy (see Fig. 10.5) indicates significant levels of transaxial polarization is a puzzle at first sight. The emitting species whose dipole moments are aligned radially would show little polarization when viewed from the side. The explanation may be found in the fact that due to the finite size of the imaging optics, including the spectrograph, the plume's image on the slit plane contains plasma emissions from directions other than the direct line of sight. The measured polarization reflects the polarized emissions from the part of the plasma that is off the line of sight but still within the finite-size solid angle of acceptance of the spectrograph.

Figure 10.12 presents an overview of an LPP plume in the form of a sketch of the plasma particle movement and the alignment of the electric dipole moments of the radiating species that includes the observations made above.

### Acknowledgments

The author's former doctoral students Conrad Lloyd-Knight, Jaechul Oh, and Hedok Lee contributed significantly in the development of the three LPP plume structure diagnostic methods that are applicable to a large class of weakly nonideal, nonaxisymmetric and self-absorbing plasmas. Financial support of the work by the CTU 5-2 Consortium for Laser Produced Plasmas and Lehigh University is acknowledged.

(Y.W. Kim)

Note added by the editors: Recently two other experiments on laser-produced plasmas have been performed: A.K. Sharma, R.K. Thareja, J. App. Phys. **98**, 33304 (2005); R.J. Gordon (private communication, 2007). Although the details of the experimental conditions are different, these authors observed substantial polarization of the emission lines from the plasma plume.

## 10.2 Polarization of X-Ray Laser

### 10.2.1 Introduction

The possibility of amplification of stimulated emission (ASE) in the soft X-ray region was first proposed by Gudzenko and Shelepin in 1965 [25], in which the authors showed that population inversion could be generated between excited levels of highly charged ions in low temperature plasmas; the collisional excitation–deexcitation dominates the populations of the upper lasing level whereas the fast radiative decay dominates that of the lower lasing level. This scheme is called the recombining plasma X-ray lasers. Another scheme was proposed by Zherikhin et al. [26], in which the neon-like ions in high-temperature (or ionizing) plasma was the gain medium. Figure 10.13 shows the schematic energy-level diagram of the neon-like ions. Consider the $(2p_{1/2}, 3s_{1/2})_1$ and $(2p_{1/2}, 3p_{1/2})_0$ excited levels. In a high-temperature plasma, collisional excitation from the ground state, $2p^6$ ($J = 0$), is the dominant populating mechanism of these excited levels. Since the $(2p_{1/2}, 3s_{1/2})_1$ level has a fast radiative decay probability to the ground state, whereas the transition from the $(2p_{1/2}, 3p_{1/2})_0$ to the ground state is optically forbidden, population inversion can be created between these two levels. This scheme is called the collisional-excitation scheme.

Lasers in both the schemes were verified experimentally in 1984 by the groups in Princeton University and Lowrence Livermore National Laboratory [27, 28]. Especially for the collisional-excitation scheme, the succeeding

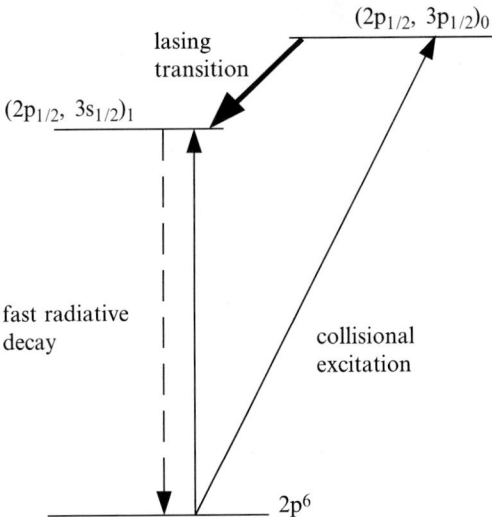

**Fig. 10.13.** Energy-level diagram of the neon-like ions. The *thin solid arrows* show collisional excitation from the ground state. The *dashed arrow* is fast radiative decay from $(2p_{1/2}, 3s_{1/2})_1$ to the ground state. *Thick solid arrow* is the lasing transition. In the case of GeXXIII (neon-like germanium ions), the lasing wavelength is 19.6 nm

experimental and theoretical studies allowed us to achieve the amplification saturation in the neon-like and nickel-like ion lasers, and substantial amplification has been obtained in the wavelength region as short as 4.3 nm [29].

In the case of the collisional-excitation laser, the solid target is illuminated by two laser pulses temporally separated by 0.2–2 ns. The first laser pulse creates a "preformed plasma" having a long scale length, which has a small density gradient in the electron density region of the amplification gain. Then the energy of the second (heating) pulse is absorbed efficiently, resulting in a high temperature plasma with sufficient abundance of neon-like or nickel-like ions which work as an X-ray laser gain medium. In the 1980s and 1990s, the typical duration and intensity of the heating pulse was $\sim$100 ps and $\sim 10^{13}\,\mathrm{W\,cm^{-2}}$, respectively. Under such heating conditions, the plasma parameter, especially the electron temperature, changes slowly as compared with the relaxation time of the excited level populations of the ions, and the quasisteady-state (QSS) approximation is valid to describe the population kinetics. This scheme is called the "QSS collsional-excitation laser." More recently, the use of the heating pulse with a ps-duration makes it possible to realize a transient high gain, resulting in the reduction of the required pumping energy for the amplification saturation. This scheme is called the "transient collisional-excitation (TCE) laser" [30].

### 10.2.2 Observation of the Polarization of QSS Collisional Excitation X-Ray Laser

Polarization study of the X-ray lasers has been done mainly on the QSS collisional-excitation laser. The early experiments were conducted jointly by Institute of Laser Engineering (ILE) [31] and Rutherford Appleton Laboratory (RAL) [32]. These experiments aimed at generating a linearly polarized X-ray laser beam; The X-ray laser beam from the gain medium was reflected by the soft X-ray mirror which selected a polarization direction, and the "polarized" X-ray laser beam was injected into the gain medium again as the seed X-ray to be amplified. In the experiment by ILE, it was found that the intensity of the obtained linearly polarized X-ray laser beam was higher than the expected intensity on the assumption that the X-ray laser beam is unpolarized. Therefore, the intrinsic polarization was suggested. In this context, the polarization property of the X-ray laser was studied [33]. Figure 10.14 shows the experimental setup.

A germanium slab target was irradiated by a linearly focused Nd:glass laser light ($\lambda = 1.053\,\mu$m). The pumping laser light consisted of two Gaussian pulses with 0.1 ns duration, separated by 0.4 ns. The X-ray laser line at the wavelength of 19.6 nm from the neon-like germanium ions (the transition from $(2p_{1/2}, 3p_{1/2})_0$ level to $(2p_{1/2}, 3s_{1/2})_1$ level) was observed in the $y$ direction by use of a grazing incidence spectrometer coupled with a reflective linear polarizer, where the $z$-axis is taken to be the quantization axis. Figure 10.15 shows, as an example, two densitometer traces of the recorded spectra of the

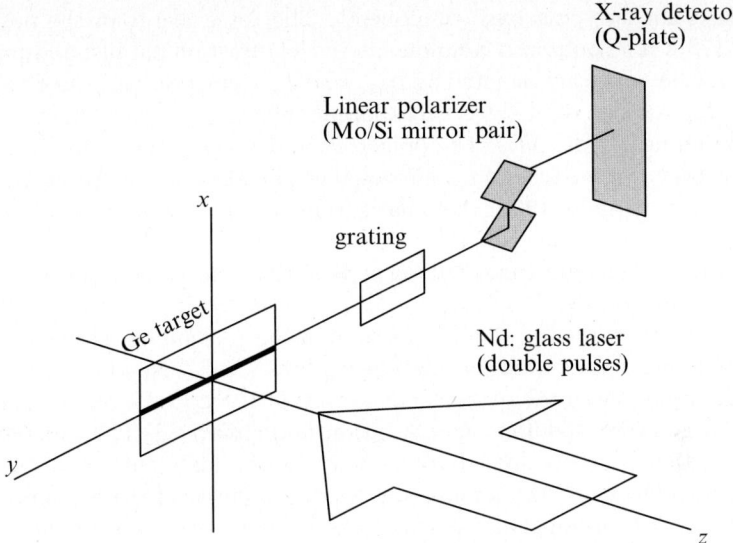

**Fig. 10.14.** Experimental setup. Nd:glass laser light travels along the $z$ direction to irradiate a slab target and the 19.6 nm laser light is amplified along the $y$ direction. The laser light is diffracted by the grating and the polarization is selected by the linear polarizer, which consists of two Mo/Si multilayer mirrors optimized for 19.6 nm light. The linear polarizer is fixed in a housing, and we can rotate it around the $y$-axis to obtain appropriate polarization direction. The detector is Ilford Q-plate

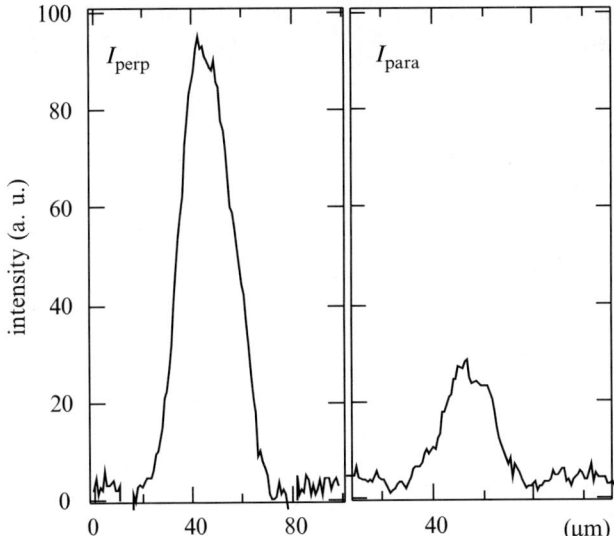

**Fig. 10.15.** Two densitometer traces of the recorded spectra of the X-ray laser for two polarized components (Quoted from [33].)

X-ray laser for two polarized components. The total energy of the pumping laser is 170 J. The polarized components in the direction parallel and perpendicular to the $z$-axis are denoted as $I_{\text{para}}$ and $I_{\text{perp}}$, respectively. As Fig. 10.15 shows $I_{\text{perp}}$ is more than three times larger than $I_{\text{para}}$, corresponding to the polarization degree of –53%. The polarization degree is quite sensitive to the pumping laser parameter, e.g., the irradiance and the form of the pulse. With the pumping energy of 190 J, the polarization degree decreased to −20%. The group of RAL conducted a similar experiment by use of single pulse irradiation with a 650 ps-duration, but no substantial polarization of the 19.6 nm line was observed [34].

Figure 10.16 shows the Kastler diagram of the 19.6 nm lasing line together with the resonance line from the lower lasing level $(2p_{1/2}, 3s_{1/2})_1$ to the ground state $2p^6$. Since the total angular momentum of the upper level of the lasing transition is $J = 0$, spontaneous transition is unpolarized. The difference in the intensities of the polarized components of the laser output is therefore due to the difference in the population inversion density of these components: A small population imbalance between the magnetic sublevels of the $(2p_{1/2}, 3s_{1/2})_1$ level is amplified exponentially by Einstein $B$ coefficient, resulting in the substantial polarization in the output beam.

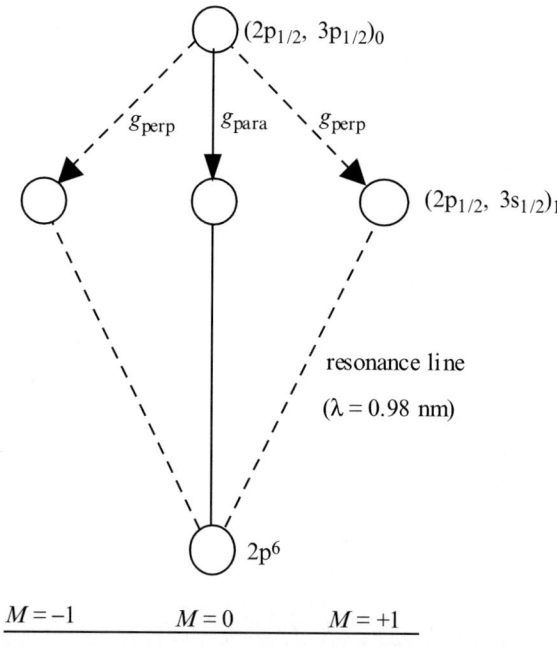

**Fig. 10.16.** Kastler diagram of the 19.6 nm laser transition and the resonance transition with the wavelength of 0.98 nm. $M$ is the magnetic quantum number. *Solid and Dotted arrows* indicate the perpendicular and parallel components of the polarized emission, respectively

To clarify the mechanism of the observed polarization of the X-ray lasers, several analyses have been made. Kawachi et al. [33] attribute this polarization to the anisotropic radiation trapping of the resonance line. The hydrodynamics simulation shows that the plasma has a streaming motion dominantly in the $z$ direction, resulting in a steep velocity gradient in this direction [33]. Under such a condition, the absorption spectrum of ions in the laboratory coordinate system is shifted according to their $z$ position owing to the Doppler effect. This could reduce the absorption of the photons of the resonance transition emitted in the $z$ direction as compared with those in the $x$ direction: i.e., the number of the dipoles oscillating in the $x$ direction ($M = \pm 1$ levels) tends to decay faster than that in the $z$ direction. This explanation is in accordance with the tendency of the experimental result: with a higher pumping energy, the increase in the ion temperature would reduce the anisotropy in the radiation trapping. Furthermore, an increase in the prepulse intensity induces a longer scale length of the preformed plasma and thus smaller velocity gradient in the $z$ direction.

Dubau et al. [35] discuss a possibility of anisotropic electron impact excitation to generate the population imbalance in the magnetic sublevels of the $(2p_{1/2}, 3s_{1/2})_1$ level. They have constructed an alignment-population collisional-radiative model for the neon-like germanium ions by taking into account the effect of beam-like electrons with the energy of 3–8 keV. The calculated result shows that, with a substantial fraction of hot electrons (5–10% of the bulk), the 19.6 nm lasing line can be polarized. It may be argued that, in the case of the QSS collisional-excitation laser like the present one, the intensity of the pumping laser may not be high enough to produce substantial anisotropic hot electrons. In the transient collisional-excitation scheme, however, the heating pulse with ∼ps-duration and ∼$10^{15}$ W cm$^{-2}$ intensity is sufficient to create an anisotropic electron velocity distribution; this would produce population imbalance in the magnetic sublevels of the $(2p_{1/2}, 3s_{1/2})_1$ level.

Since the observed polarization is from a high-density plasma (the plasma parameter is estimated to be $n_e = 5 \times 10^{26}$ cm$^{-3}$, $T_e = 500$ eV), the polarization destruction (alignment destruction) process is quite important. Benredjem et al. [36] claim that the population imbalance in the magnetic sublevels of the $(2p_{1/2}, 3s_{1/2})_1$ level is impossible owing to the fast disalignment process by electron–ion Coulomb elastic scattering, the rate of which is of the order of $10^{14}$ s$^{-1}$. This value is higher by 2–3 orders of magnitude than the estimate by Kawachi et al. on the basis of the Stark (or collision) broadening of the $(2p_{1/2}, 3s_{1/2})_1$ level. It may be argued that the Coulomb collisions do not contribute to the Stark broadening and thus to disalignment. This point has been discussed in Sect. 6.1.4.

Another polarization destruction process is suggested by Romanovsky et al. [37]: The polarization direction of an X-ray laser is modulated by the microscopic electron density fluctuation in the gain medium. This process may be effective for shorter wavelength X-ray lasers where the laser wavelength becomes comparable to the scale length of the fluctuation or the Debye length. These arguments above are good examples that polarization studies of the X-ray lasers provide a benchmark of atomic processes in dense plasmas.

The study of polarization of X-ray lasers has just started. To obtain a deep understanding of the mechanism of generating polarization, further experiments are needed. For shorter wavelength, X-ray lasers below 4 nm, the collisional-excitation may not be an appropriate scheme. One of the realistic schemes is to use inner-shell ionization to produce a gain medium. Generation of inner-shell ionized atoms requires fast pumping by a continuum intense X-ray [38] or an energetic electron beam [39], because of their short lifetimes of the order of fs. In both the cases, the pumping source makes the plasma strongly anisotropic, and plasma polarization spectroscopy may become an important tool to diagnose these plasmas.

(T. Kawachi)

## 10.3 Atomic Kinetics of Magnetic Sublevel Populations and Multipole Radiation Fields in Calculation of Polarization of Line Emissions

### 10.3.1 Introduction

As discussed in the preceding chapters, while polarization-based plasma diagnostics breaks new ground, good quality modeling for its purposes must face the same issues addressed by the more traditional radiation-based diagnostics techniques, which rely on the properties of line intensities and line shapes. Observed spectral line intensity ratios are yardsticks for measuring plasma temperature; Stark-effect-induced broadening of line profiles contains information about plasma density. Construction of synthetic spectra requires calculation of populations of the plasma ion species that are in their ground as well as excited states. In non-LTE plasmas the populations often strongly deviate from Saha–Boltzmann equilibrium values. To address this issue multi-level collisional-radiative atomic kinetics models are built. Energy-level populations are then calculated as the result of combined effects of many atomic processes (excitation/deexcitation, ionization/recombination, etc.). Many energy levels may need to be included in such models, which in turn leads to large sizes of accompanying atomic databases. The size and complexity of a particular atomic kinetic model is determined by the level of detail used in the description of energy-level structure and the number of atomic processes linking them. Since polarized line radiation emerges from collections of ions with unequal populations of magnetic sublevels within individual fine-structure levels, development of fundamental, magnetic-sublevel atomic kinetics models is warranted. Such models must be complemented with a way of calculating polarized line emissions based on magnetic-sublevel populations. Previous work in this direction has been done by Inal and Dubau using the photon density-matrix formalism [40]. In this chapter, another approach based on the properties of multipole radiation fields is presented. This technique agrees with the photon density-matrix predictions and is consistent with results of the density-matrix method originally developed for nuclear physics applications [41, 42] and electron-beam ion trap measurements [43, 44]. With this technique the more traditional (line-intensity oriented) atomic kinetics modeling is naturally extended to the area of line polarization.

## 10.3.2 Development of a Magnetic-Sublevel Atomic Kinetics Model

Magnetic sublevels are quantum states characterized by parity $\pi$, energy $E$, total angular momentum $J$, and its projection $M_J$ on the axis of quantization. (The collection of the $2J + 1$ states within a given $J$ is hereby referred to as a fine-structure energy level.) These are good quantum numbers in the absence of hyperfine interaction and external fields. The magnetic-sublevel atomic kinetics approach is based on the assumption that off-diagonal elements of the ion density matrix (coherences) are not important. Since the ion density matrix remains diagonal if the electron distribution is axially symmetric and the symmetry axis is chosen as the axis of quantization [45,46], a large category of cases can be modeled only using the diagonal elements, i.e., magnetic-sublevel populations $f$. A system of kinetic rate equations of the general type

$$\frac{\mathrm{d}f}{\mathrm{d}t} = Af + b \qquad (10.17)$$

is constructed using atomic physics data describing processes linking various magnetic sublevels in the model. Elements of the rate matrix $A$ and the rate vector $b$ are given either directly (for spontaneous processes: radiative decay, autoionization), or (for collisional processes like excitation, etc.) they are obtained by averaging the appropriate sublevel-to-sublevel cross section over the electron distribution function. The system (10.17) is then solved for magnetic-sublevel populations $f$. This method is well suited for cases with complex cascade effects because it eliminates the need of a priori identification of the dominant atomic processes. This is a particularly useful feature in modeling transient plasmas where different atomic processes may be important at different times.

## 10.3.3 Calculation of Polarization-Dependent Spectral Line Intensities

Given the magnetic-sublevel populations $f$ the polarization-dependent line intensities can now be calculated. The intensity of a spectral line is proportional to the flux of photons originating from the associated atomic line transition. This photon flux is given by the number of radiative decays per unit time, which in turn is determined by the product of the population of the transition's upper energy state and the spontaneous radiative decay rate (or transition probability) of that state. This basic principle, which is valid for the calculation of unpolarized line emissions, can be extended to polarization-based spectroscopic modeling in the following way. Radiative decay transitions between magnetic sublevels produce pure multipole radiation emissions that have distinct angular and polarization characteristics [47]. Therefore, the expression for the intensity of a sublevel-to-sublevel transition picks up another factor called a *relative multipole intensity* in addition to the upper state's population and the radiative decay rate. For example, in case of an electric dipole (E1) line transition (observed at 90° with respect to axis of quantization) whose upper and lower states have the same value of the magnetic quantum number $M_J$ the emission is linearly polarized in the direction parallel to the axis of quantization, making the E1 relative multipole intensity along the perpendicular direction equal to zero for $\Delta M = 0$. On the other hand, if the magnetic quantum number changes by one unit, an E1 emission is polarized in a direction perpendicular to the quantization axis due to the vanishing parallel relative multipole intensity. In general, the polarization of line emissions from sublevel-to-sublevel transitions is determined by the multipole type of the transition and the absolute difference in $M_J$ values of the upper and lower sublevels. This is a manifestation of angular momentum conservation of the combined ion + electromagnetic field system during ion's decay by photon emission. A fine-structure line transition $J_i \rightarrow J_f$ consists of sublevel emissions polarized in both directions whose intensities are weighted by the individual magnetic sublevel populations in the upper level. The resulting partially polarized spectral line is an incoherent superposition of sublevel-to-sublevel transitions, which overlap due to degeneracy with respect to $M_J$ [48]. In the optically thin approximation the polarization-dependent intensity is then given by

$$I_{\parallel,\perp} \propto h\nu \sum_{M_i=-J_i}^{J_i} f(M_i) \sum_{M_f=-J_f}^{J_f} A(J_i M_i \rightarrow J_f M_f) \times MI_{\parallel,\perp}(\Delta M, \theta = 90°),$$

$$(10.18)$$

where $h\nu$ is the transition energy, $f(M_i)$ are populations of upper level's magnetic sublevels, $\theta$ is the angle between the quantization axis and the line of sight, and $\Delta M = M_f - M_i$. $MI(\Delta M, \theta)$'s are the above mentioned relative multipole intensities based on the angular parts of wave-zone multipole fields that are also known as vector spherical harmonics. These fields are the basic solutions of the vector Helmholtz equation for free electromagnetic fields in spherical coordinates. They are constructed as common eigenfunctions of orbital, spin, and total angular momenta $L^2$, $S^2$, and $J^2$ (where $J = L + S$) by coupling the "ordinary" spherical harmonics $Y_{lm}(\theta, \varphi)$ (eigenstates of $L^2$) with the 3D complex spherical unit vectors $e_i$ (eigenstates of $S^2$, with photon spin $S = 1$) [47]. The $MI(\Delta M, \theta)$ values for dipole and quadrupole transitions are

**Table 10.1.** Relative multipole line intensities $MI_\theta$, $MI_\varphi$ ($MI_\parallel$, $MI_\perp$ at $\theta = 90°$) for dipole transitions

|  | E1 | | M1 | |
|---|---|---|---|---|
|  | $MI_\theta$ | $MI_\varphi$ | $MI_\theta$ | $MI_\varphi$ |
| $\Delta M = 0$ | $3\sin^2\theta$ | $0$ | $0$ | $3\sin^2\theta$ |
| $\Delta M = \pm 1$ | $(3/2)\cos^2\theta$ | $3/2$ | $3/2$ | $(3/2)\cos^2\theta$ |

**Table 10.2.** Relative multipole line intensities $MI_\theta$, $MI_\varphi$ ($MI_\parallel$, $MI_\perp$ at $\theta = 90°$) for quadrupole transitions

|  | E2 | | M2 | |
|---|---|---|---|---|
|  | $MI_\theta$ | $MI_\varphi$ | $MI_\theta$ | $MI_\varphi$ |
| $\Delta M = 0$ | $15\sin^2\theta\cos^2\theta$ | $0$ | $0$ | $15\sin^2\theta\cos^2\theta$ |
| $\Delta M = \pm 1$ | $5/2(2\cos^2\theta - 1)^2$ | $(5/2)\cos^2\theta$ | $(5/2)\cos^2\theta$ | $5/2(2\cos^2\theta - 1)^2$ |
| $\Delta M = \pm 2$ | $5/2\sin^2\theta\cos^2\theta$ | $(5/2)\sin^2\theta$ | $(5/2)\sin^2\theta$ | $5/2\sin^2\theta\cos^2\theta$ |

based on the electric field values in the wave zone (pp. 416–417 of [47]) and are listed in Tables 10.1 and 10.2. $A(J_iM_i \to J_fM_f)$ is the sublevel-to-sublevel spontaneous radiative decay rate, which is related to the fine-structure decay rate $A(J_i \to J_f)$ via the Wigner–Eckart theorem (see, for example [49]):

$$A(J_iM_i \to J_fM_f) = A(J_i \to J_f) \times \langle J_f\, q\, M_f - \Delta M | J_i\, M_i \rangle^2, \qquad (10.19)$$

in which the quantity in $\langle \cdots \rangle$ is a Clebsch–Gordan coefficient, and $q$ is the multipolarity of the transition. The two polarization-dependent line intensities (10.17) observed at $\theta = 90°$ then define the linear polarization:

$$P = \frac{I_\parallel - I_\perp}{I_\parallel + I_\perp}. \qquad (10.20)$$

For example, the combination of (10.18)–(10.20) for an E1 transition with $J_i = 1 \to J_f = 0$, where $f(M_J)$ denote the populations of the upper magnetic sublevels, yields

$$P = -\frac{f(-1) - 2f(0) + f(+1)}{f(-1) + 2f(0) + f(+1)}; \qquad (10.21)$$

for an M1 transition in which $J_i = 1 \to J_f = 0$ the result is

$$P = \frac{f(-1) - 2f(0) + f(+1)}{f(-1) + 2f(0) + f(+1)}; \qquad (10.22)$$

the polarization of an E2 transition where $J_i = 2 \rightarrow J_f = 0$ is

$$P = -\frac{f(-2) - f(-1) - f(+1) + f(+2)}{f(-2) + f(-1) + f(-1) + f(+2)};$$ (10.23)

and finally the M2 polarization result for $J_i = 2 \rightarrow J_f = 0$ is

$$P = \frac{f(-2) - f(-1) - f(+1) + f(+2)}{f(-2) + f(-1) + f(-1) + f(+2)}.$$ (10.24)

Using the assumption of axial symmetry the above expressions can be simplified using the relation $f(M_J) = f(-M_J)$. Furthermore, under isotropic conditions (Maxwellian plasmas, for instance) the populations of all magnetic sublevels within a fine-structure level are the same, which results in unpolarized (i.e., isotropic) line emissions. Polarization may therefore arise only from lines whose upper levels are aligned, i.e., the population is distributed unevenly among their sublevels. In turn, alignment can be created by anisotropic processes such as collisional excitation or electron capture due to a beam of energetic electrons.

### 10.3.4 Results

Figure 10.17 presents some results calculated by the described magnetic-sublevel atomic kinetics technique combined with multipole radiation field approach. These are results of a modeling study of polarization properties of helium-like satellite line emissions in Si plasmas driven by high-intensity, ultrashort-duration laser pulses [50, 51]. These types of emissions (see Table 10.3) have been observed in experiments and are typically optically thin, which eliminates the need to consider opacity effects. Furthermore, Si nuclei have zero spin, therefore potentially depolarizing hyperfine interaction [52] is also a nonfactor. In this application the magnetic-sublevel atomic kinetics model includes the two magnetic sublevels of the ground state in hydrogen-like Si, and the 81 magnetic sublevels in helium-like Si arising from the configurations $1s^2$, $1s2l$, $1s3l''$, and $2l2l'$ where $l$ and $l'$ can be s or p, and $l''$ can be s, p, or d. The $1s2l$ and $2l2l'$ states give rise to the lower and upper levels of the transitions of interest (Table 10.3). The magnetic-sublevel atomic kinetics model was driven by time histories of electron temperature and density calculated by a hydrodynamic code and the nonthermal part of the electron distribution function (which is responsible for the polarization effect) was calculated using a particle-in-cell code. The simulation covered the time interval of roughly 1 ps. Figure 10.17 shows how two spectral lines (from transitions connecting spin singlet states) show a polarization effect indicative of the time evolution of this electron distribution, which contains an anisotropic component. The remaining lines (related to spin triplet states) remain largely unpolarized, providing an unpolarized reference for the two singlet polarization markers of plasma anisotropy.

(P. Hakel)

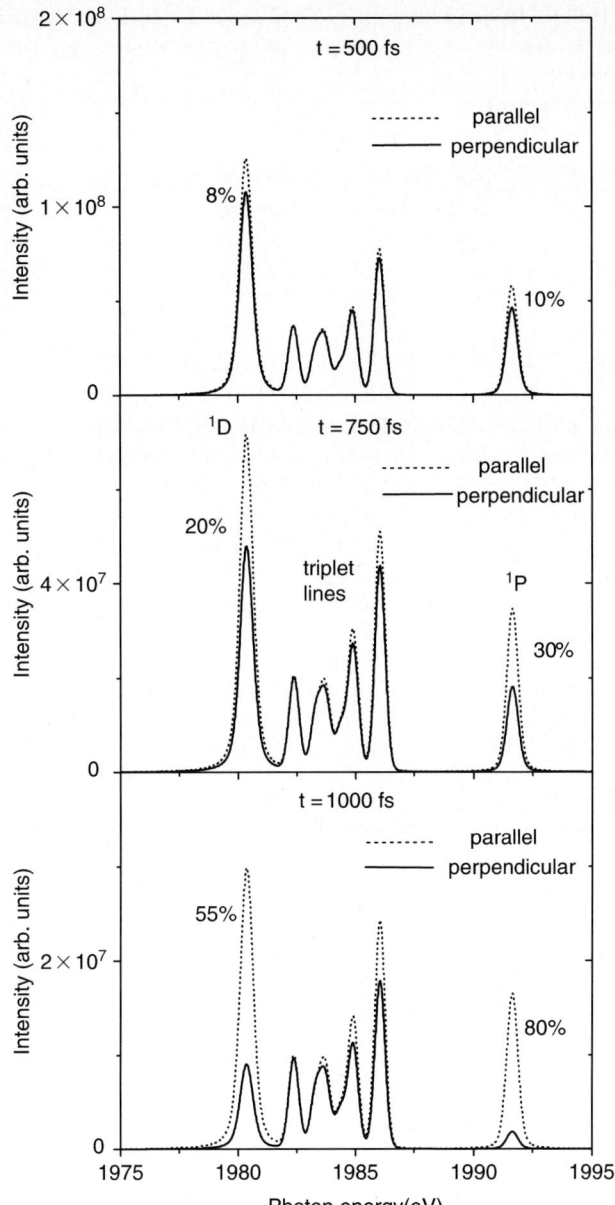

**Fig. 10.17.** Polarization-dependent helium-like Si satellite spectra at various time instants as presented in [50,51]. The percentage values refer to degrees of polarization $P$ at the given time instants for the two singlet lines

**Table 10.3.** Helium-like satellites ($n = 2$) of the Si Ly-$\alpha$. Line transitions 1 and 2 connect singlet states, the rest link triplets. Line 3 is always unpolarized since its upper level has $J = 0$

| Label | Line transition | Energy (eV) |
|-------|-----------------|-------------|
| 1 | $2p^2\ ^1D_2 \quad \rightarrow \quad 1s2p\ ^1P_1$ | 1980 |
| 2 | $2s\ 2p\ ^1P_1 \quad \rightarrow \quad 1s2s\ ^1S_0$ | 1992 |
| 3–5 | $2s\ 2p\ ^3P_{0,1,2} \quad \rightarrow \quad 1s2s\ ^3S_1$ | 1984–1986 |
| 6–8 | $2p^2\ ^3P_1 \quad \rightarrow \quad 1s2p\ ^3P_{0,1,2}$ | 1982–1984 |

# References

1. Y.W. Kim, C. Lloyd-Knight, Rev. Sci. Instr. **72**, 944 (2001)
2. C.D. Lloyd-Knight, Ph.D. Dissertation, Lehigh University, 2000
3. Y.W. Kim, J.-C. Oh, Rev. Sci. Instr. **72**, 948 (2001)
4. J.-C. Oh, Ph.D. Dissertation, Lehigh University, 1999
5. Y.W. Kim, H.-D. Lee, Rev. Sci. Instr. **75**, 3953 (2004)
6. R.B. Mohanti, J.G. Gilligan, J. Appl. Phys. **68**, 504 (1990)
7. B.V. Zelener, G.E. Norman, V.S. Filinov, Sov. Phys. – High Temp. **10**, 1160 (1972)
8. Y.W. Kim, in *Laser-Induced Plasmas and Applications*, ed. by L.J. Radziemski and D.A. Cremers. (Dekker, New York, 1989), Chap. 8
9. Y.W. Kim, High Temp. Sci. **26**, 57–70 (1990)
10. Y.W. Kim, K.-S. Lyu, J.C. Kralik, in *Current Topics in Shock Waves*, ed. by Y.W. Kim (American Institute of Physics, 1990), p. 353
11. M.R. Zaghloul, J. Phys. D: Appl. Phys. **36**, 2249 (2003)
12. L. Spitzer Jr., R. Haerm, Phys. Rev. 89, 977 (1953)
13. H. Hora, *Physics of Laser Driven Plasmas* (Wiley, New York, 1981)
14. H.H. Barrett, W. Swindell, *Radiological Imaging*, Vols. 1 and 2 (Academic Press, New York, 1981)
15. Y.W. Kim, H.-D. Lee, Rev. Sci. Instr. **74**, 2123 (2003)
16. Y.W. Kim, in *Proceedings of the Japan-US Workshop on Plasma Polarization Spectroscopy, NIFS-PROC-57* ed. by T. Fujimoto, P. Beiersdorfer, 2004, p. 57
17. H.J. Kuhl, Phys. Rep. **206**, 197 (1991)
18. M. Brouillette, Ann. Rev. Fluid Mech. **34**, 45 (2002)
19. H.S. Robertson, Statistical Thermophysics (Prentice-Hall, Englewood Cli.s, NJ, 1992)
20. T. Ditmire, T. Donnelly, A.M. Rubenchick, R.W. Falcone, M.D. Perry, Phys. Rev. A **53**, 3379 (1996)
21. M. Lezius, S. Dobosz, D. Normand, M. Schmidt, Phys. Rev. Lett. **80**, 261 (1998)
22. Y.W. Kim, Int. J. Thermophys. **23**, 1103 (2002)
23. Y.W. Kim: Int. J. Thermophys. **25**, 575 (2004)
24. Y.W. Kim: Int. J. Thermophys. **26**, 1051 (2005)
25. L.I. Gudzenko, L.A. Shelepin, Sov. Phys. Doki. **10**, 147 (1965)
26. A. Zherikhin, K. Koshelev, V. Letokhov, Sov. J. Quant. Mech. **6**, 82 (1976)
27. S. Suckewer, C.H. Skinner, H. Milchberg, C. Keane, D. Voorhees, Phys. Rev. Lett. **55**, 1753 (1985)

28. D.L. Matthews, P.L. Hagelstein, M.D. Rosen et al: Phys. Rev. Lett. **54**, 110 (1985)
29. B.J. MacGowan, L.B. Da Silva, D.J. Fields et al., Phys. Fluids B **4**, 2326 (1992)
30. P.V. Nickles, V.N. Shlyaptsev, M. Kalachnikov et al., Phys. Rev. Lett. **80**, 2825 (1998)
31. K. Murai, G. Yuan, R. Kodama, H. Daido, Y. Kato, S. Nakai, D. Neely, A. MacPhee, Jpn. J. Appl. Phys. **33**, L600 (1994)
32. B. Rus, C.L.S. Lewis, G.F. Cairns et al., Phys. Rev. A **51**, 2316 (1995)
33. T. Kawachi, K. Murai, G. Yuan, S. Ninomiya, R. Kodama, H. Daido, Y. Kato, T. Fujimoto, Phys. Rev. Lett. **75**, 3826 (1995)
34. C.L.S. Lewis, R. Keenan, A.G. MacPhee et al., in *Proceedings of "X-ray Lasers 1998"*, ed. by Y. Kato, H. Takuma, IOP 159, 1999
35. J. Dubau, M.K. Inal, D. Benredjem, M. Cornille, J. Phys. IV **2**, 277 (2001)
36. D. Benredjem, A. Sureau, B. Rus, C. Möller, Phys. Rev. A **56**, 5152 (1997)
37. M.Yu. Romanovsky, J. Ortner, V.V. Korobkin, J. Phys. IV **2**, 293 (2001)
38. S.J. Moon, D.C. Eder, Phys. Rev. A **57**, 1391 (1998)
39. D. Kim, S.H. Son, J.H. Kim, C. Toth, C.P. Barty, Phys. Rev. A **63**, 23806 (2001)
40. M.K. Inal, J. Dubau, J. Phys. B, **20**, 4221 (1987)
41. R.M. Steffen, K. Alder, in *The Electromagnetic Interaction in Nuclear Spectroscopy*, ed. by W.D. Hamilton (North-Holland, Amsterdam, 1975), p. 505
42. P. Beiersdorfer, D.A. Vogel, K.J. Reed et al., Phys. Rev. A **53**, 3974 (1996)
43. A.S. Shlyaptseva, R.C. Mancini, P. Neill et al., Phys. Rev. A **57**, 888 (1998)
44. A.S. Shlyaptseva, R.C. Mancini, P. Neill, P. Beiersdorfer, J. Phys. B **32**, 1041 (1999)
45. K. Blum, *Density Matrix Theory and Applications*, 2nd edn. (Plenum, New York, 1996)
46. J. Dubau, M.K. Inal, O.Z. Zabaydullinm, in *Proceedings of the 12th Topical Conference on Atomic Processes in Plasmas*, ed. by R.C. Mancini, R.A. Phaneuf, 2000, p. 217
47. B.W. Shore, D.H. Menzel, *Principles of Atomic Spectra* (Wiley, New York, 1968)
48. U. Fano, J. Opt. Soc. Am. **39**, 859 (1949)
49. E. Takács, E.S. Meyer, J.D. Gillaspy et al., Phys. Rev. A **54**, 1342 (1996)
50. P. Hakel, Ph.D. Dissertation, University of Nevada, Reno, 2001
51. P. Hakel, R.C. Mancini, J.C. Gauthier, E. Mínguez, J. Dubau, M. Cornille, Phys. Rev. E **69**, 056405 (2004)
52. J.C. Kieffer, J.P. Matte, H. Pépin, M. Chaker, Y. Beaudoin, T.W. Johnston, C.Y. Chien, S. Coe, G. Mourou, J. Dubau, Phys. Rev. Lett. **68**, 480 (1992)

# Polarized Atomic Radiative Emission in the Presence of Electric and Magnetic Fields

V.L. Jacobs

A reduced-density-matrix approach provides a very general framework for the theoretical description of polarized radiative emission during single-photon transitions from bound and autoionizing states of atomic systems in the presence of a general arrangement of static (or quasistatic) electric and magnetic fields. Polarized radiative emission from partially ionized atomic systems in plasmas can occur as a result of the excitation of the radiating atomic states by electrons with an anisotropic velocity distribution, which can be produced in an electron–ion beam experiment, in a nonequilibrium plasma environment, and in an electromagnetic field that is sufficiently intense in the relevant spectral region. Polarized radiative emission can also be produced or modified during the excitation of the atomic system in the presence of electric and magnetic fields and electromagnetic fields. In electric and magnetic fields, the normally overlapping angular-momentum-projection components of atomic spectral lines can be substantially shifted from their field-free positions and split into spectroscopically resolvable (and inherently polarized) features. Due to the breakdown of the field-free angular-momentum and parity selection rules, otherwise forbidden components of atomic spectral lines can be generated. Spectral patterns and polarizations produced by either a uniform static electric or magnetic field have been extensively investigated, and the individual Stark and Zeeman effects have been treated in both the weak-field (perturbative) and strong-field (nonperturbative) regimes. The more complex Stark–Zeeman patterns have been investigated primarily for simple atomic systems, including hydrogen-like or high-Rydberg atoms, and for special applied-field configurations, including parallel and perpendicular electric and magnetic fields. In the theoretical description of the polarized atomic radiative emission, it is advantageous to consider the coherent excitation of a particular subspace of the initial atomic bound or autoionizing states, which can occur as a result of sufficiently intense electromagnetic interactions. A general expression for the matrix elements of the detected-photon density operator provides a unified framework for the analysis of the spectral intensity, angular distribution, and polarization of the Stark–Zeeman patterns. From a unified development of time-domain (equation-of-motion) and frequency-domain (resolvent-operator) formulations of the more comprehensive reduced-density-matrix approach, the nonequilibrium atomic-state kinetics and the homogeneous spectral-line shapes can be systematically and self-consistently described.

## 11.1 Introduction

Polarized atomic radiative emission can be produced by a nonuniform (or nonstatistical) distribution of the population densities among the degenerate (or nearly degenerate) angular-momentum-projection or magnetic ($M$) substates of the excited atoms (or ions). A nonuniform distribution of $M$-substate populations can be created by directed-electron excitations, electric and magnetic fields, and electromagnetic fields. In the presence of electric and magnetic fields, the normally overlapping angular-momentum-projection components of atomic spectral lines can be substantially shifted from their field-free positions and split into spectroscopically resolvable features that are inherently polarized. Polarized radiative emission is anisotropic, giving rise to an angular dependence of the photon emission. The angular distribution and polarization measurements can provide detailed information on the elementary collisional interactions involving the individual magnetic substates of the atomic system, which would usually be unobtainable in an ordinary plasma spectroscopy observation of only the total intensity. The nearly monoenergetic electron-beam distribution that can be achievable in electron–ion beam experiments enables the investigation of otherwise-unobservable weak radiative-emission processes involving forbidden transitions or narrow autoionizing resonances. External electric and magnetic fields can be expected to play an important role, even for multiply charged ions, in processes involving high-Rydberg atomic states. The measurement and analysis of polarized radiative emission from bound–bound atomic transitions has been widely exploited to investigate the various anisotropic (nonequilibrium) excitation mechanisms and the strengths of the electric and magnetic fields. The theoretical analysis of the total intensities, angular distributions, and polarizations can be more extensively developed for applications to atomic radiative transitions from both bound and autoionizing states in the presence of electric and magnetic fields. For a very general theoretical description, it is necessary to consider the influence of environmental decoherence and relaxation processes, together with the corresponding spectral-line broadening mechanisms.

By means of a density-matrix approach, a general quantum-mechanical description of polarized atomic radiative emission from both bound and autoionizing states in the presence of electric and magnetic fields can be developed. Account can thereby be taken of the quantum-mechanical interference phenomena that are usually associated with the coherent excitation of these atomic states, which is of special importance in the case of the autoionizing states. Accordingly, it is advantageous to extend a previously developed density-matrix description for atomic radiative emission in electric and magnetic fields [1], for which the primary emphasis was on bound–bound atomic radiative transitions. The electric-field induced enhancement of the radiation emitted in the dielectronic-recombination processes, which can be understood in terms of the electric-field induced modification of the radiationless electron capture and autoionizing rates, has been theoretically predicted [2–4] and experimentally observed [5–7]. Recently, the modification of the electric-field enhanced dielectronic recombination by a perpendicular magnetic field has been theoretically investigated [8–10] and experimentally identified [7,11,12]. In the density-matrix description of polarized atomic radiative emission from bound and autoionizing states, account can be taken of a general set of steady-state (possibly coherent) atomic-excitation processes in the presence of an arbitrary arrangement of static (or quasistatic) electric and magnetic fields.

In the theoretical description of polarized atomic radiative emission, it is convenient to distinguish between a restricted polarization-density-matrix description [1], which is rigorously applicable only to an isolated atomic system combined (or entangled) with the relevant (observable) mode of the electromagnetic field, and a more comprehensive quantum-open-systems reduced-density-matrix formulation [13], in which the influence of a larger system (environment) of charged particles and photons can be incorporated. In the reduced-density-matrix formulation, the environmental interactions are treated in terms of decoherence and relaxation processes, together with the associated spectral-line broadening mechanisms. The ordinary Hilbert-space quantum theory of polarized radiative emission, following directed-electron collisional excitation of an isolated atomic system (in the absence of electric and magnetic fields), was first presented by Oppenheimer [14] and subsequently refined by Percival and Seaton [15]. A polarization-density-matrix approach to this theory of radiative emission has been presented by Inal and Dubau for ordinary bound–bound atomic transitions [16] and subsequently extended to dielectronic recombination radiation [17]. A polarization-density-matrix description for dielectronic recombination radiation (in the absence of electric and magnetic fields) has been developed by Shlyaptseva, Urnov, and Vinogradov [18,19] and applied by Shlyaptseva et al. [20] to spectroscopic observations on the electron beam ion trap EBIT at the Lawrence Livermore National Laboratory. Radiative emission from atomic transitions excited by electrons spirally in magnetic fields, where the electrons with velocity components perpendicular to the common electron-beam and magnetic-field direction acquire a helical trajectory, has been treated by Gu, Savin, and Beiersddorfer [21], using a polarization-density-matrix approach. In a previously developed polarization-density-matrix description, which was primarily directed at bound–bound radiative transitions [1], it has been found advantageous to exploit the angular-momentum methods and techniques advanced in the earlier density-matrix descriptions of the angular distribution and polarization in single-photon and multiphoton ionization processes [22,23], as well as those developed in the density-matrix analyses presented by Inal and Dubau [16,17]. A more comprehensive reduced-density-matrix formulation is based on a fundamental quantum-open-systems approach in the Liouville-space representation. This formulation provides a detailed theoretical description of polarized atomic radiative emission for a general set of steady-state nonequilibrium (possibly coherent) excitation and de-excitation processes involving the atomic bound and autoionizing states, in the presence of an arbitrary arrangement of static (or quasistatic) electric and magnetic fields and under the influence of environmental collisional and radiative decoherence and relaxation processes.

To provide precise interpretations and theoretical predictions of a wide class of spectroscopic observations, it is necessary to carry out reliable calculations for the relevant Stark–Zeeman spectral patterns and intensities. Separate calculations should be performed for different assumptions regarding the basic atomic excitation and de-excitation mechanisms, including collisional transitions induced by both electrons and ions. Transition probabilities (or cross sections) for elementary atomic radiative and collisional interactions can be substantially altered in the presence of electric and magnetic fields. In the density-matrix description of collisional and radiative interactions in the presence of electric and magnetic fields, the atomic-state populations (corresponding to the diagonal density-matrix elements) and the atomic-state coherences (represented by the nondiagonal elements) can be determined systematically and self-consistently. In the traditional collisional-radiative model for atomic processes in a plasma, only the population densities are usually determined and only the field-free atomic transition rates are normally employed. In the conventional description of spectral-line broadening in plasmas, the electron densities are often assumed to be sufficiently high for the establishment of local-thermodynamic-equilibrium (LTE) population densities, and the atomic-state coherences are usually assumed to be destroyed on a very rapid timescale. In plasma kinetic theory, electric and magnetic fields are frequently included in either the collisional (Boltzmann) or the collisionless (Vlasov) equation for the nonequilibrium single-particle electron-velocity distribution function. For a comprehensive description of atomic processes in a plasma, it is necessary to provide a systematic and self-consistent treatment of the effects of electric and magnetic fields on both the atomic-state and the plasma-electron kinetics.

## 11.2 Polarization-Density-Matrix Description

The polarization-density-matrix formalism has been discussed by Fano [24], Jacobs [22,23], Blum [25], and Kazantsev and Hénoux [26]. In the density-matrix description of polarized atomic radiative emission, it is convenient to assume that the matrix of the total Hamiltonian operator, describing the many-electron atomic system in the presence of an arbitrary arrangement of static (or quasistatic) electric and magnetic fields, has been diagonalized in a basis set of field-free (isolated-atom) eigenstates. The complete atomic basis set should ideally consist of the entire sets of discrete bound states, discrete autoionizing resonances, and nonresonant continuum (or electron–ion scattering) states of the isolated atomic system. The bound excited-state (or the autoionizing-state) excitation processes and the spontaneous radiative emission processes will be treated as independent events. However, in a generalized collisional-radiative model, which is based on the reduced-density-matrix formulation, all important excitation and de-excitation processes must be systematically and self-consistently taken into account in the determination of the initial bound excited-state (or autoionizing-state) density-matrix elements. The initial excitation process is accordingly treated in terms of a density matrix, whose diagonal elements give the familiar population densities of the initial atomic states and whose nondiagonal elements correspond to the initial atomic-state coherences.

The steady-state spectral intensity, angular distribution, and polarization of the radiation that is emitted in the atomic transitions $\gamma_i \to \gamma_f$ can be systematically determined from a knowledge of the photon-polarization density operator $\rho^R$. In terms of the transition operator $T$, whose lowest-order contribution is given simply by the electromagnetic-interaction operator $\mathcal{V}$, the matrix elements of the photon-polarization density operator $\rho^R$ can be expressed in the form

$$\langle \lambda \,|\, \rho^R \,|\, \lambda' \rangle = \sum_{f,\,i,\,i'}^{\bullet} \langle \gamma_f, \boldsymbol{k}\lambda | T | \gamma_i, 0 \rangle \langle \gamma_i | \rho^A | \gamma_{i'} \rangle \langle \gamma_{i'}, 0 | T | \gamma_f, \boldsymbol{k}\lambda' \rangle \,. \qquad (11.1)$$

The angular frequency $\omega$ of the observed electromagnetic radiation is given in terms of the magnitude $k$ of the photon wave vector $\boldsymbol{k}$ by means of the free-space relation $\omega = kc$. The photon-helicity quantum numbers, which represent the projections of the intrinsic spin of the spin-1 quanta along the propagation direction, may have only the numerical values $\lambda, \lambda' = \pm 1$, corresponding to the right ($\lambda = -1$) and left ($\lambda = +1$) circular polarization. The summations over f, i, and i' include the quantum numbers specifying degenerate or nearly degenerate substates of the field-dependent final and initial atomic states in the radiative transitions $\gamma_i \to \gamma_f$. These restricted summations are indicated by the dot above the summation symbol in (11.1). When field effects are neglected, the degenerate substates can be specified in terms of the usual angular-momentum projection (or magnetic) quantum numbers. In the presence of perpendicular (crossed) electric and magnetic fields, for which there is only a twofold degeneracy associated with a plane of reflection symmetry, the magnetic substates are no longer exactly degenerate. $\rho^A$ is the atomic density operator representing the initial bound exited or autoionizing states of the atomic system in the presence of the electric and magnetic fields. The quantum-mechanical interference between radiative and dielectronic recombination can be incorporated by taking into account the initial atomic-state coherences involving the autoionizing states, which correspond to the nondiagonal elements of $\rho^{\mathrm{A}}$. The description of single-photon emission processes based on (11.1), retaining only the lowest-order contribution in the perturbation expansion for the transition operator $T$, is expected to be adequate for narrow, isolated emission lines or for blended emission features, whose individual spectral-line profiles may not be resolvable. To determine the precise spectral distribution of the possibly overlapping Stark and Zeeman components, it is necessary to retain the high-order contributions. These high-order contributions give rise to the frequency shifts and the spectral widths of the bound–bound emission lines and the dielectronic satellite lines. The more comprehensive reduced-density-matrix formulation provides a systematic approach to the spectral line shape problem in the presence of electric and magnetic fields, including the environmental collisional and radiative interactions.

The initial-state atomic density matrix $\rho^{\mathrm{A}}$ could be determined on the basis of a (possibly steady-state) Master equation. This equation, which will be discussed in our more comprehensive reduced-density-matrix formulation, incorporates the influence of the multitude of elementary collisional and radiative interactions in the presence of electric and magnetic fields. In the lowest-order perturbation-theory approximation for these collisional and radiative interactions, the equation for the diagonal matrix elements of $\rho^{\mathrm{A}}$ (corresponding to the atomic-state populations) can be expressed in terms of the familiar (time-independent) rates for all possible collisional and radiative transitions between pairs of field-dependent atomic states. The electron collision rates can be evaluated in terms of the electron-impact excitation or de-excitation cross sections and the single-electron velocity distribution function, while the radiative contributions may include both the spontaneous radiative emission rates and the radiation-intensity-dependent absorption or induced emission rates. The collisional and radiative relaxation processes, together with the decoherence processes, can be most systematically and self-consistently treated within the framework of a reduced-density-matrix description, starting with the equation of motion for the density operator representing the entire, enlarged system consisting of the relevant atomic system together with the quantized-electromagnetic-field and charged-particle environment.

### 11.2.1 Field-Free Atomic Eigenstate Representation

Since polarization is intimately related to angular momentum, it is advantageous to employ the angular-momentum representation for the discrete bound states, discrete autoionizing resonances, and nonresonant continuum atomic states. The electromagnetic-multipole expansion for the quantized radiation field will also be introduced. Accordingly, the initial atomic eigenstates in the presence of the electric and magnetic fields will be expanded in a basis set of field-free angular-momentum eigenstates as

$$|\gamma_i\rangle = \sum_{\Delta_i J_i M_i} |\Delta_i J_i M_i\rangle \langle \Delta_i J_i M_i|\gamma_i\rangle . \tag{11.2}$$

Here $J_i$ is the total electronic angular momentum, $M_i$ is the projection (or component) along a suitably chosen atomic quantization axis, and $\Delta_i$ denotes the set of remaining quantum numbers. Hyperfine structure will be ignored in the present theoretical analysis. The final atomic eigenstates in the presence of the electric and magnetic fields can be represented by an expansion in the same form as (11.2). We emphasize that the complete basis set of field-free atomic states should ideally include the entire sets of the discrete bound states, the autoionizing resonances, and the nonresonance electron–ion-scattering (continuum-channel) states. In weak-field atomic-structure calculations, the field-induced mixing of only a relatively small set of low-lying field-free bound eigenstates is normally considered. For an accurate determination of the shifts and splittings of the spectral patterns in a strong field, highly excited bound and continuum field-free atomic eigenstates must be taken into account. It is also well known that the field-free unperturbed-eigenstate expansion, while formally complete, may not provide an economical representation for atomic systems under the influence of very strong fields.

The Hamiltonian operator $H^A$ for the many-electron atomic system can be expressed in the form $H^A = H^A(0) + H^A(E) + H^A(B)$, where $H^A(0)$ describes the isolated, field-free atomic system and $H^A(E)$ and $H^A(B)$ represent the electric-field and magnetic-field interactions, respectively. The field-dependent eigenvalues and eigenstates of $H^A$ can be determined as expansions in the field-free angular-momentum representation, employing a relativistic multiconfiguration atomic-structure calculation. For sufficiently weak fields, a perturbation-theory analysis may provide a preliminary approximation for the complex Stark–Zeeman spectral patterns. The Zeeman-effect spectral patterns, which are produced by a weak magnetic field acting alone, are characterized by the (conserved) total electronic angular-momentum projections $M_i$ and $M_f$. For a magnetic field that is sufficiently strong to uncouple the orbital and spin angular momenta, the alternative Paschen–Back representation may be more appropriate. In either magnetic-field regime, a preliminary perturbation-theory analysis may be used to investigate the influence of a relatively weak electric field on the Zeeman spectral patterns.

### 11.2.2 Multipole Expansion of the Electromagnetic Interaction

The photon-polarization parameters are most naturally defined with respect to the direction of spectroscopic observation. In the theoretical analysis, it is more convenient to relate these parameters to the fundamental electromagnetic-transitions amplitudes, which are usually defined with respect to the atomic quantization axis. The desired transformation can be introduced by means of the expansion of the radiative-transition matrix elements, in terms of the matrix elements of effective electric and magnetic multipole operators, in the form [27]

$$\langle \Delta_f J_f M_f, \boldsymbol{k}\lambda|T|\Delta_i J_i M_i, 0\rangle = \sum_j \sum_m (-1)^m \left(\frac{2j+1}{4\pi}\right)^{1/2}$$
$$\times A(j) D_{\lambda m}^{(j)}(\widehat{\boldsymbol{k}}) \langle \Delta_f J_f M_f|\tilde{Q}_{-m}^{(j)}|\Delta_i J_i M_i\rangle.$$
$$(11.3)$$

Here $\tilde{Q}_{-m}^{(j)}$ denotes the irreducible spherical-tensor form of the effective electromagnetic-multipole-moment operator for the many-electron atomic system. The lowest-order perturbation-theory component is the usual electromagnetic-multipole-moment tensor operator $Q_{-m}^{(j)}$. The higher-order components may be viewed as corresponding to the radiative corrections that are predicted by quantum-electrodynamical perturbation theory, and these higher-order components represent the contributions from nonlinear electromagnetic processes arising from the absorption and emission of virtual photons and the creation and destruction of intermediate atomic states. The quantities $D_{\lambda m}^{(j)}(\widehat{\boldsymbol{k}})$ designate the matrix elements $D_{\lambda m}^{(j)}(\phi, \theta, 0)$ of the Wigner rotation operator corresponding to the desired coordinate rotation. The multiplying factors $A(j)$ are defined individually for the various electromagnetic-multipole components. The photon eigenstate representation characterized by linear momentum and intrinsic spin is thereby replaced by the alternative representation based on angular momentum and parity. The matrix elements of the effective electromagnetic-multipole-moment operator can be evaluated in the angular-momentum representation, in terms of Wigner 3-$j$ symbols and reduced electromagnetic-multipole-moment matrix elements, using the Wigner–Eckart theorem:

$$\langle \Delta_f J_f M_f|\tilde{Q}_m^{(j)}|\Delta_i J_i M_i\rangle = (-1)^{J_f - M_f} \begin{pmatrix} J_f & j & J_i \\ -M_f & m & M_i \end{pmatrix} \langle \Delta_f J_f||\tilde{Q}^{(j)}||\Delta_i J_i\rangle.$$
$$(11.4)$$

To take into account both the electric and magnetic multipole contributions associated with a given value of $j$, the effective electromagnetic-multipole-moment operator $\tilde{Q}_{-m}^{(j)}$ should be defined to include the components associated with all permissible values of the photon parity. In contrast with the photon helicity $\lambda$, which can have only the values $\pm 1$, the angular-momentum projection $m$ is not limited to the values $\pm 1$. The symmetry (or angular-momentum conservation) information is completely represented by the 3-$j$ symbols, while the dynamical information is entirely contained in the reduced matrix elements of the effective electromagnetic-multipole-moment operators.

The spectroscopic observation of polarized atomic radiative emission in perpendicular (crossed) electric and magnetic fields is illustrated in Fig. 11.1. The special case of perpendicular electric and magnetic fields is encountered in numerous experimental arrangements. In their rest frame of reference, an atomic system moving in a magnetic field will experience a Lorentz (or motional) electric field, which is perpendicular to both the magnetic field and the direction of motion. In any reference frame, an electromagnetic field can be treated as composed of perpendicular electric and magnetic field components,

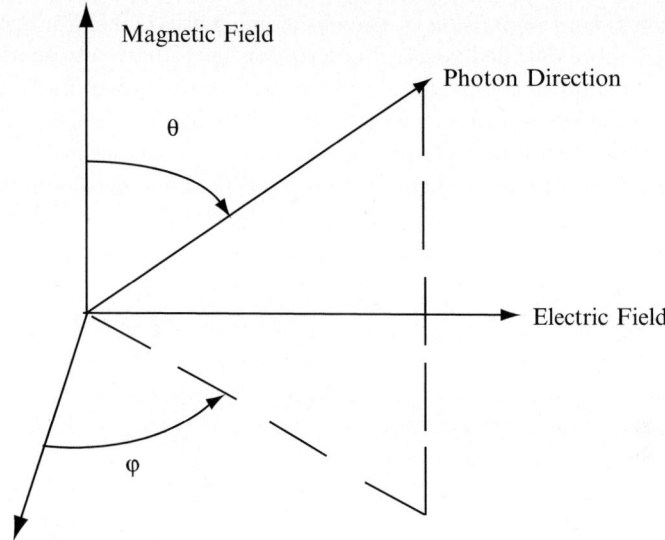

**Fig. 11.1.** Viewing angles appropriate for the spectroscopic observation of polarized atomic radiative emission in perpendicular (crossed) electric and magnetic fields

which are also perpendicular to the photon propagation direction. In both charged-particle beam and plasma environments, the total electric and magnetic fields acting on the radiating atomic system should be treated as composed of an external (applied) field and an internal (dynamical) field. In a high-density plasma, the action of the relatively slowly moving ions has been customarily treated in the quasistatic approximation, in terms of an isotropic (nearly equilibrium) statistical distribution of static electric fields. However, it is often necessary to consider an anisotropic (nonequilibrium) component corresponding to a distribution of turbulent electric fields. In a tokamak plasma, a dynamical (poloidal) magnetic field is generated perpendicular to the externally applied (toroidal) magnetic field. In a tokamak plasma, the viewing angle $\theta$ may be selected to be perpendicular to the known toroidal magnetic field. In an electron–ion beam experiment, the most convenient angle of observation is usually at a direction perpendicular to the electron beam. However, it has been recognized that more detailed spectroscopic investigations can be made by a variation of the angle of observation away from the perpendicular direction.

### 11.2.3 Photon-Polarization Density Matrix Allowing for Coherent Excitation Processes in a General Arrangement of Electric and Magnetic Fields

For ordinary bound–bound radiative transitions, an expression has been obtained for the photon-polarization density matrix [1]. In the derivation of this expression, only the lowest-order contribution was retained in the perturbation-theory expansion of the electromagnetic-transition operator $T$. To provide a more detailed spectral description for radiative transitions from bound excited and autoionizing states, as well as to incorporate quantum-mechanical interference phenomena, it is necessary to consider the entire electromagnetic-transition operator and to introduce the associated effective electromagnetic-multiple-moment operators $\tilde{Q}^{(j)}_{-m}$. The corresponding generalized expression for the photon-polarization density matrix elements may be presented in the form

$$
\langle \lambda | \rho^{R} | \lambda' \rangle = \overset{\bullet}{\sum_{i,i',f}} \langle \gamma_i | \rho^{A} | \gamma_{i'} \rangle \sum_{\Delta_f J_f M_f} \sum_{\Delta_f' J_f' M_f'} \langle \gamma_f | \Delta_f J_f M_f \rangle \langle \gamma_f | \Delta_f' J_f' M_f' \rangle^*
$$

$$
\times \sum_{\Delta_i J_i M_i} \sum_{\Delta_{i'} J_{i'} M_{i'}} \langle \Delta_i J_i M_i | \gamma_i \rangle \langle \Delta_{i'} J_{i'} M_{i'} | \gamma_{i'} \rangle^*
$$

$$
\times \sum_{j,m} \sum_{j',m'} \sum_{J,M,M'} \langle \Delta_f J_f || \tilde{Q}^{(j)} || \Delta_i J_i \rangle \langle \Delta_f' J_f' || \tilde{Q}^{(j')} || \Delta_{i'} J_{i'} \rangle^*
$$

$$
\times \left( \frac{1}{4\pi} \right) (2j+1)^{1/2} (2j'+1)^{1/2} (2J+1) A(j) A(j')^*
$$

$$
\times (-1)^{J_f + J_f' - M_f - M_f' - m - m' + \lambda' - m' + 2j - 2j' - M - M'} \begin{pmatrix} J_f & j & J_i \\ -M_f & -m & M_i \end{pmatrix}
$$

$$
\times \begin{pmatrix} J_f' & j' & J_{i'} \\ -M_f' & -m' & M_{i'} \end{pmatrix} \begin{pmatrix} j & j' & J \\ \lambda & -\lambda' & -M \end{pmatrix} \begin{pmatrix} j & j' & J \\ m & -m' & -M \end{pmatrix} D^{(J)}_{MM'}(\widehat{\boldsymbol{k}}) .
$$

$$(11.5)$$

The dot above the summation symbol is used to indicate that the summations over f, i, and i′ are to be taken over quantum numbers specifying degenerate or nearly degenerate substates of the many-electron atomic system in the presence of the electric and magnetic fields, as in (11.1). In the derivation of this expression, the product of two Wigner rotation matrices, with ranks $j$ and $j'$, has been expanded as a summation of Wigner rotation matrices corresponding to the total angular momentum $J$, each of which is multiplied by two additional 3-$j$ symbols. Although the primary interest is usually in electric-dipole transitions from the bound excited and autoionizing states, it is advantageous to retain the generality for arbitrary electromagnetic-multipole interactions, including the interference between different multipole amplitudes. The general, nonperturbative expression for the effective electromagnetic-multiple-moment operators $\tilde{Q}^{(j)}_{-m}$ is derived using the entire electromagnetic-transition operator $T$.

It should also be emphasized that the general expression given by (11.5) can be employed for an arbitrary arrangement of (static or quasistatic) electric and magnetic fields and for a general set of steady-state (or time-dependent) nonequilibrium excitation processes. The dependence on the electric and magnetic fields, which may be quite complex and not expressible in a simple

analytical form, is incorporated by the field transformation or mixing co-
efficients $\langle \Delta_i J_i M_i | \gamma_i \rangle$ and $\langle \gamma_f | \Delta_f J_f M_f \rangle$ and also by the initial atomic-state
density-matrix elements $\langle \gamma_i | \rho^A | \gamma_{i'} \rangle$. The nondiagonal elements of $\rho^A$ describe
the coherent excitation of the field-dependent initial atomic substates. Coher-
ences involving atomic states with different energy eigenvalues, which can be
produced by short-pulse laser-photon excitation, generate quantum beats in a
time-resolved photon-detection process. With the incorporation of relaxation
and decoherence processes, the time-dependent density-matrix description
provides a rigorous starting point for the introduction of the time-integrated,
reduced photon density matrix. The coherences for excited atomic states with
different energy eigenvalues are expected to be important when the energy-
level separations are not large in comparison with the spectral-line widths.

If the field-induced mixing of the atomic eigenstates is neglected, as well
as the coherences corresponding to the nondiagonal matrix elements of the
initial atomic-state density operator $\rho^A$, the general expression for the photon-
polarization density matrix may be reduced to the result

$$
\langle \lambda | \rho^R | \lambda' \rangle = \sum_{M_i} N(\Delta_i J_i M_i) \sum_{j, j', J} (-1)^{J_f + j + j' + \lambda' - M_i}
$$
$$
\times \left[ (2j+1)(2j'+1)(2J+1)/4\pi \right]^{1/2} A(j) A(j')^*
$$
$$
\times \langle \Delta_f J_f || \tilde{Q}^{(j)} || \Delta_i J_i \rangle \langle \Delta_f J_f || \tilde{Q}^{(j')} || \Delta_i J_i \rangle^*
$$
$$
\times \begin{pmatrix} J & j & j' \\ \lambda' - \lambda & \lambda & -\lambda' \end{pmatrix} \begin{pmatrix} J & J_i & J_i \\ 0 & -M_i & M_i \end{pmatrix} \begin{Bmatrix} J & J_i & J_i \\ J_f & j' & j \end{Bmatrix} Y_{\lambda' - \lambda}^{(J)}(\hat{\boldsymbol{k}}) .
$$

$$(11.6)$$

The diagonal matrix elements of $\rho^A$, which correspond to the usual level population densities, are now denoted by $N(\Delta_i J_i M_i)$. The condition $M' = 0$ follows
from the neglect of the nondiagonal matrix elements of $\rho^A$, together with the
neglect of the field-induced mixing of the upper, initial substates. The Wigner
rotation matrix elements $D_{MM'}^{(J)}(\hat{\boldsymbol{k}})$ can then be reduced to the spherical harmonic functions $Y_{\lambda' - \lambda}^{(J)}(\hat{\boldsymbol{k}})$, for which the angular momentum $J$ can assume
only integer values. The 6-$j$ symbol is introduced with the neglect of the field-
induced mixing of the lower, final substates, which is expected to be valid for
low-lying bound states. The dominant radiative emission process is usually
assumed to involve only a single multipole component of the electromagnetic
field, in which case $j = j'$. Equations (11.5) and (11.6) are valid for inter-
fering electromagnetic-multipole components, corresponding to $j \neq j'$. This
interference can have a more important effect on the angular distribution and
polarization than on the total photon intensity.

In the lowest-order perturbation-theory approximation for the effective
electromagnetic-multiple-moment operators $\tilde{Q}_{-m}^{(j)}$, the result given by (11.6)
is found to be in agreement with that obtained by Inal and Dubau [17],
who investigated the directed electron excitation of polarized atomic radiative

emission during dielectronic-recombination satellite transitions in the absence of electric and magnetic fields. Only even values of $J$ can contribute for excitation by a beam of unpolarized electrons, due to the reflection symmetry in the plane perpendicular to the atomic quantization axis (which is taken as the electron-beam propagation direction). The usual axial symmetry can be broken in the coherent excitation of the magnetic substates by a spin-polarized electron beam for which the spin projection is defined at an angle with respect to the atomic quantization axis. The axial symmetry can also be broken during atomic excitation in perpendicular (crossed) electric and magnetic fields, due to the coherent excitation of the magnetic substates. It should be pointed out that the density-matrix formulation that has been introduced in this investigation is based on a boundary condition in which the initial atomic states are taken as the autoionizing resonances. This leads to a different treatment of the autoionizing-state population kinetics than that presented by Inal and Dubau [17], which is based on the alternative boundary condition involving the initial electron–ion continuum states. In the application of the present density-matrix formulation to electron–ion beam experiments, the initial atomic-state populations and coherences (together with their implicit dependences on the electron-beam direction) would be introduced in the course of the generalized collisional-radiative-model determination of the initial bound-state and autoionizing-state density-matrix elements.

### 11.2.4 Irreducible Spherical-Tensor Representation of the Density Operators

By means of (11.5), the matrix elements of photon-polarization density operator have been expressed in a nonperturbative form, in terms of the reduced matrix elements of the effective electromagnetic-multiple-moment operators. These matrix elements are explicitly given as functions of the photon-emission direction, in terms of which the photon-helicity quantum number $\lambda$ is defined. Following Fano and Racah [28], Happer [29,30], Omont [31], and Baylis [32], this photon-polarization density operator may be presented, as an expansion in terms of the irreducible spherical-tensor operators $T_M^{(J)}(j, j')$, as

$$\rho^{\mathrm{R}} = \sum_{j,j'} \sum_{J,M,M'} \rho^{\mathrm{R}}(j, j'; J, M') T_M^{(J)}(j, j') \, D_{MM'}^{(J)}(\widehat{\boldsymbol{k}}) \,. \tag{11.7}$$

In the Liouville-space Dirac notation, which will be adopted in the more comprehensive reduced-density-operator formulation, this expansion corresponds to a transformation from the representation of the uncoupled states $|jm, j'm'\rangle\rangle$ to the alternative representation of the coupled states $|j, j'; J, M\rangle\rangle$. The coefficients $\rho^{\mathrm{R}}(j, j'; J, M')$ are referred to as the irreducible spherical-tensor components of the photon-polarization density operator. They are also known as state multipoles or statistical tensors. The photon-helicity quantum numbers are defined, with respect to the photon propagation direction, by means of the Wigner rotation matrices $D^{(J)}_{M\,M'}(\hat{\boldsymbol{k}})$. The ordinary Hilbert-space matrix elements of the irreducible spherical-tensor operator $T^{(J)}_M(j,j')$ can be evaluated using the Wigner–Eckart theorem as

$$\langle j\,\lambda|T^{(J)}_M(j,j')|j'\,\lambda'\rangle = (-1)^{j-\lambda}(2J+1)^{1/2}\begin{pmatrix} j & J & j' \\ -\lambda & M & \lambda' \end{pmatrix}. \qquad (11.8)$$

The reduced matrix element of $T^{(J)}_M(j,j')$ is accordingly given by the factor $(2J+1)^{1/2}$. The general expression for the irreducible spherical-tensor components $\rho^{\mathrm{R}}(j,j';J,M')$ can be obtained by comparing (11.5) and (11.7), employing (11.8) and the symmetry properties of the 3-$j$ symbols. This irreducible spherical-tensor representation is often advantageous for the photon density operator, because only a few multipole components of the electromagnetic field are usually required to provide an adequate description of an atomic radiative-emission process.

In contrast to the photon density operator, the irreducible spherical-tensor representation of the field-dependent atomic bound excited-state or autoionizing-state density operator, which may be expressed in the form,

$$\rho^{\mathrm{A}} = \sum_{\Delta_{\mathrm{i}}\,\Delta_{\mathrm{i}'}} \sum_{J_{\mathrm{i}}\,J_{\mathrm{i}'}} \sum_{K\,N} \rho^{\mathrm{A}}(\Delta_{\mathrm{i}}\,\Delta_{\mathrm{i}'}\,J_{\mathrm{i}}\,J_{\mathrm{i}'};K,N)T^{(K)}_N(J_{\mathrm{i}}\,J_{\mathrm{i}'})\,, \qquad (11.9)$$

involves two separate expansions that ideally should be taken over the complete basis set of unperturbed (field-free) atomic eigenstates. Consequently, the irreducible spherical-tensor representation of the atomic density operator could be advantageous for weak fields or perhaps for parallel electric and magnetic fields. For atomic systems with axial symmetry, in which case $N=0$, it is useful to introduce a representation of the atomic bound excited-state or autoionizing-state density operator in terms of components describing orientation (corresponding to $K=1$) and alignment (corresponding to $K=2$), and possibly higher-rank components.

To express the irreducible spherical-tensor components of the photon-polarization density operator in terms of the atomic irreducible spherical-tensor components $\rho^{\mathrm{A}}(\Delta_{\mathrm{i}}\Delta_{\mathrm{i}'}J_{\mathrm{i}}J_{\mathrm{i}'};K,N)$, it is necessary employ the following transformation:

$$\langle\gamma_{\mathrm{i}}|\rho^{\mathrm{A}}|\gamma_{\mathrm{i}'}\rangle = \sum_{\Delta_{\mathrm{i}}J_{\mathrm{i}}M_{\mathrm{i}}\,\Delta_{\mathrm{i}'}J_{\mathrm{i}'}M_{\mathrm{i}'}} \langle\Delta_{\mathrm{i}}J_{\mathrm{i}}M_{\mathrm{i}}|\gamma_{\mathrm{i}}\rangle^*\langle\Delta_{\mathrm{i}'}J_{\mathrm{i}'}M_{\mathrm{i}'}|\gamma_{\mathrm{i}'}\rangle$$

$$\times\sum_{K,N}\rho^{\mathrm{A}}(\Delta_{\mathrm{i}}\Delta_{\mathrm{i}'}J_{\mathrm{i}}J_{\mathrm{i}'};K,N)(-1)^{J_{\mathrm{i}}-M_{\mathrm{i}}}(2K+1)^{1/2}\begin{pmatrix} J_{\mathrm{i}} & K & J_{\mathrm{i}'} \\ -M_{\mathrm{i}} & N & M_{\mathrm{i}'} \end{pmatrix}.$$

$$(11.10)$$

By means of (11.5), (11.7), (11.9), and (11.10), the irreducible spherical-tensor expansion coefficients (or components) of the photon-polarization density operator are expressed in terms of the irreducible spherical-tensor components of the atomic density operator and the reduced matrix elements describing the atomic electromagnetic transition. A natural separation between the geometrical (or symmetry) and the dynamical factors is thereby achieved.

A complex dependence on the electric and magnetic fields is implicitly incorporated by the occurrence of the coefficients $\langle \Delta_i J_i M_i | \gamma_i \rangle$ and $\langle \Delta_f J_f M_f | \gamma_f \rangle$. These coefficients must be determined by a diagonalization, in the field-free angular-momentum basis representation, of the Hamiltonian $H^A$ describing the many-electron atomic system in the presence of the electric and magnetic fields. In applications to dielectronic recombination, the electric-field induced mixing of the autoionizing states can significantly alter the radiationless electron capture and autoionization rates [2], which occur in the generalized collisional-radiative model used in the determination of the autoionizing-state populations and coherences. Physically, the enhancement of the dielectronic-recombination radiation in the presence of an electric field may be understood as a redistribution of population among the outer-electron $nl$ states in favor of higher-$l$ states, which have negligible field-free autoionization rates but exhibit radiative emission rates that are essentially independent of $l$.

If field-induced mixing of the atomic eigenstates and the atomic-state coherences are neglected, the irreducible spherical-tensor components of the photon-polarization density operator can be simply related to the irreducible spherical-tensor components of the initial atomic-state density operator as

$$\rho^R(j, j'; J, 0) = \rho^A(\Delta_i \Delta_i J_i J_i; J, 0)$$
$$\times (-1)^{J_i + J_f + J + 2j + j'} (1/4\pi) \left[(2j + 1)(2j' + 1)\right]^{1/2} A(j) A(j')^*$$
$$\times \langle \Delta_f J_f || \tilde{Q}^{(j)} || \Delta_i J_i \rangle \langle \Delta_f J_f || \tilde{Q}^{(j')} || \Delta_i J_i \rangle^* \begin{Bmatrix} J & J_i & J_i \\ J_f & j' & j \end{Bmatrix} , \quad (11.11)$$

which is equivalent to (11.6). Coherences between initial atomic states with different angular-momentum-projection quantum numbers can be taken into account only by including the irreducible spherical-tensor components $\rho^A(\Delta_i \Delta_i J_i J_i; K, N)$ with $N \neq 0$.

For systems with spherical or axial symmetry, the steady-state or time-dependent Master equation governing the atomic density operator (which will be introduced in the more comprehensive reduced-density-operator formulation) can be advantageously re-expressed in the irreducible spherical-tensor representation. These equations may be solved, taking into account the relevant set of elementary atomic collisional and radiative relaxation and decoherence processes in the presence of the electric and magnetic fields. If the influence of the electric and magnetic fields on the relaxation and decoherence rates can be neglected, the spherical-tensor components $\rho^A(\Delta_i \Delta_{i'} J_i J_{i'}; K, N)$ thereby obtained could then be simply transformed into the field-dependent representation by means of (11.10). However, the spherical-tensor representation may not retain its characteristic advantage in the presence of a general arrangement of electric and magnetic fields, for which the field-free angular-momentum and parity selections rules may no longer be rigorously valid.

## 11.2.5 Stokes-Parameter Representation of the Photon Density Operator

The photon-polarization density operator is most commonly presented in a representation based on the Stokes parameters. This representation can be expressed as [25]

$$\rho^R = \left(\frac{I}{2}\right)\begin{pmatrix} 1+V & -Q+iU \\ -Q-iU & 1-V \end{pmatrix}. \tag{11.12}$$

The total spectral intensity $I$, summed over all photon-polarization states, is determined by the normalization condition on $\rho^R$:

$$I = \mathrm{Tr}\,\rho = \langle 1|\rho^R|1\rangle + \langle -1|\rho^R|-1\rangle. \tag{11.13}$$

The parameters $Q$ and $U$ specify linear polarization, while $V$ represents circular polarization. Using the nonperturbative approach, involving the entire electromagnetic-transition operator $T$ rather than the lowest-order contribution $\mathcal{V}$, the four photon-polarization parameters can be determined, as functions of the photon frequency and propagation direction, by means of the polarization-density-matrix formulation.

In the directed excitation of an atomic system (in the absence of electric and magnetic fields) by a beam of electrons that are not spin polarized, axial symmetry is preserved and reflection symmetry is maintained in the plane perpendicular to the quantization axis, which is taken to be the electron-beam direction. Under these conditions, the atomic magnetic substates cannot be coherently excited. It then follows that the Stokes parameters $U$ and $V$ must vanish. The only nonzero polarization parameter $Q$, which is one of the two linear-polarization parameters, can be conveniently defined by means of the relationship

$$Q = (I_\| - I_\perp)/(I_\| + I_\perp), \tag{11.14}$$

where $I_\|$ and $I_\perp$ refer to the emitted photon intensities with polarization directions parallel and perpendicular, respectively, to the plane that is defined by the photon propagation direction and the atomic quantization axis. In terms of the helicity matrix elements of the photon-polarization density operator $\rho^R$, this linear-polarization parameter may be expressed in the form

$$Q = \frac{-\langle 1|\rho^R|-1\rangle + \langle -1|\rho^R|1\rangle}{\langle 1|\rho^R|1\rangle + \langle -1|\rho^R|-1\rangle}. \tag{11.15}$$

The four Stokes parameters $I$, $Q$, $U$, and $V$ can be evaluated, as functions of the emitted photon angles, by means of our general expression for the matrix elements of the photon-polarization density operator presented by (11.5), which is applicable to the steady-state (or time-dependent) coherent excitation of the initial atomic states in an arbitrary arrangement of static (or quasistatic) electric and magnetic fields. Alternatively, by utilizing the expansion given by (11.7), the Stokes parameters may be expressed in terms of the irreducible spherical-tensor components of the photon-polarization density operator. However, the irreducible spherical-tensor components of the photon-polarization density operator can be conveniently related to the irreducible spherical-tensor components of the initial atomic-state density operator only for special high-symmetry situations or for sufficiently weak fields.

## 11.3 Polarization of Radiative Emission Along the Magnetic-Field Direction

Due to a limited line-of-sight access in some experimental arrangements, the polarized atomic radiative emission is often observed only in the direction of the external magnetic field. However, in tokamak plasmas, the polarized atomic radiative emission has been measured in a direction perpendicular to the known toroidal magnetic field and parallel to the poloidal magnetic field of interest. For observation along the magnetic-field direction, the main emphasis has been on the determination of the circular-polarization parameter $V$, which can be defined as

$$V = (I_r - I_l)/(I_r + I_l) \,. \tag{11.16}$$

Here $I_r$ and $I_l$ denote the intensities of the right and left circularly polarized radiative emissions. The circular-polarization parameter $V$ may be conveniently expressed, in terms of the diagonal matrix elements of the photon-polarization density operator $\rho^R$, as

$$V = \frac{\langle 1|\rho^R|1\rangle - \langle -1|\rho^R|-1\rangle}{\langle 1|\rho^R|1\rangle + \langle -1|\rho^R|-1\rangle} \,. \tag{11.17}$$

### 11.3.1 Polarization of Radiative Emission in the Presence of Perpendicular (crossed) Electric and Magnetic Fields and Coherent Excitation Processes

Spectroscopic observations in the magnetic-field direction, which will be taken as the direction of the atomic quantization axis, can be treated by setting $\theta = 0$ in the general expression for the matrix elements of the photon-polarization density operator $\rho^R$, which is given by (11.5). This expression can then be substantially simplified by taking advantage of the relationship $D^{(J)}_{MM'}(\phi, 0, 0) = \delta(M, M')$. In addition, only a single electromagnetic-multipole contribution (which will be specified by $j$) will be assumed to be dominant, and the interference between different electromagnetic-multipole components will accordingly be ignored. Using the orthogonality properties of the final two 3-$j$ symbols, the general expression can then be reduced to the result

$$\langle\lambda|\rho^{\mathrm{R}}|\lambda'\rangle = \overset{\bullet}{\sum_{i,i',f}} \langle\gamma_i|\rho^{\mathrm{A}}|\gamma_{i'}\rangle \sum_{\Delta_f J_f M_f} \sum_{\Delta'_f J'_f M'_f} \langle\gamma_f|\Delta_f J_f M_f\rangle\langle\gamma_f|\Delta'_f J'_f M'_f\rangle^*$$

$$\times \sum_{\Delta_i J_i M_i} \sum_{\Delta_{i'} J_{i'} M_{i'}} \langle\Delta_i J_i M_i|\gamma_i\rangle\langle\Delta_{i'} J_{i'} M_{i'}|\gamma_{i'}\rangle^* \left(\frac{2j+1}{4\pi}\right) A(j)\,A(j)^*$$

$$\times (-1)^{J_f+J'_f-M_f-M'_f-\lambda-\lambda'} \langle\Delta_f J_f||\tilde{Q}^{(j)}||\Delta_i J_i\rangle\langle\Delta'_f J'_f||\tilde{Q}^{(j)}||\Delta_{i'} J_{i'}\rangle^*$$

$$\times \begin{pmatrix} J_f & j & J_i \\ -M_f & -\lambda & M_i \end{pmatrix} \begin{pmatrix} J'_f & j & J_{i'} \\ -M'_f & -\lambda' & M_{i'} \end{pmatrix}. \tag{11.18}$$

This simplified expression can be used to evaluate both the linear-polarization and the circular-polarization parameters for observation along the magnetic-field direction of atomic radiative emission corresponding to a single electromagnetic-multipole component.

## 11.3.2 Circular Polarization of Radiative Emission in the Absence of a Perpendicular Electric Field and a Coherent Excitation Process

In the absence of a perpendicular component of the electric field, axial symmetry may be present and the $z$-components of the total electronic angular momenta can then be treated as conserved quantities. It then follows that the final-state angular-momentum-projection (or magnetic) quantum numbers must be equal, i.e., it is permissible to set $M_f = M'_f$ in (11.18). Coherent excitation of the initial atomic magnetic substates, which are designated by $M_i$, can still occur as a result of a nonparallel directed-excitation process. In the absence of a coherent excitation process, the initial atomic-state density operator $\rho^{\mathrm{A}}$ must be diagonal in the $M_i$-representation, i.e., it is permissible to assume that $M_i = M_{i'}$ in (11.18). In the absence of a perpendicular component of the electric field and a coherent excitation of the initial magnetic substates, the selection rules governing the remaining two 3-$j$ symbols in (11.18) can be exploited to obtain the result that the photon-polarization density operator $\rho^{\mathrm{R}}$ can have only diagonal elements, corresponding to $\lambda = \lambda' = \pm 1$. Consequently, only the circularly polarized radiative emission will be observable in the direction of the magnetic field. According to (11.17), the circularly polarized radiative emission along the magnetic-field direction can be described, in terms of the diagonal photon-polarization density-matrix elements, by means of the following expression:

$$\langle\lambda|\rho^{\mathrm{R}}|\lambda\rangle = \overset{\bullet}{\sum_{M_i M_f}} \langle\gamma_i M_i|\rho^{\mathrm{A}}|\gamma_i M_i\rangle \sum_{\Delta_f J_f} \sum_{\Delta_{f'} J'_f} \langle\gamma_f M_f|\Delta_f J_f M_f\rangle\langle\gamma_f M_f|\Delta'_f J'_f M_f\rangle^*$$

$$\times \sum_{\Delta_i J_i} \sum_{\Delta_{i'} J_{i'}} \langle\Delta_i J_i M_i|\gamma_i M_i\rangle\langle\Delta_{i'} J_{i'} M_i|\gamma_i M_i\rangle^* \left(\frac{2j+1}{4\pi}\right)$$

$$\times A(j)\, A(j)^*\, (-1)^{J_f + J'_f - 2M_f} \langle \Delta_f J_f || \tilde{Q}^{(j)} || \Delta_i J_i \rangle \langle \Delta'_f J'_f || \tilde{Q}^{(j)} || \Delta_{i'} J_{i'} \rangle^*$$

$$\times \begin{pmatrix} J_f & j & J_i \\ -M_f & -\lambda & M_i \end{pmatrix} \begin{pmatrix} J'_f & j & J_{i'} \\ -M_f & -\lambda & M_i \end{pmatrix}. \tag{11.19}$$

The indices $\gamma_i$ and $\gamma_f$ now represent the quantum numbers for the magnetic-field-dependent many-electron atomic eigenstates, excluding the conserved $z$-components of the total electronic angular momenta. The observation of linearly polarized radiative emission in the magnetic-field direction, which is described in terms of the nondiagonal matrix elements of the photon-polarization density operator $\rho^R$, could reveal the presence of a perpendicular component of the electric field or of a coherent excitation mechanism.

### 11.3.3 Radiative Emission in the Absence of Electric and Magnetic Fields and Coherent Excitation Processes

In the conventional description of atomic radiative emission, initial atomic-state coherences and field-induced mixing of initial and final states are ignored. In addition, the initial magnetic substates are often assumed to be uniformly (statistically) populated. This assumption can be expressed as

$$\langle \gamma_i M_i | \rho^A | \gamma_i M_i \rangle = \frac{\rho^A(\Delta_i \Delta_i J_i J_i; 0, 0)}{(2J_i + 1)^{1/2}} = \frac{N(\Delta_i J_i)}{(2J_i + 1)}. \tag{11.20}$$

The summations over the magnetic quantum numbers $M_i$ and $M_f$ in (11.19) can now be performed. The total intensity of the right or left circularly polarized radiative emission in the magnetic-field direction can then be reduced to the following result:

$$\langle \lambda | \rho^R | \lambda \rangle = \left( \frac{1}{4\pi} \right) \left[ \frac{N(\Delta_i J_i)}{2J_i + 1} \right] |A(j)|^2 |\langle \Delta_f J_f || \tilde{Q}^{(j)} || \Delta_i J_i \rangle|^2, \tag{11.21}$$

which is independent of the photon helicity $\lambda$. With these approximations, the relationship between the irreducible spherical-tensor components of the photon-polarization and many-electron atomic-system density operators may be simply expressed as

$$\rho^R(j, j; 0, 0) = \rho^A(\Delta_i \Delta_i J_i J_i; 0, 0)$$

$$\times \left( \frac{1}{4\pi} \right) \left( \frac{1}{2j + 1} \right)^{1/2} |A(j)|^2 |\langle \Delta_f J_f || \tilde{Q}^{(j)} || \Delta_i J_i \rangle|^2, \tag{11.22}$$

and the irreducible spherical-tensor components $\rho^R(j, j; J, M)$ of the photon-polarization density operator $\rho^R$ corresponding to $J \neq 0$ must then vanish. For anisotropic initial atomic-state population distributions, represented by an atomic-system density operator $\rho^A$ with nonvanishing irreducible spherical-tensor components $\rho^A(\Delta_i \Delta_i J_i J_i; K, N)$ corresponding to $K \neq 0$, the photon-polarization density operator $\rho^R$ can have nonvanishing irreducible spherical-tensor components with $J \neq 0$. This is in accord with the conservation of total angular momentum in the field-free spontaneous radiative-emission process.

### 11.3.4 Electric-Dipole Transitions

Although the importance of forbidden radiative transitions from certain excited atomic states is widely recognized, atomic systems are customarily assumed to undergo spontaneous radiative decay predominantly by electric-dipole transitions. Moreover, the quantum-mechanical interference of different electromagnetic-multipole components is usually ignored. For electric-dipole transitions in the absence of electric and magnetic fields, the matrix elements of the electromagnetic interaction are given, in accord with the standard approximation, as

$$\langle \Delta_f J_f M_f, \, \boldsymbol{k}\lambda|\mathcal{V}|\Delta_i J_i M_i, 0\rangle = -\mathrm{i}(2\pi\hbar\omega)^{1/2}\langle \Delta_f J_f M_f|\boldsymbol{D}_\lambda|\Delta_i J_i M_i\rangle \,, \quad (11.23)$$

where $\boldsymbol{D}_\lambda$ denotes the component of the many-electron electric-dipole-moment operator corresponding to photon helicity $\lambda$. The familiar (lowest-order perturbation theory) expression for the spontaneous electric-dipole emission rate may be evaluated using the Fermi golden-rule formula, which contains the factor $2\pi/\hbar^2$ and the density-of-final-states (per unit frequency) factor $\omega^2/(2\pi c)^3$. The diagonal matrix elements of the photon-polarization density operator describing spontaneous electric-dipole emission are then given as

$$\langle \lambda|\rho^R|\lambda\rangle = \left(\frac{\omega^3}{2\pi\hbar c^3}\right)\left[\frac{N(\Delta_i J_i)}{2J_i + 1}\right]|\langle \Delta_f J_f||Q^{(1)}||\Delta_i J_i\rangle|^2 \,. \quad (11.24)$$

The conventional spontaneous radiative-emission rate (or Einstein $A$ coefficient) is defined in terms of the summation over final magnetic substates $M_f$ and photon polarizations $\lambda$, together with the integration over the photon emission angles. The result thereby obtained can be expressed in the familiar form

$$A_r(\Delta_i J_i \to \Delta_f J_f) = \left(\frac{4}{3}\right)\left(\frac{\omega^3}{\hbar c^3}\right)\left(\frac{1}{2J_i + 1}\right)|\langle \Delta_f J_f||Q^{(1)}||\Delta_i J_i\rangle|^2 \,. \quad (11.25)$$

Equations (11.24) and (11.25) have been recovered by introducing the electric-dipole approximation for the electromagnetic interaction, assuming a uniform (statistical) distribution of the initial magnetic-substate populations, and ignoring field-induced mixing of the unperturbed initial and final many-electron atomic eigenstates.

### 11.3.5 Directed Excitation Processes

In the anisotropic excitation of the many-electron atomic system, which can occur by means of an electron beam or a laser source, a nonuniform distribution of initial magnetic-substate population densities can be established.

Transition rates between pairs of individual magnetic substates can be expressed, in terms of the usual $M$-averaged Einstein $A$ coefficients, as

$$A_r(\gamma_i \rightarrow \gamma_f) = \left| \begin{pmatrix} J_f & j & J_i \\ -M_f & m & M_i \end{pmatrix} \right|^2 (2J_i + 1) A_r(\Delta_i J_i \rightarrow \Delta_f J_f) . \qquad (11.26)$$

It should be emphasized that this expression is valid for any spontaneous radiative emission process for which only a single electromagnetic-multipole components is taken into account. To evaluate the matrix elements of the photon-polarization density operator, the initial atomic-state population densities $N(\gamma_i) = N(\Delta_i J_i M_i)$ must be determined from an $M$-resolved collisional-radiative model. A steady-state (or time-dependent) radiative-cascade model may provide an adequate description at sufficiently low particle densities for which the effects of collisional de-excitation processes can be neglected. If the total rate due to electron and photon excitations is assumed to be balanced only by the rates for spontaneous radiative transitions, the steady-state equations can be expressed in the form

$$N(g)W_{ex}(g \rightarrow \gamma_i) + \sum_{\gamma_j > \gamma_i} N(\gamma_j) A_r(\gamma_j \rightarrow \gamma_i) = N(\gamma_i) A_r(\gamma_i) . \qquad (11.27)$$

The quantity $W_{ex}(g \rightarrow \gamma_i)$ denotes the total excitation rate from the ground state g and $A_r(\gamma_i)$ represents the sum of the spontaneous radiative-decay rates. The number of important higher excited states $\gamma_j > \gamma_i$ is expected to grow rapidly with increasing electron or photon energy, especially in an $M$-resolved atomic-kinetics description. The steady-state radiative-cascade model has been adopted in an analysis of X-ray emission from electron–ion beam interactions [33]. Radiative emission following inner-shell-electron ionization and dielectronic-recombination radiation can be treated by extending the conventional bound-state radiative-cascade model to include the appropriate class of autoionizing states [34]. In an $M$-resolved collisional-radiative model, it may be necessary to take into account collisional and radiative transitions involving a large set of bound excited and autoionizing states. Fujimoto and Kawachi [35] have employed a collisional-radiative model based on the density-matrix approach for axially symmetric excitation in the absence of electric and magnetic fields. The most general time-independent (or time-dependent) atomic-kinetics description would be based on the Master equations for all elements of the initial atomic-state density operator, allowing for a general set of steady-state (or time-varying) excitation processes in an arbitrary arrangement of static (or quasistatic) electric and magnetic fields.

## 11.3.6 Spectral Patterns Due to the Circularly Polarized Radiative Emissions

If the initial magnetic substates are assumed to be uniformly populated and the electric fields are neglected, the weak-field Zeeman-effect spectral patterns can be characterized by equal intensities of symmetrically shifted left- and right-circularly polarized radiative emissions in the magnetic-field direction. The photon-energy shift is defined as the difference between the atomic-energy-level shifts produced by the magnetic-field perturbation $H^A(B) = -\boldsymbol{M} \cdot \boldsymbol{B} = \mu_B g_e \boldsymbol{J} \cdot \boldsymbol{B}$, where $g_e$ denotes the electron gyromagnetic ratio and $\mu_B$ is the Bohr magneton. Using first-order perturbation theory, the photon-energy shift may be expressed in the linear-Zeeman-effect form [36, 37]:

$$\Delta\hbar\omega(B) = \langle \Delta_f J_f M_f | H^A(B) | \Delta_f J_f M_f \rangle - \langle \Delta_i J_i M_i | H^A(B) | \Delta_i J_i M_i \rangle$$
$$= [M_f g(\Delta_f J_f) - M_i g(\Delta_i J_i)] \,\mu_B B \ . \tag{11.28}$$

The quantities $g(\Delta_i J_i)$ and $g(\Delta_f J_f)$ denote the gyromagnetic ratios or $g$-factors for initial and final many-electron atomic states, respectively. The weak-field linear-Zeeman-effect spectral pattern for $^3P_2 \rightarrow {}^3P_1$ magnetic-dipole transitions is illustrated in Fig. 11.2. Due to Doppler broadening and other broadening mechanisms, one usually observes two blended spectral features, corresponding to the unresolved components for each of the two circular polarizations. These patterns can be conveniently analyzed in terms of the intensity-weighted average of the shifts for the unresolved right- or left-circularly polarized spectral components [38].

The more complex Stark–Zeeman spectral patterns may be describable in terms of additional Stark shifts of the Zeeman-effect patterns. Since the electric-dipole interaction $H^A(E) = -\boldsymbol{D} \cdot \boldsymbol{E}$ has nonvanishing matrix elements only between unperturbed (field-free) atomic eigenstates with different parities, the Stark shifts for many-electron (nonhydrogenic) atomic systems must be evaluated at least in second-order perturbation theory. Ignoring the diamagnetic (quadratic-Zeeman-effect) contribution, the additional photon-energy shift can be expressed in the resolvent-operator form

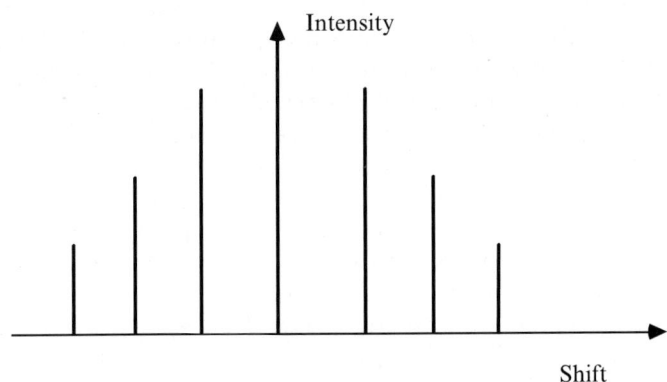

**Fig. 11.2.** The weak-field Zeeman pattern describing a $^3P_2 \rightarrow {}^3P_1$ magnetic-dipole emission from a many-electron atomic system. The symmetrically shifted right- and left-circularly polarized emissions are observed in the magnetic-field direction in the absence of electric fields and for uniformly populated magnetic substates

$$\Delta\hbar\omega(E) = \left\langle \Delta_f J_f M_f \left| \boldsymbol{D} \cdot \boldsymbol{E} \frac{1}{E(0) - H^A(0)} \boldsymbol{D} \cdot \boldsymbol{E} \right| \Delta_f J_f M_f \right\rangle$$
$$- \left\langle \Delta_i J_i M_i \left| \boldsymbol{D} \cdot \boldsymbol{E} \frac{1}{E(0) - H^A(0)} \boldsymbol{D} \cdot \boldsymbol{E} \right| \Delta_i J_i M_i \right\rangle . \quad (11.29)$$

If the magnetic field is sufficiently strong to uncouple the orbital and spin angular momenta, the alternative, uncoupled magnetic-substate (Paschen–Back-effect) representation may be more appropriate and the diamagnetic contribution should then be included. An extension of the average-shift analysis could be useful for perpendicular (crossed) electric field and magnetic fields. The general density-matrix description should be applied starting with initial and final eigenstate expansions that have been obtained from a diagonalization of the field-dependent many-electron atomic Hamiltonian matrix.

## 11.4 Reduced-Density-Matrix Formulation

The reduced-density-matrix approach can provide the fundamental basis for a nonperturbative and nonequilibrium quantum-statistical description of electromagnetic interactions involving many-electron atomic systems in electron–ion beam interactions and in high-temperature laboratory and astrophysical plasmas. The reduced-density-operator formulation, which has been described by Fano [24] and by Cohen-Tannoudji [39], is a convenient framework for the incorporation of a microscopic description of the collisional and radiative decoherence and relaxation processes (together with the associated spectral-line broadening mechanisms) that arise from the influence of a much larger system (consisting of charged-particles and photons), which is referred to as the environment. In the conventional reservoir approximation, which is illustrated in Fig. 11.3, the environment is assumed to be essentially unaffected by its interactions with the much smaller quantum system of interest, which is referred to as the relevant system. Consequently, the environment can be represented by a time-independent density operator. The kinetics and spectral phenomena due to the environmental interactions can be systematically and self-consistently investigated in terms of the self-energy corrections that arise in the complementary time-domain and frequency-domain formulations of the reduced-density-operator approach. In this manner, the influence of the environment is described in terms of nonequilibrium kinetics (decoherence and relaxation) processes together with spectral-line broadening mechanisms associated with these processes, as illustrated in Fig. 11.4.

The partition of the entire, interacting quantum system into a relevant system (consisting of the atomic system and the observable photon modes) and an environment is by no means unique. In addition, the appropriate decomposition may strongly depend on the particular type of measurement that is desired. In the ordinary Hilbert-space quantum-mechanical perturbation theory, different partitions of the total Hamiltonian operator $H$ into an unperturbed Hamiltonian operator $H_0$ and an interaction operator $\mathcal{V}$ would be

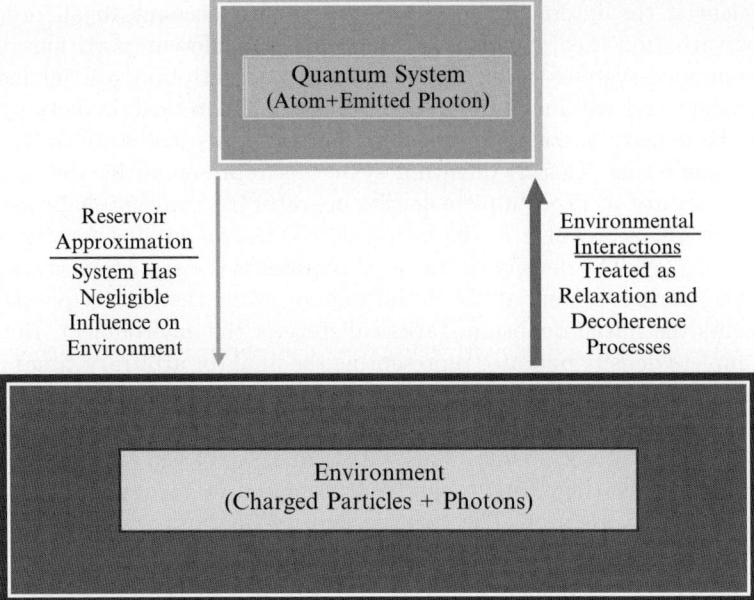

**Fig. 11.3.** The reduced-density-operator description, illustrating the partition of the entire, interacting system into a "relevant" quantum system (which may consist of a many-electron atomic system and the emitted or observable photon) and an environment (consisting of charged particles and photons), which is treated using the reservoir approximation

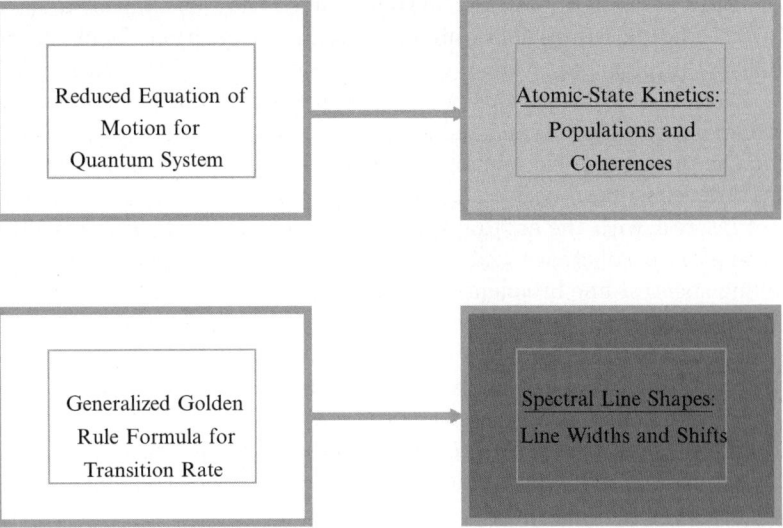

**Fig. 11.4.** Time-domain (equation-of-motion) and frequency-domain (resolvent-operator) formulations of the reduced-density-operator-approach, illustrating the unified treatment of atomic-state kinetics and spectral-line shapes

equivalent if the interaction could be taken into account to all orders in the perturbation-theory expansion. In contrast, different partitions in the quantum-open-systems reduced-density-matrix description are intrinsically inequivalent and will inevitably lead to different theoretical predictions.

In the density-operator approach, a general statistical state of the combined, interacting (closed) quantum system is represented by the complete density operator $\rho$. The complete density operator $\rho$ is conventionally assumed to be initially expressible in the uncorrelated, tensor-product form given by $\rho(t_0) = \rho^S \otimes \rho^E$. The density operator $\rho^S$ represents the statistical state of the relevant quantum system at the initial time $t_0$, while the density operator $\rho^E$ represents the initial quantum-statistical state of the environment. However, the complete density operator representing the final (or arbitrary-time) statistical state of the combined, interacting quantum system cannot be expressed in the simple uncorrelated, tensor-product form that is conventionally adopted for the initial statistical state. The statistical state of the relevant quantum system, at an arbitrary time $t$, can be represented by means of the reduced, relevant density operator. The reduced density operator $\rho^R(t)$ is defined in terms of the average (partial trace) operation $\rho^R(t) = \mathrm{Tr}_E\{\rho(t)\}$, which is taken over the large set of quantum numbers specifying the environmental degrees of freedom.

For the reduced-density-matrix description of radiative emission, the relevant quantum system will consist of the many-electron atomic system (which is initially in a set of bound excited or autoionizing states) combined (or entangled) with the relevant (observable) modes of the quantized electromagnetic field. The final state of the entire, interacting quantum system corresponds to the final state of the atomic system (which may encompass a set of final atomic bound states following the radiative transitions from the bound excited or autoionizing states), combined with that of the observable photons. The radiative transitions occur under the simultaneous influence of the environmental collisional and radiative decoherence and relaxation processes and in the presence of electric and magnetic fields. We emphasize that, in the reduced-density-operator description, the interactions of the quantum subsystem of interest with the environment are treated stochastically, as collisional and radiative decoherence and relaxation processes together with the corresponding spectral-line broadening mechanisms. For the treatment of coherent interactions, the reduced, relevant system must obviously be defined to include all coherently coupled (or correlated) subsystems.

It is advantageous to formally introduce the reduced description by adopting an abstract procedure based on the Zwanzig Liouville-space projection operators [40, 41]. The Liouville-space operators (which will be denoted by overbars) are defined within a generalized Hilbert space (Liouville space), in which ordinary Hilbert-space operators, such as density operators, play the role of state vectors. The elements of the Liouville-space basis set $\{|\alpha\rangle\langle\beta|\}$ may be denoted, in the Liouville-space double-bracket (Dirac) notation, by $|\alpha, \beta\rangle\rangle$. The complete Liouville-space basis set $\{|\alpha, \beta\rangle\rangle\}$ can be constructed from the complete Hilbert-space basis sets $\{|\alpha\rangle\}$ and $\{\langle\beta|\}$. Finally, the complex inner product $\langle\langle\rho_1|\rho_2\rangle\rangle$ of two Liouville-space state vectors $\rho_1$ and $\rho_2$ is represented by the trace operation $\mathrm{Tr}(\rho_1^+\rho_2)$, where the superscript $+$ indicates the adjoint.

## 11.4.1 Frequency-Domain (Resolvent-Operator) Formulation

In the time-independent (frequency-domain) formulation of the reduced-density-operator description, the radiative-transition rate (transition probability per unit time) may be expressed in the generalized (Liouville-space) Fermi golden-rule form [13,42]:

$$A_R(i \rightarrow f) = -i \lim_{\varepsilon\to 0}\langle\langle P_f^r|\bar{T}^r(+i\varepsilon)|\rho_i^r\rangle\rangle \, . \tag{11.30}$$

The reduced, relevant Liouville-space transition operator $\bar{T}^r(+i\varepsilon)$ is a generalization of the ordinary Hilbert-space transition operator, and the conventional asymptotic (scattering-theory) boundary condition is indicated by the $\varepsilon \rightarrow 0$ limit. The reduced, relevant Liouville-space transition operator can be evaluated using the Lippmann–Schwinger relationship $\bar{T}^r(+i\varepsilon) = \bar{V}^r + \bar{V}^r\bar{G}^r(+i\varepsilon)\bar{V}^r$. The reduced, relevant Liouville-space resolvent (or Green) operator is expressed by $\bar{G}^r(+i\varepsilon) = [+i\varepsilon - \bar{L}^r - \bar{\Sigma}(+i\varepsilon)]^{-1}$, where $\bar{L}^r$ is the relevant Liouvillian operator and $\bar{\Sigma}(+i\varepsilon)$ is the Liouville-space self-energy operator. The relevant Liouvillian operator $\bar{L}^r$ is defined in terms of a commutator involving the relevant system Hamiltonian operator $H^r$, which describes the many-electron atomic system together (or entangled) with the restricted set of observable photons. The electromagnetic field can be treated as a quantized field by introducing the Liouville-space basis set composed of the tensor products of the individual electronic and photon basis states. The definition of the Liouville-space operator $\bar{L}^r$ may be expressed by means of the following relationships [43]:

$$\bar{L}^r \rho^r = \left(\frac{1}{\hbar}\right)[H^r, \rho^r] = \left(\frac{1}{\hbar}\right)(H^r \otimes I^{r*} - I^r \otimes H^{r*})\, \rho^r \, , \tag{11.31}$$

where the second form is given in terms of tensor products involving the relevant Hamiltonian operator $H^r$ and the relevant identity operator $I^r$. The relevant Liouville-space operator $\bar{L}^r$ can be decomposed as $\bar{L}^r = \bar{L}_0^r + \bar{V}^r$, where $\bar{V}^r$ includes the electromagnetic interaction of the many-electron atom system, together with the interaction responsible for autoionization.

The initial statistical state of the relevant quantum system is represented by the reduced density operator $\rho_i^r$. In this investigation, the initial-state density operator represents the bound excited-state or autoionizing-state populations together with the coherences. The final state of interest in the electromagnetic transition is formally represented by means of the relevant final-state projection operator $P_f^r$. For the description of the single-photon spontaneous-emission process, the initial-state reduced density operator represents the electronic system with no photons and the relevant final-state projection operator projects onto the subspace corresponding to the electronic system in some final state and a single photon in an observable mode. The polarization of the emitted photons can be described by introducing a final-state projection operator pertaining only to the observable photon mode but defining the desired reduced photon-polarization density matrix. Natural generalizations can be introduced for the description of single-photon absorption and multiphoton (nonlinear optical) processes. It is clear that the quantum-field-theory description of nonlinear optical phenomena, in terms of multiphoton components of the initial-state reduced density operators and the relevant final-state projection operators, is practical only for not-too-intense electromagnetic fields, for which the number of photons in the observable field modes is not arbitrarily large.

The Liouville-space self-energy operator $\bar{\Sigma}(+i\varepsilon)$ represents the spectral effects of the environmental electron–electron, electron–ion, and electron–photon decoherence and relaxation processes. The Liouville-space self-energy operator $\bar{\Sigma}(+i\varepsilon)$ can be formally expressed in terms of the Zwanzig Liouville-space projection operators $\bar{P}$ and $\bar{Q}$. The Zwanzig Liouville-space projection operator $\bar{P}$ has the explicit form $|\rho^E\rangle\rangle\langle\langle I^E|$, where $I^E$ denotes the environmental identity operator. As a result of the average over the environmental degrees of freedom, the projection operator $\bar{P}$ projects a Liouville-space element onto the subspace of the relevant-system degrees of freedom (uncorrelated with the environmental degrees of freedom). The complementary projection operator $\bar{Q} = 1 - \bar{P}$ projects a Liouville-space element onto the orthogonal subspace of the irrelevant (environmental) degrees of freedom (taking into account the system-environment correlations). With these definitions, the Liouville-space self-energy operator $\bar{\Sigma}(z)$ can be expressed, as a function of the complex variable $z$, by means of the relationships [13, 42]

$$\bar{\Sigma}(z) = \bar{P}\bar{\mathcal{V}}^{ir}\bar{P} + \bar{P}\bar{\mathcal{V}}\bar{Q}\frac{1}{z - \bar{Q}\bar{L}\bar{Q}}\bar{Q}\bar{\mathcal{V}}\bar{P}$$

$$= \mathrm{Tr}_E\left[\left(\bar{\mathcal{V}}^{ir} + \bar{\mathcal{V}}\bar{Q}\frac{1}{z - \bar{Q}\bar{L}\bar{Q}}\bar{Q}\bar{\mathcal{V}}\right)\rho^E\right].$$

(11.32)

The complete Liouville-space operator $\bar{L}$ is defined in terms of the total Hamiltonian operator for the entire (closed) interacting quantum system. The total Liouville-space interaction operator $\bar{\mathcal{V}}$ is partitioned as $\bar{\mathcal{V}} = \bar{\mathcal{V}}^r + \bar{\mathcal{V}}^{ir}$, where the irrelevant interaction operator $\bar{\mathcal{V}}^{ir}$ includes the environmental interactions. The Zwanzig Liouville-space projection operators, which occur in the explicit expression for the Liouville-space self-energy operator $\bar{\Sigma}(z)$, introduce (quantum-statistical) averages over the environmental degrees of freedom. Although these averages are often carried out assuming Maxwellian and Boltzmann equilibrium distributions for the free and bound electrons, respectively, and Bose–Einstein equilibrium distributions for the photons, our general formulation is applicable to nonequilibrium distributions. The Liouville-space projection-operator formalism has the advantage that the analysis can be carried out in considerable generality, without specifying the physical nature of the environmental interactions.

The general (tetradic-matrix) expression given by (11.30) can be used to determine the overall spectral-line shape for an arbitrary array of (possibly overlapping) spectral lines due to radiative transitions among the substates from two groups of closely spaced upper (bound excited or autoionizing) and lower (bound) atomic states. In the widely used isolated-line approximation, the expressions for the standard Lorentzian spectral-line shape parameters (the line shifts and widths) can be recovered from the diagonal tetradic matrix elements of the general Liouville-space self-energy operator $\bar{\Sigma}(+i\varepsilon)$ given by (11.32). When the system–environment interactions are sufficiently weak, the Liouville-space self-energy operator $\bar{\Sigma}(+i\varepsilon)$ may be expanded in a perturbation-theory series involving increasing powers of the total Liouville-space interaction operator $\bar{\mathcal{V}}$. If only the lowest-order nonvanishing contribution is retained in this perturbation-theory expansion (which corresponds to the Born approximation), the total (isolated-line) shift and width can be expressed simply as the sums of the partial contributions from the elementary decoherence and relaxation processes acting alone. Quantum-mechanical interference between the individual transition amplitudes can occur in the high-order contributions to the isolated-line width and shift, as well as in the more general tetradic-matrix forms of the spectral-profile expressions describing overlapping lines. To include Stark and Zeeman broadening, the spectral-line-shape formula must be evaluated in a basis of electric-field- and magnetic-field-dependent atomic eigenstates, and this procedure has been adopted in the general theory of spectral line broadening by plasmas [43]. The atomic states could be determined taking into account a quasistatic (ion-produced) plasma electric microfield and a (poloidal + toroidal) magnetic field or an arbitrary arrangement of externally applied static (or quasistatic) electric and magnetic fields. For the description of intense-field (nonlinear optical) phenomena, the many-electron atomic eigenstates can be taken to be "dressed" states, which are determined including the interaction with a classical (possibly multiple-mode) electromagnetic field. To make comparisons with experimental observations, it is usually necessary to include either the equilibrium or nonequilibrium Doppler effect, which is a major source of inhomogeneous spectral-line broadening.

### 11.4.2 Time-Domain (Equation-of-Motion) Formulation

The time-dependent (time-domain) formulation of the reduced-density-operator description is based on the equation of motion for the reduced, relevant density operator $\rho^r(t)$, which describes the dynamics of the many-electron atomic system combined (entangled) with the observable modes of a quantized electromagnetic field. The reduced equation of motion can be derived from the more general quantum Liouville (or Liouville–von Neumann) equation for the combined, complete density operator $\rho(t)$. The reduced equation of motion can be expressed in the generalized-Master-equation form [13, 44]

$$i\frac{\partial}{\partial t}\rho^{\mathrm{R}}(t) = \bar{L}^{\mathrm{R}}(t)\rho^{\mathrm{R}}(t) + \int_{t_0}^{t} dt'\, \bar{\Sigma}(t,t')\, \rho^{\mathrm{R}}(t') \,. \tag{11.33}$$

This is an integral–differential equation, which is expressed in terms of the Liouville-space self-energy operator kernel $\bar{\Sigma}(t, t')$. The equation-of-motion (time-domain) formulation provides a general framework for the evaluation of the electromagnetic-response functions describing linear and nonlinear optical interactions. These electromagnetic-response functions can be defined within the context of a simpler semiclassical description, in which the dynamics of the many-electron atomic system is investigated using the quantum-open-systems equation of motion while the electromagnetic field is treated according to classical electrodynamics (i.e., the Maxwell Equations).

The reduction procedure provides a closed-form set of equations for the various matrix elements of the reduced, relevant density operator $\rho^{\mathrm{r}}(t) = \bar{P}\rho(t)$, which represents (at an arbitrary time $t$) the quantum-statistical state of the relevant system. The closed-form result has been obtained by neglecting the initial-state correlations, which are described by an omitted contribution involving the irrelevant projection $\bar{Q}\rho(t_0)$ (evaluated at the initial time $t_0$). The initial-state correlations are excluded by the assumption that the entire initial-state density operator $\rho(t_0)$ is expressible as an uncorrelated, tensor product of individual density operators for the separate, isolated systems. Initial-state correlations are expected to be negligible for times $t$ such that $t - t_0$ is long compared with the characteristic "correlation" time.

The Liouville-space self-energy operator kernel $\bar{\Sigma}(t, t')$ can be formally expressed by means of the relationships [42, 45]

$$\bar{\Sigma}(t,t') = -i\bar{P}\bar{V}(t)\bar{Q}\bar{g}_Q(t,t')\bar{Q}\bar{V}(t')\bar{P} = -i\mathrm{Tr}_{\mathrm{E}}\left[\bar{V}(t)\bar{Q}\bar{g}_Q(t,t')\bar{Q}\bar{V}(t')\rho^{\mathrm{E}}\right] \,. \tag{11.34}$$

The Q-subspace projection $\bar{g}_Q(t, t')$ of the Liouville-space propagator is defined, in terms of the time-ordering operator $\mathcal{T}$, as

$$\bar{g}_Q(t,t') = \mathcal{T}\exp\left[-i\int_{t'}^{t} dt''\bar{Q}\bar{L}(t'')\bar{Q}\right] \,. \tag{11.35}$$

Since the Liouville-space operator $\bar{Q}\bar{L}(t)\bar{Q}$ is a non-Hermitian operator, the Liouville-space propagator defined by (11.35) will be a nonunitary operator. The Liouville-space self-energy operator kernel $\bar{\Sigma}(t,t')$, which appears in the time-domain (equation-of-motion) formulation, can be related to the time-independent Liouville-space self-energy operator $\bar{\Sigma}(z)$, occurring in the frequency-domain (resolvent-operator) formulation. This relationship can be most directly derived by invoking the Fourier-transform relationship between the time-domain propagator and corresponding frequency-domain resolvent operator.

In the commonly adopted Markov (short-memory-time) approximation, the Liouville-space self-energy operator kernel $\bar{\Sigma}(t, t')$ is assumed to be independent of time. The Markov approximation may be introduced into the equation of motion for the reduced, relevant density operator $\rho^{\mathrm{r}}(t)$ by utilizing the relationship

$$\bar{\Sigma}(t, t') = \lim_{z \to i0} \bar{\Sigma}(z)\delta(t - t') . \tag{11.36}$$

In this approximation, the corresponding frequency-domain Liouville-space self-energy operator $\bar{\Sigma}(i0)$, which will subsequently be denoted by $\bar{\Sigma}$, is independent of the frequency. However, if a "dressed" electronic basis set is adopted, taking into account the influence of a classical electromagnetic field, the Markov form of the frequency-domain Liouville-space self-energy operator will acquire an implicit dependence on the amplitude and frequency of the external (applied) time-dependent field. For the treatment of ultrashort-pulse electromagnetic interactions, the Markov approximation may not be valid.

A set of (further reduced) quantum-kinetics equations for the atomic-state population densities (corresponding to the diagonal reduced-density-matrix elements), together with the atomic-state coherences (represented by the nondiagonal reduced-density-matrix elements), can be derived from (11.33) by performing the additional average (trace) operation over the (relevant) degrees of freedom specifying the observed photon states. The result may be described as providing the foundation for a generalized collisional-radiative model, which can be used to determine the populations of the bound excited and autoionizing states and as well as the atomic-state coherences. In addition, the corresponding quantum-kinetics equation for the reduced density matrix describing the observable modes of the quantized electromagnetic field can be obtained from (11.33) by carrying out the complimentary additional average (trace) operation over the relevant atomic states. If the electromagnetic-field coherences are neglected, the quantum-kinetics equation for the spectral intensity of the radiation field can be used to derive the familiar equation of radiation transport. Casini [45] and Casini and Degl'Innocenti (in Chap. 12) have developed a closely related reduced-density-matrix formulation for polarized atomic radiative emission in the presence of electric and magnetic fields. They have emphasized the importance of a self-consistent solution of the coupled set of equations consisting of the atomic-state kinetics equations and the radiation-transport equation.

As in the time-independent (resolvent-operator) formulation, it is convenient to expand the self-energy operator kernel $\bar{\Sigma}(t, t')$ in a perturbation-theory expansion in powers of the full Liouville-space interaction operator. The time-dependent perturbation-theory analysis is usually carried out in the Liouville-space interaction representation, in place of the Schrödinger representation adopted in the derivation of (11.33). Since the full Liouville-space interaction operator is the sum of electron–electron, electron–ion, and electron–photon interaction operators, the (tetradic) matrix elements of the self-energy operator kernel can involve quantum-mechanical interference terms. Taking into account only the lowest-order nonvanishing perturbation-theory (or Born-approximation) contributions to the self-energy operator kernel, the equation of motion for the atomic-state population densities can be expressed in terms of the familiar collisional and radiative transition rates that are obtained from an evaluation of the standard (lowest-order) golden-rule formula of ordinary Hilbert-space perturbation theory. For the precise description of polarized radiative emission, allowing for the possible coherent excitation of the initial bound or autoionizing states of a many-electron atomic system in the presence of electric and magnetic fields, it is necessary to employ a generalized collisional-radiative model, taking into account the atomic-state coherences.

## Acknowledgment

This work has been supported by the U. S. Department of Energy and by the U. S. Office of Naval Research.

# References

1. V.L. Jacobs, A.B. Filuk, Phys. Rev. A **60**, 1975 (1999)
2. V.L. Jacobs, J. Davis, P.C. Kepple, Phys. Rev. Lett. **37**, 1390 (1976)
3. K. LaGattuta, Y. Hahn, Phys. Rev. Lett. **51**, 558 (1983)
4. D.C. Griffin, M.S. Pindzola, C. Bottcher, Phys. Rev. A **33**, 3124 (1986)
5. A. Muller, D.S. Belic, B.D. DePaola, N. Djuric, G.H. Dunn, D.W. Mueller, C. Trimmer, Phys. Rev. Lett. **56**, 127 (1986)
6. P.F. Dittner, S. Datz, P.D. Miller, P.L. Pepmiller, C.M. Fou, Phys. Rev. A **33**, 124 (1986)
7. D.W. Savin, L.D. Gardner, D.B. Reisenfeld, A.R. Young, J.L. Kohl, Phys. Rev. A **53**, 280 (1996)
8. F. Robicheaux, M.S. Pindzola, Phys. Rev. Lett. **79**, 2237 (1997)
9. D.C. Griffin, F. Robicheauz, M.S. Pindzola, Phys. Rev. A **57**, 2708 (1998)
10. K. LaGattuta, G. Borca, J. Phys. B **31**, 4781 (1998)
11. T. Bartsch et al., Phys. Rev. Lett. **99**, 3779 (1999)
12. V. Klimenko, T.F. Gallagher, Phys. Rev. Lett. **85**, 3357 (2000)
13. V.L. Jacobs, J. Cooper, S.L. Haan, Phys. Rev. A **50**, 3005 (1994)
14. J.R. Oppenheimer, Z. Phys. **43**, 27 (1927)
15. I.C. Percival, M.J. Seaton, Phil. Trans. R. Soc. A **251**, 113 (1958)
16. M.K. Inal, J. Dubau, J. Phys. B **20**, 4221 (1987)
17. M.K. Inal, J. Dubau, J. Phys. B **22**, 3329 (1989)
18. A.S. Shlyaptseva, A.M. Urnov, A.V. Vinogradov, P.N. Lebedev Physical Institute of the U. S. S. R. Academy of Sciences Report No. **194**, 1981
19. A.V. Vinogradov, A.M. Urnov, A.S. Shlyaptseva, in *Atomic and Ionic Spectra and Elementary Processes in Plasmas*, Proceedings of the P. N. Lebedev Physics Institute, Academy of Sciences of Russia, Vol. 912, ed. by I.I. Sobelman (Nova Science, Commack, New, York, 1992) p. 93
20. A.S. Shlyaptseva et al., Phys. Rev. A **57**, 888 (1998)

21. M.F. Gu, D.W. Savin, P. Beiersddorfer, J. Phys. B **32**, 5371 (1999)
22. V.L. Jacobs, J. Phys. B **5**, 2257 (1972)
23. V.L. Jacobs, J. Phys. B **6**, 1461 (1973)
24. U. Fano, Rev. Mod. Phys. **29**, 74 (1957)
25. K. Blum, *Density Matrix Theory and Applications*, 2nd ed. (Plenum, New York, 1996)
26. A. Kazantsev, J.-C. Hénoux, *Polarization Spectroscopy of Ionized Gases*, (Kluwer, Dordrecht, 1995)
27. V.B. Berestetskii, E.M. Lifshitz, L.P. Pitaevskii, *Quantum Electrodynamics* (Pergamon, Oxford, 1982)
28. U. Fano, G. Racah, *Irreducible Tensorial Sets* (Academic, New York, 1959)
29. W. Happer, Ann. Phys. **48**, 579 (1968)
30. W. Happer, Rev. Mod. Phys. **44**, 169 (1972)
31. A. Omont, Prog. Quant. Electron. **5**, 69 (1977)
32. W.E. Baylis, in *Progress in Atomic Spectroscopy* ed. by W. Hanle, H. Kleinpoppen (Plenum, New York, 1979)
33. B.M. Brown, U. Feldman, G.A. Doschek, J.F. Seely, R.E. LaVilla, V.L. Jacobs, J.R. Henderson, D.A. Knapp, R.E. Marrs, P. Beiersdorfer, M.A. Levine, Phys. Rev. A **40**, 4089 (1989)
34. V.L. Jacobs, B.F. Rozsnyai, Phys. Rev. A **34**, 216 (1986)
35. T. Fujimoto, T. Kawachi, in *Atomic Processes in Plasmas*, 9th APS Topical Conference, ed. by W.R. Rowan, AIP Conference Proceedings **322** (AIP, New York, 1995)
36. R.D. Cowan, *The Theory of Atomic Structure and Spectra* (University of California Press, Berkeley, 1981)
37. B.W. Shore, D.H. Menzel, *Principles of Atomic Spectra* (Wiley, New York, 1968)
38. V.L. Jacobs, J.F. Seely, Phys. Rev. A **36**, 3267 (1987)
39. C. Cohen-Tannoudji, J. Dupont-Roc, G. Grynberg, *Atom–Photon Interactions* (Interscience, New York, 1992)
40. R. Zwanzig, in *Lectures in Theoretical Physics III* ed. by W.E. Brittin, D.W. Downs, J. Downs (Interscience, New York, 1961)
41. U. Fano, Phys. Rev. **131**, 259 (1963)
42. A. Ben-Reuven, Y. Rabin, Phys. Rev. A **19**, 2056 (1979)
43. H.R. Griem, *Spectral Line Broadening by Plasmas* (Academic, New York, 1974)
44. K. Burnett, J. Cooper, R.J. Ballagh, E.W. Smith, Phys. Rev. A **22**, 2005 (1980)
45. R. Casini, Phys. Rev. **A** 71, 062505 (2005)

# Astrophysical Plasmas

R. Casini and E. Landi Degl'Innocenti

In this chapter, we discuss the application of spectro-polarimetry diagnostics to the investigation of astrophysical plasmas. We first present an overview of why polarization is expected in the spectral-line radiation that we receive from a large variety of cosmic objects, and then treat in some detail specific atomic models (e.g., the 0–1 and 1–0 two-level atoms), which illustrate how physical and electro-dynamical properties of the emitting plasma can be inferred by studying the polarized radiation in the corresponding spectral lines. The practical applications described in this chapter are taken exclusively from the realm of solar physics, mainly for two reasons: (a) from a historical point of view, the Sun was the first cosmic object to which polarization analysis of radiation was successfully applied, proving the existence of solar magnetic fields, and demonstrating the diagnostic potential of radiation phenomena involving resonance polarization, and (b) because spectro-polarimetric signals are generally very weak, their detection with a sufficient signal-to-noise ratio is possible only for strong radiation sources. In particular, a plethora of atomic-polarization effects (magnetic and collisional depolarization, alignment-to-orientation conversion, level crossing and anti-crossing interferences) could be detected in the polarized light from the Sun only because of the high sensitivity that can be attained in solar observations. As the light-collecting capabilities of night-time astronomical instrumentation keep growing, it is expected that the diagnostic techniques illustrated in this chapter will become increasingly available for the investigation of plasma properties all over the universe.

## 12.1 Introduction

The importance of understanding the origin of polarized radiation in astrophysical plasmas cannot be overestimated. Unlike laboratory plasmas – for which many physical parameters can be directly controlled or measured – in the case of astrophysical plasmas, radiation is the only carrier of information available to the observer for understanding the physical processes that determine the dynamics of the object under study.

Radiation polarization is expected whenever the interaction of photons with atoms (or molecules) is affected by symmetry-breaking processes. This symmetry breaking can be intrinsic to the process of photon–atom interaction itself (e.g., anisotropic illumination of atoms, leading to scattering polarization), or be determined by the presence of external causes (e.g., anisotropic excitation by collisions; interaction of the atom with deterministic magnetic and/or electric fields). Examples of processes leading to radiation polarization in laboratory plasmas have been discussed in other chapters of this book. In this chapter, we will discuss polarization phenomena observed in astrophysics, and illustrate how from them we can gain an understanding of the undergoing physical processes.

Historically, solar physics has led the way in exploiting polarization information contained in the radiation coming from the Sun in order to understand solar magnetism. This privileged role of our star among polarization studies of cosmic objects has a very natural explanation: both spectral and polarimetric analysis are photon starved techniques, and therefore they are more difficult to apply to faint objects. Even for the Sun, the detection of very low levels of polarization – characterizing, for instance, the resonance scattered radiation from quiet (i.e., magnetically "inactive") regions in the solar atmosphere [1] – poses to the present day significant challenges, which have only partially been overcome by the availability of increasingly more efficient detectors.

Magnetic field strengths in the solar atmosphere range from the very weak fields permeating the quiet-sun corona ($B \lesssim 10^{-3}\,\mathrm{T}$) to the strong fields characteristic of the umbral regions of sunspots ($B \gtrsim 10^{-1}\,\mathrm{T}$).[1] Because of such diverse regimes of magnetic fields, different phenomena affecting the polarization of the emitted radiation can occur in solar plasmas (Zeeman effect, Hanle effect, alignment-to-orientation conversion, level crossing and anti-crossing interferences), whose signatures can in principle be modeled in order to infer the magnetic topology of the emitting plasma.

We refer the reader to other chapters of this book for an exhaustive treatment of collision-induced polarization. Here we will focus exclusively on the effects of anisotropic illumination and external fields on the polarization of spectral lines.

---

[1] Here the comparative terms "weak" and "strong" are used in connection with the relative size of magnetic splitting to line broadening of nonmagnetic origin (typically, the Doppler thermal linewidth). However, the same terminology can also be used to classify the regime of interaction of the atom with the external fields, so that a field can be said to be "weak" if the corresponding magnetic splitting is small compared with some characteristic energy scale of the atomic structure (e.g., the energy separation between atomic levels due to the presence of fine structure and/or hyperfine structure).

## 12.2 Origin of Polarized Radiation

The origin of spectral-line polarization can be traced back to the intrinsic properties of atomic transitions. Let us consider the simplest atomic transition, between an upper level $J_u = 1$ and a lower level $J_l = 0$. Quantum mechanics tells us that the atoms in the upper level can occupy any of the three possible $M$ substates ($M = -1, 0, +1$). In the language of the vector model of atoms [2], these substates correspond to the three possible projections of the vector $\boldsymbol{J}$ along the quantization axis (see Fig. 12.1). The electric-dipole transitions, from these three substates to the one substate $M = 0$ of the lower level, have different polarization properties. The transition with $\Delta M = 0$ is linearly polarized along the $z$-axis ($\pi$ transition), whereas the two transitions with $|\Delta M| = 1$ are circularly polarized, with opposite signs, around the $z$-axis, and linearly polarized perpendicularly to the same axis ($\sigma$ transitions).

If the excitation processes are completely isotropic, then the excited atoms have equal probability of populating any of the three upper $M$ substates. The three components of the atomic transition will then contribute to the polarized emission with well defined weights. Figure 12.2 illustrates the two cases of longitudinal and transversal radiation emitted by an atom that has been isotropically illuminated (the quantization axis is defined by the direction of the magnetic field, $\boldsymbol{B}$), with the respective polarization properties of the three transition components. Since the three substates of $J = 1$ are isoenergetic in the absence of external fields ($\omega_B = 0$, in the case of Fig. 12.2), positive and

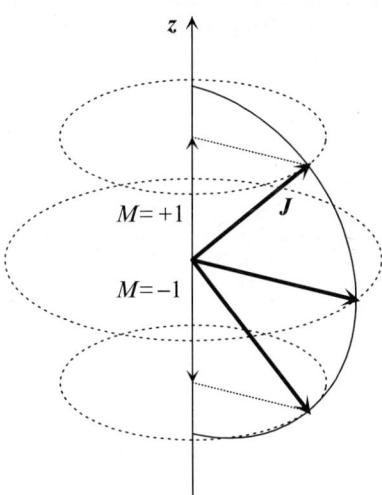

**Fig. 12.1.** Vector-model description of the $J = 1$ atomic state. The norm of the total angular momentum is $|\boldsymbol{J}| = \sqrt{J(J+1)}\hbar = \sqrt{2}\hbar$. The $z$-axis defines the direction of quantization of the atomic system

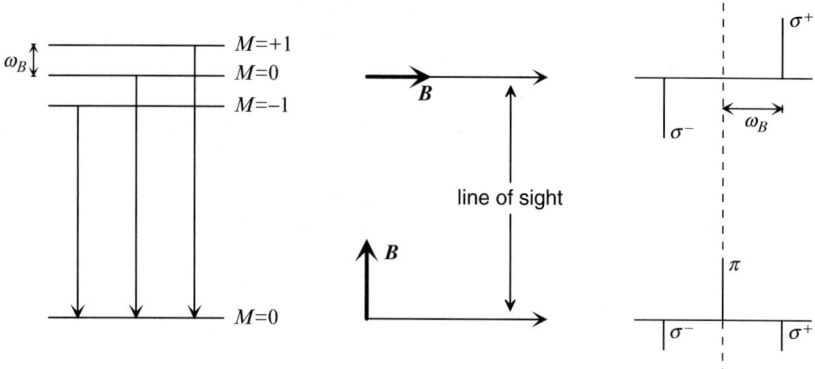

**Fig. 12.2.** Definition of $\pi$ and $\sigma$ components of polarized radiation for the transition $1 \rightarrow 0$. $B$ defines the direction of the quantization axis. The case at the top shows the circular-polarization signal, and the one at the bottom the linear-polarization signal, for the respective geometries of $B$ with respect to the line-of-sight

negative polarizations exactly cancel out in this case and so the emitted radiation appears to be completely unpolarized.

One can imagine two different scenarios (not mutually exclusive), where the different polarization properties of the three transition components can be made manifest in the emitted radiation. The first scenario corresponds to the possibility that the upper $M$ substates may be unevenly populated, and so the transition components no longer contribute to the emitted radiation with the very particular weights illustrated in Fig. 12.2, which make the total polarization vanish. This condition typically occurs when the excitation processes are anisotropic (e.g., a collimated beam of radiation illuminating the atom). The second scenario corresponds to the possibility that the upper $M$ substates may be separated in energy, so that a spectral analysis of the emitted radiation would reveal its varying polarization properties with wavelength, even if the upper $M$ substates were equally populated. This condition typically occurs when an external magnetic or electric field is interacting with the atom, so that the energy structure of the atom is modified, by the additional interaction energy, with respect to the zero-field case. Figure 12.2 illustrates this last case, when a magnetic field is present (*Zeeman effect* [2]).

In astrophysical plasmas, both scenarios can generally be present at the same time, and so it is possible in principle to investigate the excitation conditions of a plasma, and its electro-dynamic properties, by studying the polarization signature of spectral lines formed in the plasma.

### 12.2.1 Description of Polarized Radiation

Before considering the problem of plasma diagnostics via polarization measurements, we must define a set of observable parameters from the physical quantities that describe polarized radiation. Let us consider for simplicity a wave train of monochromatic radiation propagating along the $z$-axis of a given reference frame. It is well known that the polarization information of the wave train is completely contained in the (complex) Cartesian components, $E_x$ and $E_y$, of the radiation electric field. In the physical description of polarized radiation, it is customary to introduce the *coherency matrix* (or *polarization tensor*) of the radiation field (averaged over the acquisition time and elemental surface of the light detector)

$$C \equiv \begin{pmatrix} \langle E_x^* E_x \rangle & \langle E_x^* E_y \rangle \\ \langle E_y^* E_x \rangle & \langle E_y^* E_y \rangle \end{pmatrix} , \tag{12.1}$$

from which it is evident that four independent parameters are needed in order to describe polarized radiation. In observational polarimetry, this description is somewhat inconvenient (although not at all abandoned), because it is not possible to define a direct operational procedure that gives all four complex entries of the coherency matrix as its outcome. For this reason, it has become customary to describe polarization signals in terms of the four *Stokes parameters*, $I, Q, U, V$, where $I$ is the radiation intensity, $Q$ and $U$ are the two independent parameters needed to describe linear polarization on the $xy$ plane, and finally $V$ is the circular-polarization parameter. See Appendix A.

The Stokes parameters have the advantage of being real quantities that can be directly related to detectable signals of polarization measurements. Figure 12.3 describes a typical polarimetric setup for the determination of the four Stokes parameters of a radiation beam, consisting of a linear polarizer and

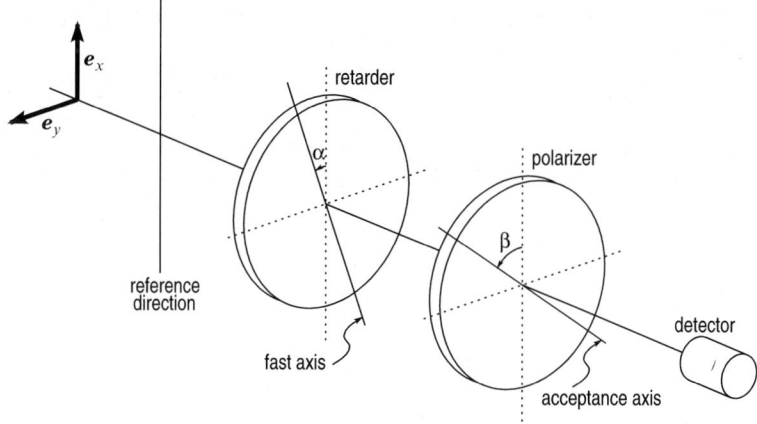

**Fig. 12.3.** Typical configuration of a polarimeter. Different polarization states of the radiation passing through the instrument are measured by suitable combinations of the retarder and polarizer position angles, $\alpha$ and $\beta$

a wave retarder.[2] For a $\lambda/4$ retarder, the intensity of the radiation reaching the detector is

$$S(\alpha, \beta) = \frac{1}{2} \left[ I + (Q \cos 2\alpha + U \sin 2\alpha) \cos 2(\beta - \alpha) + V \sin 2(\beta - \alpha) \right],$$
(12.2)

so it is possible to establish a series of measurements involving different values of the position angles $\alpha$ and $\beta$, which allows the independent measurements of the four Stokes parameters.

Of course, the two descriptions of polarized radiation, in terms of the coherency matrix or the Stokes parameters of the radiation field, must be equivalent, and in fact, it is possible to rewrite (12.1) in the form

$$C = \frac{1}{2} \begin{pmatrix} I + Q & U - iV \\ U + iV & I - Q \end{pmatrix},$$
(12.3)

apart from an inessential multiplicative factor.

## 12.3 Quantum Theory of Photon–Atom Processes

A proper formulation of the problem of polarized line formation in complex atoms must rely on a quantum-mechanical description of the interaction of radiation with atoms. The evolution of the total system atom + radiation is governed by the quantum-mechanical *Liouville equation* for the statistical operator of the system [3],

$$\frac{d}{dt} \rho(t) = \frac{1}{i\hbar} [H(t), \rho(t)] ,$$
(12.4)

where $H(t)$ is the sum of the atomic Hamiltonian, $H_A$, the radiation Hamiltonian, $H_R$, and the interaction Hamiltonian, $H_I(t)$, which is responsible for all photon–atom processes. If we assume that the interaction is switched on at a time $t = t_0$, then for $t \leq t_0$ the atomic system and the radiation field evolve independently, and the statistical operator of the total system can be written as the direct product of the two statistical operators for the atom and the radiation field,

$$\rho(t) = \rho_A(t) \otimes \rho_R(t) , \quad \forall t \leq t_0 .$$
(12.5)

Equation (12.4) has the formal integral solution

$$\rho(t) = \rho(t_0) + \frac{1}{i\hbar} \int_{t_0}^{t} dt' [H(t'), \rho(t')] ,$$
(12.6)

---

[2] A wave retarder is a device that separates the incoming beam into two beams with orthogonal states of polarization, and then recombines them after introducing a phase retardation between them.

which can be rewritten as a perturbation series by means of recursive substitution of the same solution under the sign of integral,

$$\rho(t) = \rho(t_0) + \sum_{n=1}^{\infty} \frac{1}{(i\hbar)^n} \int_{t_0}^{t} dt_n \int_{t_0}^{t_n} dt_{n-1} \cdots \int_{t_0}^{t_2} dt_1$$

$$\times \left[ H(t_n), \left[ H(t_{n-1}), \cdots, [H(t_1), \rho(t_0)] \cdots \right] \right]. \tag{12.7}$$

It is possible to show that (12.7) is equivalent to

$$\rho(t) = U(t, t_0)\, \rho(t_0)\, U^{\dagger}(t, t_0)\,, \tag{12.8}$$

where $U(t, t_0)$ is the evolution operator corresponding to the total Hamiltonian, $H(t)$, which also can be written as a perturbation series,

$$U(t, t_0) = 1 + \sum_{n=1}^{\infty} \frac{(i\hbar)^{-n}}{n!} \int_{t_0}^{t} dt_n \cdots \int_{t_0}^{t} dt_1 \, T\{H(t_n) \cdots H(t_1)\}\,, \tag{12.9}$$

where $T\{\cdots\}$ is Dyson's time-ordered product [4].

It is well known that (12.9) can be expressed as a series expansion of Feynman diagrams describing all possible photon–atom processes. For example, Figs. 12.4 and 12.5 show all Feynman diagrams corresponding to processes of 1st and 2nd order, respectively. In this chapter, we limit ourselves to consider 1st-order processes only, so as to be able to properly treat the individual mechanisms of radiation absorption and emission. Resonance scattering, on the other hand, is intrinsically a 2nd-order process (described by the first two diagrams of Fig. 12.5), where frequency correlations between the incoming and outgoing photons can occur (*partial redistribution* in frequency, *PRD*).

It is still possible to treat consistently the phenomenon of resonance scattering to 1st order, if we can assume that the coherence of the scattering process is destroyed because of the particular plasma conditions. Then, resonance scattering can be described as the temporal succession of 1st-order absorption and re-emission, interpreted as statistically independent processes (*complete redistribution* in frequency; *CRD*). This situation has also been

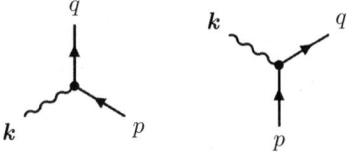

**Fig. 12.4.** Feynman diagrams for photon–atom processes of 1st order. $p$ and $q$ are the initial and final atomic states, respectively. $k$ is the incoming or outgoing photon

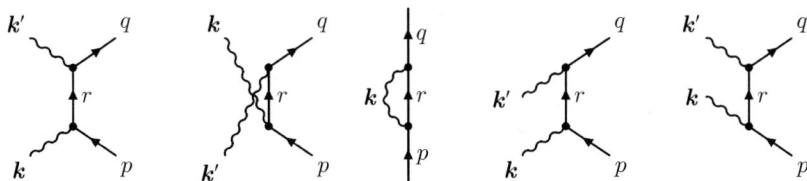

**Fig. 12.5.** Feynman diagrams for photon–atom processes of 2nd order. $p$ and $q$ are the initial and final atomic states, respectively, while $r$ is an intermediate (virtual) state. $k$ and $k'$ are the incoming or outgoing photons. A summation over all possible intermediate states, $r$, is implied by these diagrams

assumed in Chap. 7. A typical scenario is the presence of collisions that are effective in perturbing the atom during the interaction with the radiation field, to the point of completely relaxing the coherence of the scattering process. In a collisionless plasma, illumination of the atom with a spectrally flat radiation also determines the physical conditions for noncoherent scattering. In the rest of this chapter, we will assume that such *flat-spectrum approximation* of atom irradiation is always valid.

From (12.7) it is possible to derive both the evolution equations for the atomic system (*statistical equilibrium equations*) and the radiation field (*radiative transfer equations*). This is accomplished by a process of partial averaging of (12.7) over the quantum states of one or the other of the two systems [5,6]. Within the 1st-order theory of polarized line formation, the solution of the statistical equilibrium equations (hereafter, SEE) determines the excitation state of the atomic system, from which the reemitted radiation can then be calculated by solving the radiative transfer equations (hereafter, RTE).

For any pair of atomic levels, $(p, q)$, the corresponding SEE for the atomic density–matrix element, $\rho_{pq}$, has the form

$$
\begin{aligned}
\frac{\mathrm{d}}{\mathrm{d}t} \rho_{pq} + \mathrm{i}\omega_{pq}\,\rho_{pq} = & -\sum_{p'q'} \Big[ R_\mathrm{E}(pq, p'q') + R_\mathrm{S}(pq, p'q') + R_\mathrm{A}(pq, p'q') \Big] \rho_{p'q'} \\
& + \sum_{p_\mathrm{u} q_\mathrm{u}} \Big[ T_\mathrm{E}(pq, p_\mathrm{u} q_\mathrm{u}) + T_\mathrm{S}(pq, p_\mathrm{u} q_\mathrm{u}) \Big] \rho_{p_\mathrm{u} q_\mathrm{u}} \\
& + \sum_{p_\mathrm{l} q_\mathrm{l}} T_\mathrm{A}(pq, p_\mathrm{l} q_\mathrm{l})\, \rho_{p_\mathrm{l} q_\mathrm{l}} ,
\end{aligned} \tag{12.10}
$$

where $\omega_{pq}$ is the Bohr frequency between the two levels. The physical interpretation of the various radiative rates in (12.10) is illustrated by the diagram of Fig. 12.6 (the subscripts u and l stand for "upper" and "lower"; they indicate atomic levels of higher and lower energy, respectively, than the levels $p$ and $q$, to which they are radiatively connected). Explicit expressions for these rates have been given by [6], for several typologies of atomic structures (atom without fine structure, atom with fine structure in $LS$ coupling, atom with hyperfine structure).

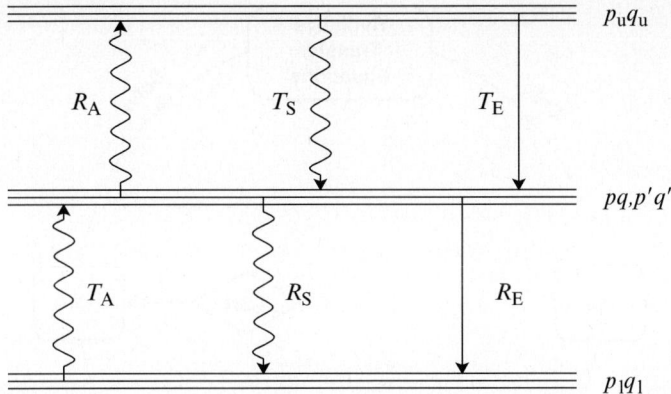

**Fig. 12.6.** Energy-level diagram illustrating the different radiative processes connecting a pair of atomic levels $(p, q)$ with others levels in the atom. *Straight arrows* indicate spontaneous emission processes, whereas *wiggly lines* indicate radiation-induced processes (absorption and stimulated emission)

The RTE can be written compactly as a vector equation,

$$\frac{1}{c}\frac{\mathrm{d}}{\mathrm{d}t}\boldsymbol{S} = -\mathsf{K}\boldsymbol{S} + \boldsymbol{J} , \tag{12.11}$$

where $\boldsymbol{S} \equiv (I, Q, U, V)$ is the Stokes vector, $\mathsf{K}$ is the absorption matrix (corrected for stimulated emission), and $\boldsymbol{J}$ is the Stokes emissivity. For the explicit expressions of $\mathsf{K}$ and $\boldsymbol{J}$, we refer to [6]. Here we only remark that these quantities depend on the solution density matrix of the atomic system, so one must first solve for the statistical equilibrium of the atom in order to calculate its absorptivity and emissivity.

In the *single scattering approximation*, radiation is scattered only once – on its way from the source to the observer – before leaving the plasma. Typically this is a very good approximation for optically thin gases, and it allows to avoid the integration of (12.11) by expressing the scattered radiation simply in terms of the Stokes emissivity vector, $\boldsymbol{J}$. In optically thick plasmas, instead, the solution of (12.11) cannot be avoided. Moreover, the scattered radiation tends to modify locally the radiation field that enters as an input of the SEE, so the radiative transfer problem in optically thick plasma is affected by nonlinearity, as well as nonlocality issues, because of the back-reaction of the radiation field on the atomic system. The solution to this problem relies on iterative schemes, of the kind illustrated in Fig. 12.7, which must be followed through until self-consistency of the two solutions of the SEE and the RTE is achieved.

The problem of the convergence of the iterative loop of Fig. 12.7 is a research subject on itself, which is still relatively new for the general case involving the presence of atomic polarization [7]. For this reason, in the following

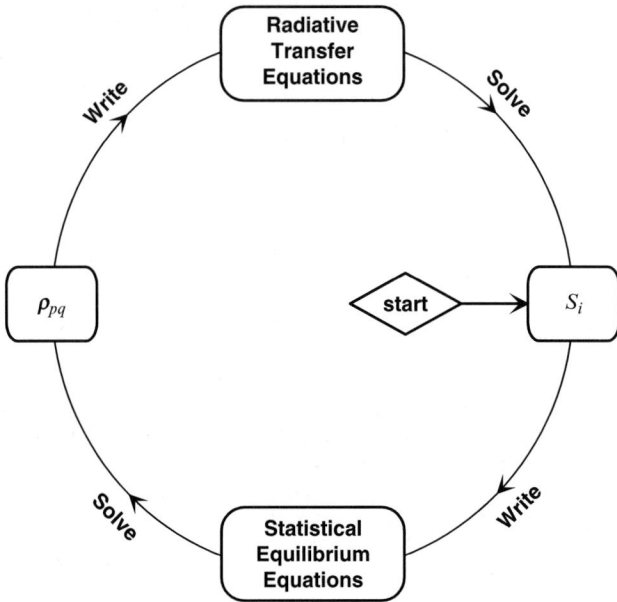

**Fig. 12.7.** Self-consistency loop representing the iterative solution of the problems of statistical equilibrium and radiative transfer in an optically thick medium

examples of this chapter, we will only consider optically thin plasmas, as well as optically thick plasmas that are homogeneous in their thermodynamical and electrodynamical properties, and for which the back-reaction of the locally scattered radiation on the excitation conditions of the plasma can be neglected. These two idealized cases are nonetheless sufficient for the purpose of demonstrating the diagnostic potential of atomic polarization – and of its modification in the presence of external fields – in the investigation of the topology of magnetic and electric fields in astrophysical plasmas.

## 12.4 The Hanle Effect in the Two-Level Atom

The simplest atomic system that we can consider to illustrate the effect of a magnetic field on the scattering polarization in a spectral line is obviously that which is composed of just two atomic levels: the lower level, $J_\mathrm{l}$, and the upper level, $J_\mathrm{u}$, of the atomic transition corresponding to the spectral line of interest.

Such model of the *two-level atom* has played an important role for the description and the understanding of polarization phenomena, for several reasons. First of all, the polarization properties of the special case with $J_\mathrm{l} = 0$ and

$J_u = 1$ (the 0–1 *atom*) can be derived completely within a classical (as opposed to quantum) approach to the electrodynamics of radiation processes [6, 8]. Thus, the 0–1 atom helped gaining fundamental insights into the process of magnetic depolarization of scattered radiation at a time when Quantum Mechanics was still being developed [9]. On the other hand, for some branches of Plasma Polarization Spectroscopy – e.g., the study of the optical pumping of atomic levels by anisotropic radiation [10] – the two-level atom can actually be a good representation of the true atomic system, and for this reason it still is a subject of active research [11, 12].

In general, however, and in particular for astrophysical applications, the two-level atom can only be regarded as a very rough – and often completely inadequate – approximation, with very few exceptions (notably, the case of the Sr I line at 4,607 Å in the solar spectrum). This is expected whenever the radiation intensity is distributed within a large interval of frequencies. Then the statistical equilibrium of the atom typically involves all levels that are connected by radiative transitions within the spectral range of the incident radiation. The complexity of the polarization phenomena in such multilevel atoms cannot be reproduced in general within the limited scheme of the two-level atom [13, 14].

In this section, we will focus on the two-level atom as a textbook case for illustrating the Hanle effect, and its potential for the diagnostics of magnetic fields in astrophysical plasmas. The Hanle effect is also treated in Appendix D. We will study two different atomic structures: (a) the 0–1 atom, and its connections to classical electrodynamics, and (b) the 1–0 atom, as the simplest case showing the role of lower-level polarization (e.g., the mechanism of *depopulation pumping* [10]).

In keeping with (12.10), we assume that the atomic system is subject only to radiative processes. We also assume that the illuminating radiation is unpolarized, and contained within a cone of half aperture $\vartheta_M \leq 90°$, with the vertex centered at the atom. Then, the illumination conditions are completely described in terms of the radiation mean intensity, $J(\omega)$, and anisotropy, $w(\omega)$. These radiation properties are easily expressed if we adopt the formalism of the irreducible spherical tensors [3], and introduce accordingly the tensor components of the radiation field, $J_Q^K(\omega)$, where $K = 0, 1, 2$ and $Q = -K, \ldots, K$.[3] For the illumination conditions described above, and also assuming that the radiation intensity, $I(\omega)$, is independent of the propagation direction within the radiation cone, we have

$$J_0^0(\omega) = \frac{1}{2}\left(1 - \cos\vartheta_M\right) I(\omega) \, , \tag{12.12a}$$

$$J_0^2(\omega) = \frac{1}{4\sqrt{2}}(1 - \cos\vartheta_M)(1 + \cos\vartheta_M)\cos\vartheta_M \, I(\omega) \, , \tag{12.12b}$$

---

[3] The general expression of the tensors $J_Q^K(\omega)$ in terms of the Stokes vector of the radiation field is given in [6].

for the only two nonvanishing components of the radiation tensors, expressed in a reference frame with the $z$-axis along the incident direction. $J_0^0(\omega)$ corresponds to the radiation mean intensity, $J(\omega)$, whereas the anisotropy of the radiation field is defined by

$$w(\omega) \equiv \sqrt{2}\,\frac{J_0^2(\omega)}{J_0^0(\omega)} = \frac{1}{2}\,(1 + \cos\vartheta_{\mathrm{M}})\cos\vartheta_{\mathrm{M}}\,. \tag{12.13}$$

We notice that $w(\omega)$ attains its maximum value of 1 for a collimated beam of radiation ($\vartheta_{\mathrm{M}} = 0°$). When the solid angle occupied by the radiation cone reaches the maximum aperture of $2\pi$ ($\vartheta_{\mathrm{M}} = 90°$) instead, the anisotropy factor vanishes.

Finally, we notice that the solution of (12.10) is expressed in a reference frame that has the $z$-axis along the quantization axis ($S'$). If this is different from the reference frame defined by the incident radiation ($S$), then the radiation tensors that enter (12.10) are given by

$$[J_0^0(\omega)]_{S'} = [J_0^0(\omega)]_S\,, \tag{12.14a}$$

$$[J_Q^2(\omega)]_{S'} = D_{0Q}^2(R_{SS'})\,[J_0^2(\omega)]_S\,, \tag{12.14b}$$

where $D_{QQ'}^K$ is the rotation matrix of order $K$, and $R_{SS'}$ is the rotation operator that transforms the original reference frame, $S$, into the frame of the quantization axis, $S'$ [15].

### 12.4.1 The 0–1 Atom in a Magnetic Field

The two-level atom with $J_l = 0$ and $J_u = 1$ is the simplest atomic system to illustrate the Hanle effect. This model is particularly instructive, because all of its radiative properties can be reproduced adopting a classical description of the atom as a three-dimensional, damped harmonic oscillator with forcing term. In this classical picture, the forcing term corresponds to the electric component, $\boldsymbol{E}(t)$, of the incident radiation field. The damping term corresponds instead to the contribution of the radiation reaction to the dynamical equation of the harmonic oscillator [16]. Rather than adopting the classical expression of the damping term, one can use more accurately the observed value of the line transition amplitude, i.e., the Einstein coefficient for spontaneous emission, $A_{10}$. The classical dynamical equation of the harmonic oscillator, in the additional presence of a stationary magnetic field of strength $B$ and direction $\boldsymbol{b}$, is then

$$\ddot{\boldsymbol{x}} + 2A_{10}\dot{\boldsymbol{x}} - 2\omega_B \boldsymbol{b} \times \dot{\boldsymbol{x}} + \omega_{10}^2 \boldsymbol{x} + \frac{e_0}{m}\,\boldsymbol{E}(t) = 0\,. \tag{12.15}$$

In this equation, $\omega_B = e_0 B/2m$ is the Larmor angular frequency, whereas $\omega_{10}$ is the resonance frequency of the oscillator, which must correspond to the frequency of the line transition of the quantum-mechanical 0–1 atom. $e_0$ and $m$ are the electron's charge and mass, respectively. The solution of the dynamical problem (12.15) can be found, e.g., in [6].

The quantum-mechanical description of the resonance scattering of radiation in the 0–1 atom, in the presence of a stationary magnetic field, requires first the solution of the SEE (12.10) for this particular problem. If we introduce the irreducible spherical components of the density matrix, $\rho_Q^K(J)$ [6] (see also Chaps. 4 and 5 in this book), with $K = 0, \ldots, 2J$ and $Q = -K, \ldots, K$, then (12.10) for the 0–1 atom becomes, neglecting stimulated emission,

$$\frac{d}{dt} \rho_Q^K(1) + i\omega_B Q \, \rho_Q^K(1) = -A_{10} \, \rho_Q^K(1) + \frac{B_{01}}{\sqrt{3}} (-1)^{K-Q} J_{-Q}^K(\omega_{10}) \, \rho_0^0(0) \,, \tag{12.16}$$

along with the normalization condition for the atomic population,

$$\rho_0^0(0) + \sqrt{3}\,\rho_0^0(1) = 1 \,. \tag{12.17}$$

In (12.16), $B_{01}$ is the Einstein coefficient for absorption. The radiation tensors in that equation are expressed in the reference frame of the magnetic field, in agreement with (12.14a) and (12.14b). In statistical equilibrium, $(d/dt)\rho_Q^K(1) = 0$, (12.16) gives at once [17]

$$\rho_Q^K(1) = \frac{1}{\sqrt{3}} \frac{B_{01}}{A_{10}} \frac{(-1)^{K-Q} J_{-Q}^K(\omega_{10})}{1 + iQ(\omega_B/A_{10})} \, \rho_0^0(0) \,. \tag{12.18}$$

In particular, for the population of the upper level, we find the well-known result

$$\sqrt{3}\,\rho_0^0(1) = \left[1 + \frac{A_{10}}{B_{01} J_0^0(\omega_{10})}\right]^{-1} \,. \tag{12.19}$$

We see that the polarization properties of the excited state of the atom are determined directly by the polarization properties of the incident radiation field, as it is expected on the basis of the classical picture presented earlier. From (12.18) we also conclude that the components with $Q = 0$ of the density matrix tensor are insensitive to the magnetic field strength. On the other hand, the components with $Q \neq 0$ (atomic coherences) are generally complex quantities that depend on the magnetic field, and in particular they vanish when $\omega_B \gg A_{10}$. This phenomenon of *relaxation* of the atomic coherences for increasing magnetic strengths is known as the *Hanle effect*. The critical value of $B$ for the onset of the Hanle effect is determined by the condition $\omega_B \approx A_{10}$. Figure 12.8 illustrates intuitively the mechanism of coherence relaxation in terms of the decreasing degree of overlapping of the wavefunctions associated with the Zeeman sublevels of $J_u = 1$ with increasing magnetic strength.

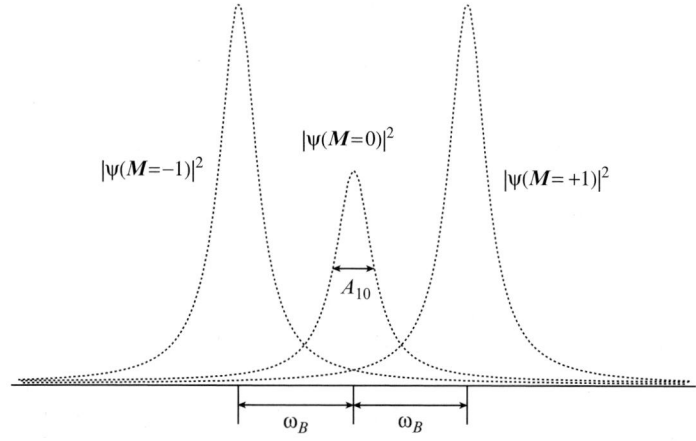

**Fig. 12.8.** In the presence of a magnetic field, the degeneracy of the atomic levels is removed because of the Zeeman splitting (proportional to $\omega_B$). Each sublevel $M$ is described by a wavefunction $\psi(M)$. In this figure, we represent the associated probability distribution for the three sublevels of $J_u = 1$, for a case of positive alignment of the level. The width $A_{10}$ of the distributions is the Einstein coefficient for spontaneous emission. If the magnetic field is such that $\omega_B \lesssim A_{10}$, then the wavefunctions of the sublevels overlap, and the effect of the quantum interferences, $\psi_i \psi_j^*$, can be observed in the scattered radiation. When $\omega_B \gg A_{10}$, instead, quantum interferences become negligible, as $|\psi_i + \psi_j|^2 \rightarrow |\psi_i|^2 + |\psi_j|^2$

If we introduce the *fractional atomic polarization*, $\sigma_Q^K(J) \equiv \rho_Q^K(J)/\rho_0^0(J)$, and also recall (12.14a) and (12.14b), we finally obtain

$$\sigma_0^0(1) \equiv 1 , \tag{12.20a}$$

$$\sigma_Q^1(1) = 0 , \tag{12.20b}$$

$$\sigma_Q^2(1) = \frac{1}{\sqrt{2}} \frac{(-1)^Q D_{0\,-Q}^2(R_{SS'})}{1 + iQ\,\dfrac{\omega_B}{A_{10}}} \, w(\omega_{10}) . \tag{12.20c}$$

We then conclude that the excited level of the 0–1 atom can only be populated ($K = 0$) and aligned ($K = 2$), but it cannot be oriented ($K = 1$), if the incident radiation field is not circularly polarized ($J_Q^1 \equiv 0$). This conclusion remains valid for any multilevel atom that can be described through density–matrix components of the form $\rho_Q^K(J)$.

Figure 12.9 shows the alignment of the upper level of the 0–1 atom as a function of the magnetic field strength, for various inclinations of the magnetic field ($0°, 30°, 60°, 90°$) from the direction of the incident radiation. This plot reproduces the result of (12.20c), on the assumption of $A_{10} = 10^8\,\mathrm{s}^{-1}$ and $w(\omega_{10}) = 10^{-2}$. We notice that the magnetic strength for which the imaginary part reaches its extremal value corresponds to the critical condition $\omega_B = A_{10}$ of Hanle effect ($B \sim 10^{-3}\,\mathrm{T}$, in this case).

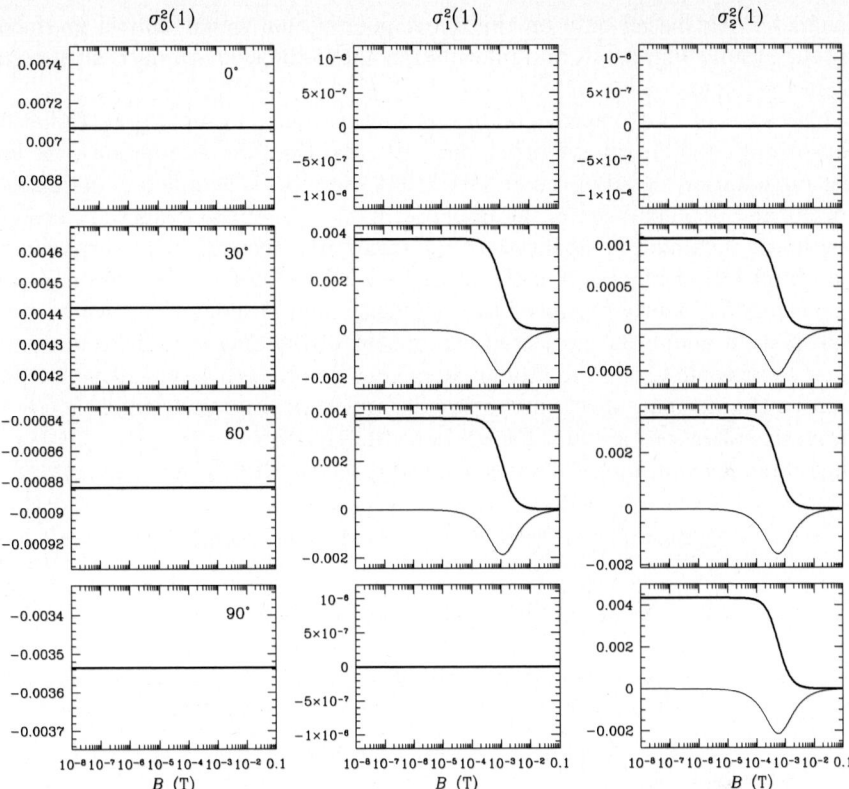

**Fig. 12.9.** Fractional alignment of the upper level, $J_u = 1$, of the 0–1 atom, as a function of the magnetic field strength, and for various inclinations $(0°, 30°, 60°, 90°)$ of the field with respect to the direction of incident radiation. For this plot, we used $A_{10} = 10^8\,\mathrm{s}^{-1}$ and $w(\omega_{10}) = 10^{-2}$. The thick and thin curves correspond to the real and imaginary parts of $\sigma_Q^2$, respectively

### Application: Polarized Coronal Emission

Despite the simplicity of the 0–1 atomic model, there are polarization phenomena in astrophysics that are suitably described by it. A noteworthy example from solar physics is the polarized emission of the forbidden (M1) transition of FeXIII $\lambda 1074.7\,\mathrm{nm}$ $(3s^2 3p^2\,{}^3P_0 - 3s^2 3p^2\,{}^3P_1)$, which is predominantly produced in the $2 \times 10^6\,\mathrm{K}$ regions of the solar corona.

Because of the very small Einstein $A$-coefficients characterizing M1 transitions (for the FeXIII $\lambda 1074.7\,\mathrm{nm}$ line, it is $A_{10} = 14\,\mathrm{s}^{-1}$), the atomic coherences are always completely relaxed, even in the presence of the very weak magnetic field of the quiet-sun corona $(B \sim 10^{-4} - 10^{-3}\,\mathrm{T})$. The atomic polarization of the first excited level of FeXIII is thus completely described by the fractional alignment $\sigma_0^2(1)$. Since this quantity is independent of the magnetic-field strength, it is impossible to achieve a complete diagnostics of the coronal magnetic field based only on the linear polarization signal that is produced by the atomic alignment, and information about the field strength must come from the Zeeman effect.

Because of the plasma conditions in the solar corona (large radiation anisotropy, small magnetic fields, large thermal Doppler broadening), the linear polarization signal of the FeXIII $\lambda1074.7$ nm line is completely dominated by the atomic alignment in the upper level, $J_u = 1$, of the transition. In fact, the linear polarization signature of the transverse Zeeman effect turns out to be completely negligible. On the other hand, the circular polarization of this line is predominantly produced by the longitudinal Zeeman effect, which gives information about the projected component of the magnetic field along the line-of-sight. However, the atomic alignment determines a systematic correction to the Zeeman effect in the circular polarization signal, which must be properly taken into account for a reliable diagnostics.

In more detail, the Stokes vector of the FeXIII $\lambda1074.7$ nm line is [6, 18]

$$I(\omega) \propto \left[1 + \frac{1}{2\sqrt{2}}\left(3\cos^2\Theta_{\boldsymbol{B}} - 1\right)\sigma_0^2(1)\right]\phi(\omega_{10} - \omega) , \quad (12.21\text{a})$$

$$Q(\omega) \propto -\frac{3}{2\sqrt{2}}\sin^2\Theta_{\boldsymbol{B}}\cos 2\Phi_{\boldsymbol{B}}\,\sigma_0^2(1)\,\phi(\omega_{10} - \omega) , \quad (12.21\text{b})$$

$$U(\omega) \propto -\frac{3}{2\sqrt{2}}\sin^2\Theta_{\boldsymbol{B}}\sin 2\Phi_{\boldsymbol{B}}\,\sigma_0^2(1)\,\phi(\omega_{10} - \omega) , \quad (12.21\text{c})$$

$$V(\omega) \propto -\frac{3}{2}\left[1 + \frac{1}{\sqrt{2}}\sigma_0^2(1)\right]\omega_{\boldsymbol{B}}\cos\Theta_{\boldsymbol{B}}\,\phi'(\omega_{10} - \omega) , \quad (12.21\text{d})$$

where $\phi(\omega_{10} - \omega)$ is a Gaussian profile with Doppler width corresponding to the coronal temperature. The scattering and magnetic geometries entering the above equations are illustrated in Fig. 12.10. We notice the "$-$" sign in front of (12.21b) and (12.21c), which is a consequence of the M1 character of the FeXIII $\lambda1074.7$ nm line [6].

Introducing the frequency-integrated Stokes parameters, $\bar{S}_i$, we note that

$$\bar{U}/\bar{Q} = \tan 2\Phi_{\boldsymbol{B}} , \quad (12.22\text{a})$$

$$\sqrt{\bar{Q}^2 + \bar{U}^2} \propto \sin^2\Theta_{\boldsymbol{B}}\,|\sigma_0^2(1)| , \quad (12.22\text{b})$$

so it is possible in principle to retrieve the full vector information of $\boldsymbol{B}$, if one takes into account both the linear and circular polarization signals of the line. However, one also needs a preliminary knowledge of $\sigma_0^2(1)$, which is determined by the location and the illumination conditions of the emitting plasma, the rates of collisional polarization and depolarization of the upper level, and the inclination of the magnetic field from the local vertical, $\vartheta_{\boldsymbol{B}}$. For example, if we assume that the atomic alignment is originated only by radiation anisotropy, from (12.20c) we find

$$\sigma_0^2(1) = \frac{1}{2\sqrt{2}}(3\cos^2\vartheta_{\boldsymbol{B}} - 1)\,w(\omega_{10}) , \quad (12.23)$$

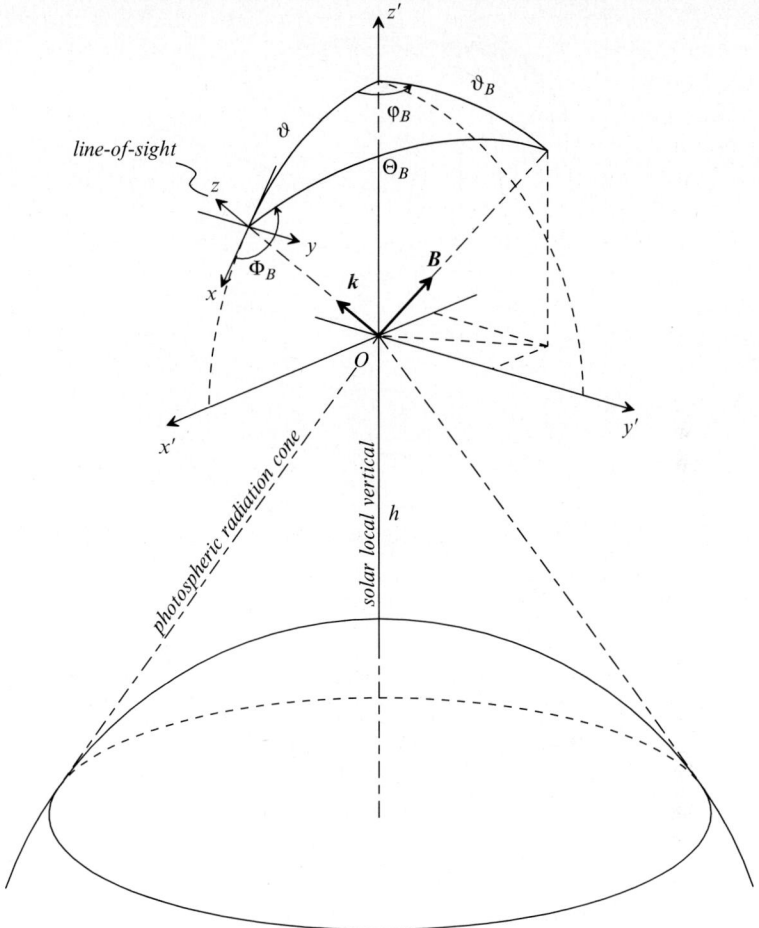

**Fig. 12.10.** Geometry for the diagnostics of coronal magnetic fields. The direction of $y \equiv y'$ identifies the parallel to the limb, which is assumed as the reference direction for positive $Q$ polarization. The direction of the magnetic field, $\boldsymbol{B}$, is identified by the angles $\vartheta_B$ and $\varphi_B$ in the reference frame of the local vertical, and by $\Theta_B$ and $\Phi_B$ in the reference frame of the scattered direction, $\boldsymbol{k}$

having recalled the expression of the rotation matrix $D_{00}^2(R_{SS'}) \equiv d_{00}^2(\vartheta_B)$ [15]. We then see that $\sigma_0^2(1)$ vanishes at the *Van Vleck angle*, defined by $\vartheta_{\mathrm{VV}} \equiv \arccos(1/\sqrt{3}) \approx 54.7°$ (see also Fig. 12.9), and at its supplementary angle, $180° - \vartheta_{\mathrm{VV}}$. The former angle is called the "magic angle" in atomic physics. In general, when the sign of $\sigma_0^2(1)$ is not known a priori (e.g., it is not known whether $\vartheta_B < \vartheta_{\mathrm{VV}}$ or $\vartheta_B > \vartheta_{\mathrm{VV}}$), from (12.21b)–(12.21c) and (12.22a), we can only determine the direction of the field on the plane of the sky with an ambiguity of 90° [18]. This is illustrated in Fig. 12.11, for a particular scattering geometry (see caption to the figure for more details).

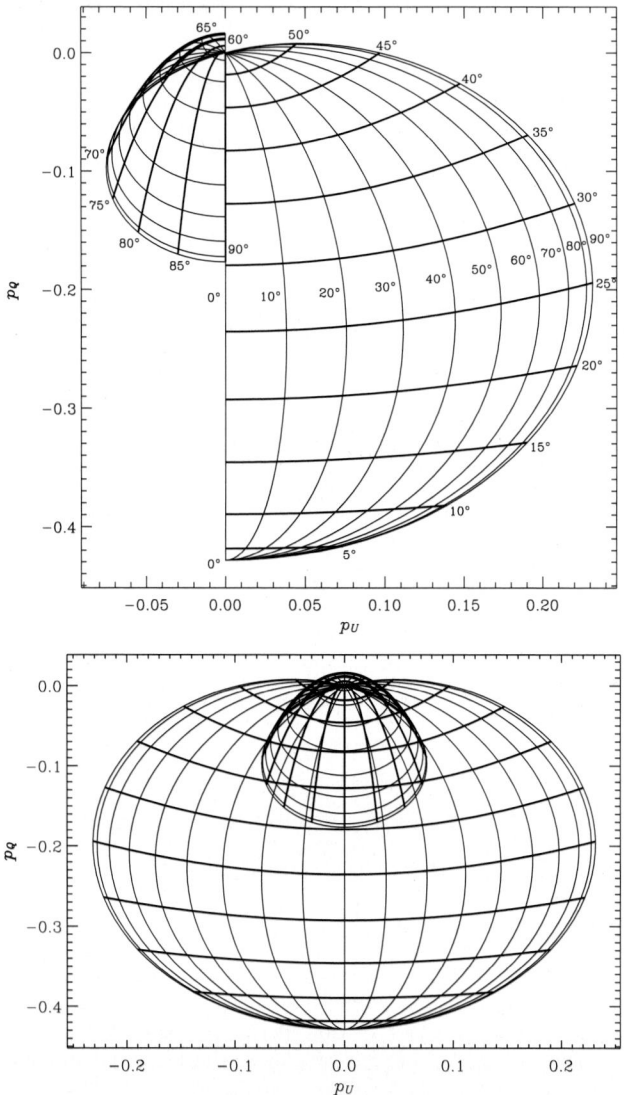

**Fig. 12.11.** Hanle diagrams of $p_Q = \bar{Q}/\bar{I}$ against $p_U = \bar{U}/\bar{I}$, for the FeXIII $\lambda 1074.7$ nm line in a 90° scattering event. The atom is illuminated by a radiation field with anisotropy factor $w(\omega_{10}) = 0.5$. The magnetic-field strength is $B = 10^{-4}$ T, and for this line it already ensures the complete relaxation of the atomic coherences. The magnetic field inclination, $\vartheta_B$, varies from 0° to 90°, and the azimuth, $\varphi_B$, from −90° to 90° (see Fig. 12.10). The *thick lines* correspond to curves of constant $\vartheta_B$, while the *thin lines* correspond to curves of constant $\varphi_B$. In the top figure, we show only the region of positive azimuth in greater detail. (The curve for $\vartheta_B = 55°$ is not shown, because it practically reduces to the point of zero polarization, being very close to the Van Vleck angle.) The 90° ambiguity can only occur where a point $(p_U, p_Q)$ in the diagram can belong to two different pairs of isocurves of $\vartheta_B$ and $\varphi_B$ (see bottom figure)

FeXIII 1074.7 Azimuth of B 4/21/05

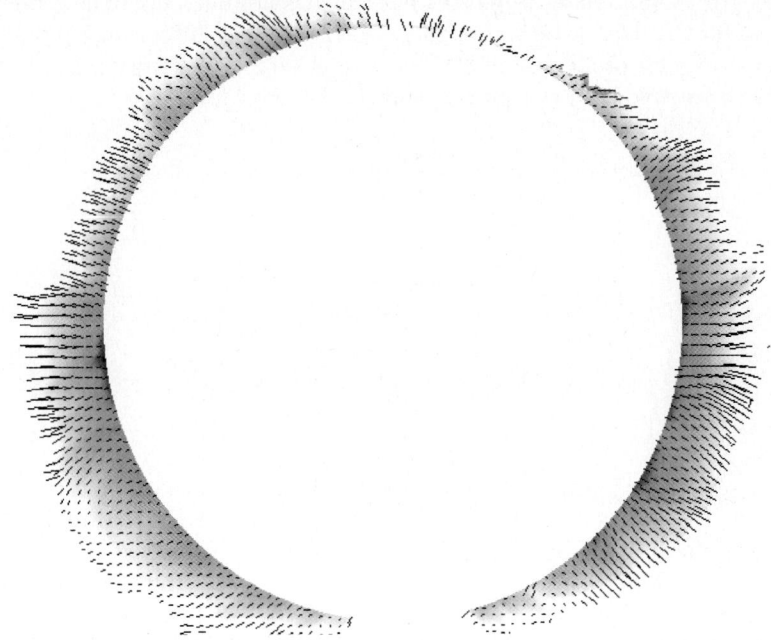

**Fig. 12.12.** Polarimetric observations of the solar corona in FeXIII $\lambda 1074.7\,\mathrm{nm}$, with the Coronal Multi-channel Polarimeter. The figure shows the direction and strength of the linear polarization, superimposed to a gray-scale image of the corona. (Courtesy of S. Tomczyk, High Altitude Observatory)

We remark that the use of all the information contained in the Stokes vector of the emitted radiation is a necessary condition (although not always sufficient) for the resolution of such $90°$ ambiguity in practical cases.

Figure 12.12 shows recent observations of the solar corona by Tomczyk (High Altitude Observatory) with the Coronal Multi-channel Polarimeter (CoMP [19]), deployed at the One-Shot coronograph of the National Solar Observatory at Sacramento Peak (Sunspot, New Mexico, USA). The figure displays the direction and strength of the linear polarization as inferred from (12.22a) and (12.22b), respectively, superimposed to a gray-scale image of the solar corona at FeXIII $\lambda 1074.7\,\mathrm{nm}$. We note the presence of regions of vanishing polarization, which indicate the possible occurrence of longitudinal fields ($\Theta_B \approx 0°, 180°$) or of the Van Vleck effect ($\vartheta_B \approx \vartheta_{\mathrm{VV}}, 180° - \vartheta_{\mathrm{VV}}$).

## 12.4.2 The 1–0 Atom in a Magnetic Field

As a second example, we discuss the two-level atom with $J_l = 1$ and $J_u = 0$. The polarizability of the lower level makes it impossible to describe this atomic model in classical terms, like we did for the 0–1 atom. The quantum approach illustrated in Sect. 12.3 is thus the only means of investigation of the properties of the 1–0 atom. These properties have in recent times become of relevant interest for the diagnostics of magnetic fields in the solar chromosphere [20].

To study the properties of the 1–0 model atom, we write again the SEE of the polarizable level (which this time is the lower level),

$$
\frac{d}{dt}\,\rho_Q^K(1) + i\omega_B\,Q\,\rho_Q^K(1) = -R_A(KQ) + \frac{1}{\sqrt{3}}\,\delta_{K0}\,\delta_{Q0}\,A_{01}\,\rho_0^0(0)\,,\qquad (12.24)
$$

where

$$
R_A(KQ) = B_{10} \sum_{K'K''} \frac{1+(-1)^{K+K'+K''}}{2}\sqrt{3(2K+1)(2K'+1)(2K''+1)}
$$

$$
\times \begin{Bmatrix} K & K' & K'' \\ 1 & 1 & 1 \end{Bmatrix} \sum_{Q'Q''}(-1)^{K''+Q'}\begin{pmatrix} K & K' & K'' \\ Q & -Q' & Q'' \end{pmatrix}\rho_{Q'}^{K'}(1)\,J_{Q''}^{K''}(\omega_{01})\,.
$$

We see immediately that, despite the simplicity of the atomic structure involved, the solution of the statistical equilibrium of the 1–0 atom is far more complicated than in the case of the 0–1 atom. In fact, the density-matrix element, $\rho_Q^K(1)$, is no longer determined directly by the illumination conditions, as in the case of the 0–1 atom (cf. (12.18)), but it now depends on all the other elements $\rho_{Q'}^{K'}(1)$ as well. In particular, this implies that the atomic polarization of the lower level, as well as its population, are in general nonlinear functions of the radiation field tensors, $J_Q^K$.

However, in the so-called limit of *weak anisotropy* ($|J_Q^2| \ll J_0^0$), which often is a valid assumption for polarization studies in solar physics, an approximate algebraic solution of (12.24) can easily be determined. In fact, neglecting all radiation tensors $J_Q^K$ for $K > 0$, at the statistical equilibrium, (12.24) gives at once

$$
\rho_0^0(1) \approx \frac{1}{\sqrt{3}}\,\frac{A_{01}}{B_{10}J_0^0(\omega_{01})}\,\rho_0^0(0)\,,
$$

whereas all other tensor components with $K > 0$ vanish identically. Recalling the normalization condition, (12.17), we then find

$$
\sqrt{3}\,\rho_0^0(1) \approx \left[1 + \frac{B_{10}J_0^0(\omega_{01})}{A_{01}}\right]^{-1}.\qquad (12.25)
$$

which is the counterpart for the 1–0 atom of (12.19). One can then go back to considering (12.24) for $K = 2$,[4] and solve for the fractional atomic alignment of the lower level, $\sigma_Q^2(1)$. In keeping with the weak-anisotropy approximation, we neglect all terms in that equation that are nonlinear in the anisotropy factor,

----

[4] Because we assumed $J_0^1 \equiv 0$, the linear system for $K = 1$ is decoupled, giving only the trivial solution $\rho_Q^1(1) \equiv 0$.

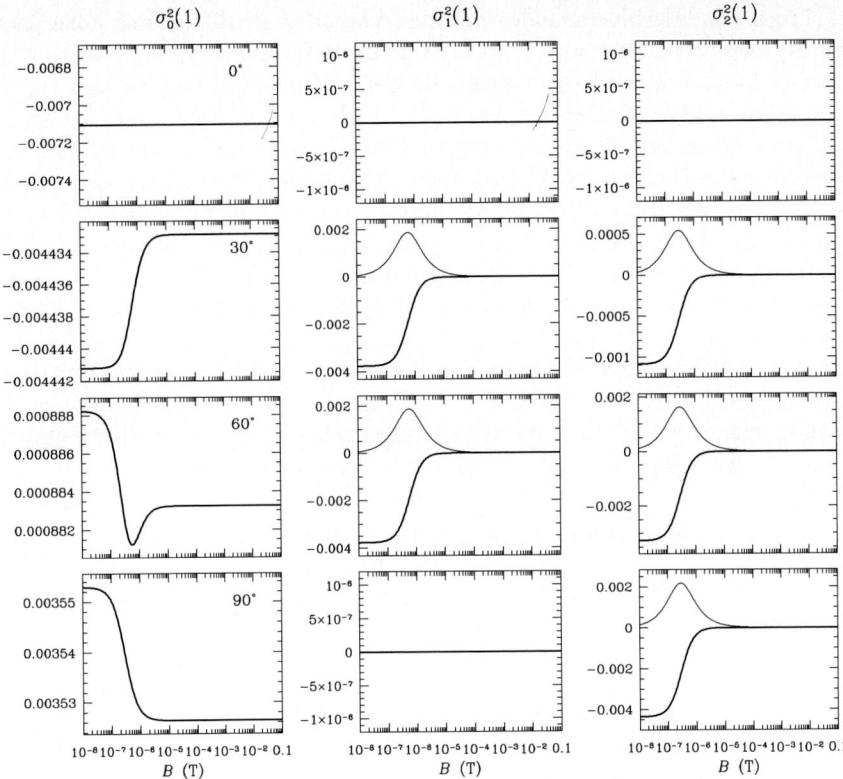

**Fig. 12.13.** Fractional alignment of the lower level, $J_l = 1$, of the 1–0 atom, as a function of the magnetic field strength, and for various inclinations $(0°, 30°, 60°, 90°)$ of the field with respect to the direction of incident radiation. For this plot, we used $B_{10} J_0^0(\omega_{01}) \approx 5.2 \times 10^4 \, \text{s}^{-1}$ and $w(\omega_{01}) = 10^{-2}$. The *thick* and *thin curves* correspond to the real and imaginary parts of $\sigma_Q^2$, respectively

$w(\omega_{01})$. It is easy to realize that such terms are those with $K' = K'' = 2$. Solving the linearized equation in $w(\omega_{01})$, and also recalling (12.14b), finally gives

$$\sigma_Q^2(1) \approx -\frac{1}{\sqrt{2}} \frac{(-1)^Q D_{0-Q}^2(R_{SS'})}{1 + iQ \dfrac{\omega_B}{B_{10} J_0^0(\omega_{01})}} w(\omega_{01}), \qquad (12.26)$$

which is the counterpart for the 1–0 atom of (12.20c).

Figure 12.13 clearly illustrates the applicability of this linearization procedure for small values of the anisotropy factor. In fact, the values of $\sigma_Q^2(1)$ given in the figure are very well represented by the analytic approximation, (12.26). The small residuals (most notably, the modulation of $\sigma_0^2(1)$ for inclined magnetic fields) are all determined by the nonlinear dependence on $w(\omega_{01})$ of the exact solution of (12.24).

From (12.26), we conclude that the relaxation of the atomic coherences in the lower level of the 1–0 atom is again determined by the ratio of the Larmor frequency with the radiative width of that level, $B_{10}J_0^0(\omega_{01})$ (which is now a function of the radiation field). However, this relaxation occurs for magnetic fields significantly smaller than in the case of the Hanle effect of the upper level of the 0–1 atom. This peculiarity of the *Hanle effect of the lower level* is well understood by comparing the two cases of Figs. 12.9 and 12.13. Because for those cases $B_{10}J_0^0(\omega_{01})/A_{10} \approx 5.2 \times 10^{-4}$, the critical value of the magnetic field for the Hanle effect in the atom 1–0 ($B \sim 5 \times 10^{-7}$ T in this case) is correspondingly smaller than the critical value for the Hanle effect in the atom 0–1.

Under illumination conditions that are typical of astrophysical plasmas, it is generally true that the atomic coherences of metastable states (like the atomic ground state) in a multilevel atom relax at much smaller magnetic strengths than the coherences of spontaneously decaying states.

### Application: Dichroism Polarization in Solar Filaments

Because the upper level of the 1–0 atom cannot carry any atomic polarization, the radiation emitted in the process of deexcitation in this model atom can only be polarized because of the Zeeman effect (through the Zeeman splitting of the lower level), if a magnetic field is present. The presence of atomic polarization in the lower level is instead made manifest in the process of radiation absorption (*dichroism polarization*). If the lower level is aligned, for example, the cross-section for absorption is different for the $\pi$ and the $\sigma$ transitions (see Sect. 12.1 in this chapter), thus the radiation transferred through an absorptive medium is polarized, even in the absence of a magnetic field.

Atomic polarization plays a very important role in the diagnostics of magnetic fields in plasmas that are optically thick (i.e., both absorptive and emissive). Figure 12.14 illustrates this point very clearly. Each column shows the emergent Stokes vector from an optically thick slab (optical depth, $\tau = 2$) immersed in a homogeneous magnetic field (see caption for more details), for different atomic structures. The first column shows the Stokes vector for *both* the 0–1 and 1–0 atoms, in the absence of atomic polarization (i.e., subject exclusively to the Zeeman effect). It is evident that in this case, those two atomic structures cannot be distinguished. The situation is completely different if we allow for the presence of atomic polarization. The second and third columns show the same emergent Stokes vector for the 0–1 and the 1–0 atoms, respectively, but allowing for both the Zeeman effect and the presence of atomic alignment. While the circular-polarization Stokes $V$ is practically the same,[5] the linear polarization profiles in the two cases differ completely. In particular, we notice the reversal of the sign of Stokes $Q$.

---

[5] We recall that the atomic alignment determines a correction to Stokes $V$ (cf. (12.21d)).

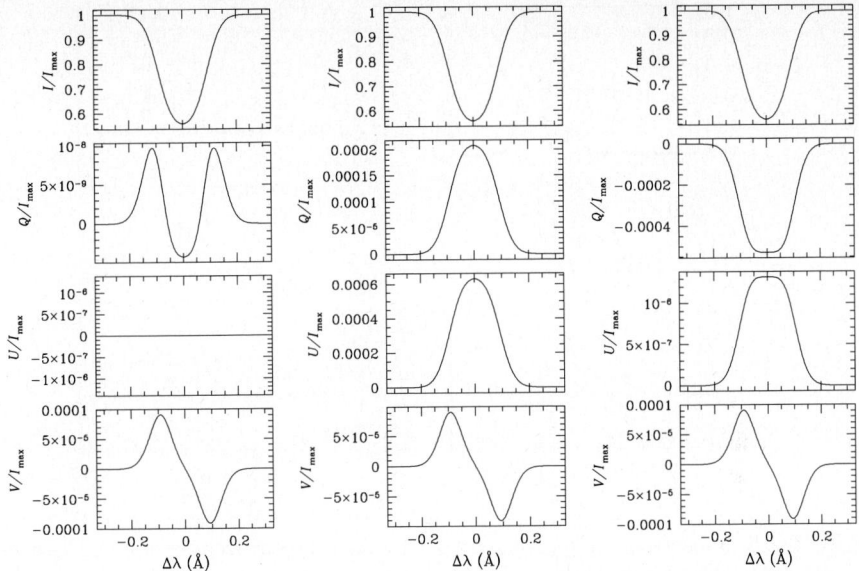

**Fig. 12.14.** Emergent Stokes vectors from a uniform, optically thick slab ($\tau = 2$) in the presence of a homogeneous magnetic field with $B = 10^{-3}$ T, and inclined of $45°$ from the vertical (which in this calculation coincides with the line-of-sight). The cases of different atomic structures are illustrated. In the first column, we show the Stokes vector for both the 0–1 and 1–0 atoms without atomic polarization (subject only to the Zeeman effect). The two Stokes vectors coincide in this case. In the second and third column, we show instead the general case in which atomic alignment induced by radiation anisotropy ($w = 10^{-2}$) is also present, in the 0–1 and 1–0 atoms, respectively

This signature of the presence of atomic polarization in metastable levels has indeed been observed in the polarized radiation from quiescent solar *filaments* [20, 21]. These structures are visible (e.g., in hydrogen H$\alpha$ and HeI $\lambda$1083.0 nm images of the Sun), as dark filamentary structures against the solar disk. They consist of cool gas that is trapped by magnetic fields at coronal heights. When they are seen off disk, extending above the solar limb, they are called *prominences*. In a first, very rough approximation, a filament or prominence can be thought of as a homogeneous slab of gas that absorbs and reemits the radiation coming from the underlying photosphere.

Figures 12.15 and 12.16 show two different observations of the Stokes vector of HeI $\lambda$1083.0 nm in a prominence and a filament, respectively [20]. This multiplet originates in the transition between the metastable state, $2\,^3S_1$, and the first excited term, $2\,^3P_{2,1,0}$, of the triplet species of HeI. The weaker, blue component visible in those figures corresponds to the 1–0 transition of the multiplet. We note that it practically does not contribute to the linearly polarized scattered radiation from a prominence (Fig. 12.15), because the upper level $J_u = 0$ cannot be aligned. On the contrary, the signature of

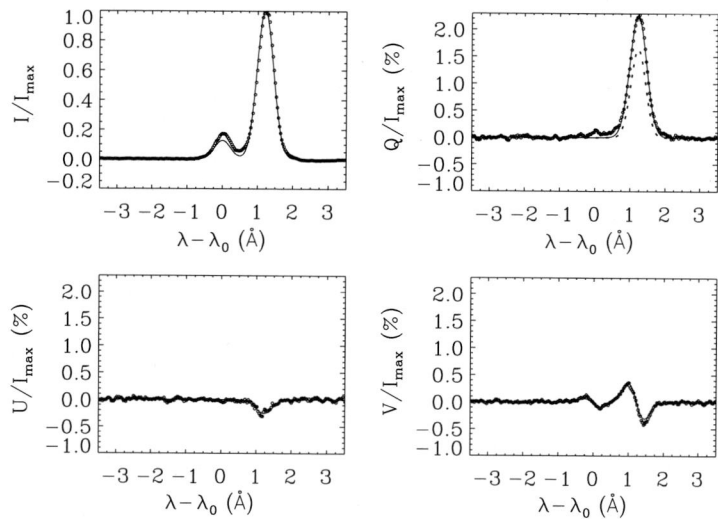

**Fig. 12.15.** Observed Stokes vectors of HeI $\lambda 1083.0\,\mathrm{nm}$ from a solar quiescent prominence (courtesy of [20])

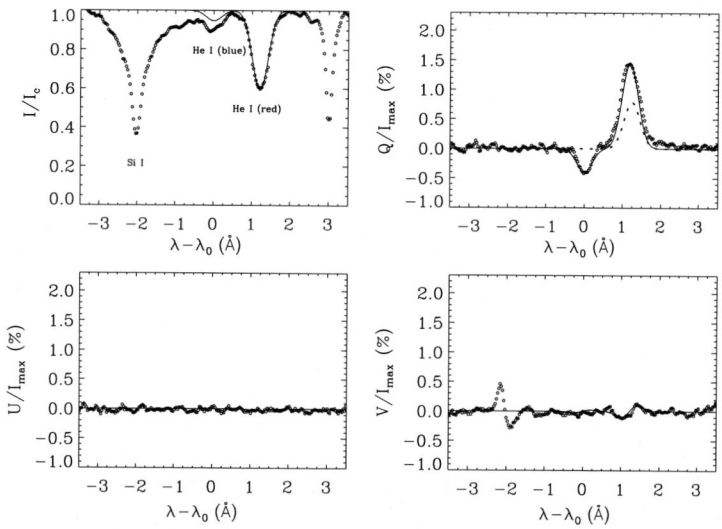

**Fig. 12.16.** Observed Stokes vectors of HeI $\lambda 1083.0\,\mathrm{nm}$ from a solar quiescent filament (courtesy of [20])

the atomic alignment in the metastable level $J_1 = 1$ is well visible in the Stokes $Q$ polarization of the radiation that is absorbed and reemitted by a filament in that same transition (Fig. 12.16). We also note that this polarization signal has opposite sign with respect to the signal from the other transitions in the multiplet, which have instead a polarizable upper level, as it was predicted on the basis of the simple model of Fig. 12.14.

The modeling of the polarimetric signature of HeI $\lambda1083.0$ nm has been applied with success to the diagnostics of magnetic fields in prominences and filaments [20], and more recently also to the investigation of the magnetism of chromospheric spicules [22]. The dichroism polarization of the 1–0 component of this multiplet also represents an important diagnostic of the plasma density, since the radiative-transfer modeling of such component is directly dependent on the optical depth of the absorptive plasma.

Finally, we point out that the Hanle-effect critical field for the upper state of HeI $\lambda1083.0$ nm is of the order of $10^{-4}$ T. Therefore this multiplet is generally formed in the saturated regime of the Hanle effect, in prominences and spicules, because the average field in these structures is of the order of $10^{-3}$ T. The Stokes polarization of HeI $\lambda1083.0$ nm in pure emission can thus be described through approximate formulae similar in structure to (12.21a)–(12.21d) (with the due changes for the atomic quantum numbers and the electric-dipole character of this transition), and its interpretation is thus generally affected by the problem of the 90° ambiguity illustrated at the end of Sect. 12.4.1 [22, 23].

In concluding this section on the two-level atom, we would like to comment briefly on the relevant properties of an atomic system where both the upper and lower levels are polarizable (e.g., the 1–1 atom). The SE problem of such a system is clearly much more complicated than in the case of the 0–1 and 1–0 systems presented above. For example, the possibility of polarization transfer from one level to the other introduces a whole new set of phenomena (e.g., the mechanism of *repopulation pumping* [10]). Nonetheless, the previous general conclusions concerning the onset of the Hanle effect in the presence of a magnetic field are still applicable in this case. In particular, under the condition $B_{lu} J_0^0(\omega_{ul}) \ll A_{ul}$, typical of permitted transitions in the visible spectrum, one must expect the appearance of two well-distinct magnetic regimes for the Hanle effects of the lower and the upper level, respectively (see Fig. 12.17).

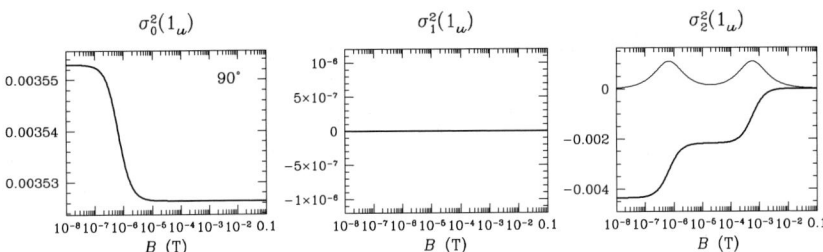

**Fig. 12.17.** Fractional alignment of the upper level, $J_u = 1$, of the 1–1 atom, as a function of the strength of a horizontal magnetic field. The illumination conditions are the same as in the computation of Figs. 12.9 and 12.13. In the rightmost panel, the two magnetic-field regimes for the Hanle effect of the lower level and the upper level are clearly distinguishable. The *thick* and *thin curves* correspond to the real and imaginary parts of $\sigma_Q^2$, respectively

# 12.5 Scattering Polarization from Complex Atoms: The Role of Level-Crossing Physics

The examples of two-level atoms illustrated in the previous section are very useful to understand the mechanism of depolarization of resonance radiation by a magnetic field. However, their applicability to real diagnostics of magnetic fields in astrophysical plasmas is very limited, precisely because of the intrinsic simplicity of the atomic systems involved. When we take into account the number of variables that contribute to the polarization signal of the resonance radiation that we collect from a distant source, we soon realize that the amount of information contained in the full Stokes vector of simple atomic transitions is often insufficient for even the simplest applications. The 90° ambiguity discussed in the case of coronal emission lines is an example of this limitation.

For this reason, plasma polarization diagnostics in solar physics often relies on multiline spectropolarimetry, as well as on the use of transitions from atoms with fine structure and hyperfine structure, where the complexity of the atomic structure gives rise to several new phenomena, induced by the presence of an external field. The inclusion of these phenomena in the modeling of resonance scattering polarization greatly improves the reliability of the inference of the magnetic properties of the emitting plasma.

In this section, we will consider two such phenomena: (1) the alignment-to-orientation conversion mechanism, which is capable of introducing a net amount of circular polarization in the scattered radiation, even when the atom is illuminated by unpolarized radiation and (2) the effect of electric-field-induced level interferences on the scattering polarization of hydrogen lines formed in a magnetized plasma.

## 12.5.1 The Alignment-to-Orientation Conversion Mechanism

We consider the case of an atom with fine structure in the approximation of $L$–$S$ coupling. The structure of such an atom is organized according to atomic terms $(\beta LS)$, where $\beta$ identifies the atomic configuration, and $L$ and $S$ are the quantum numbers for the orbital and spin angular momenta, respectively. Each atomic term is composed of a number of fine-structure levels, distinguished by the value of the total angular momentum, $J$, where $J = |L - S|, \ldots, L + S$.

In the presence of a magnetic field, the energy structure of the atom is modified by the effect of the magnetic Hamiltonian, $H_B = \mu_0 \boldsymbol{B} \cdot (\boldsymbol{J} + \boldsymbol{S})$. The best known among these magnetic effects is undoubtedly the Zeeman effect [2]. This is the lifting of the $(2J + 1)$-fold degeneracy of a $J$ level, because of the energy separation (Zeeman splitting) of its magnetic sublevels, proportionally to the azimuthal quantum number $M = -J, \ldots, J$, under the action of the magnetic field (see also Sect. 12.2).

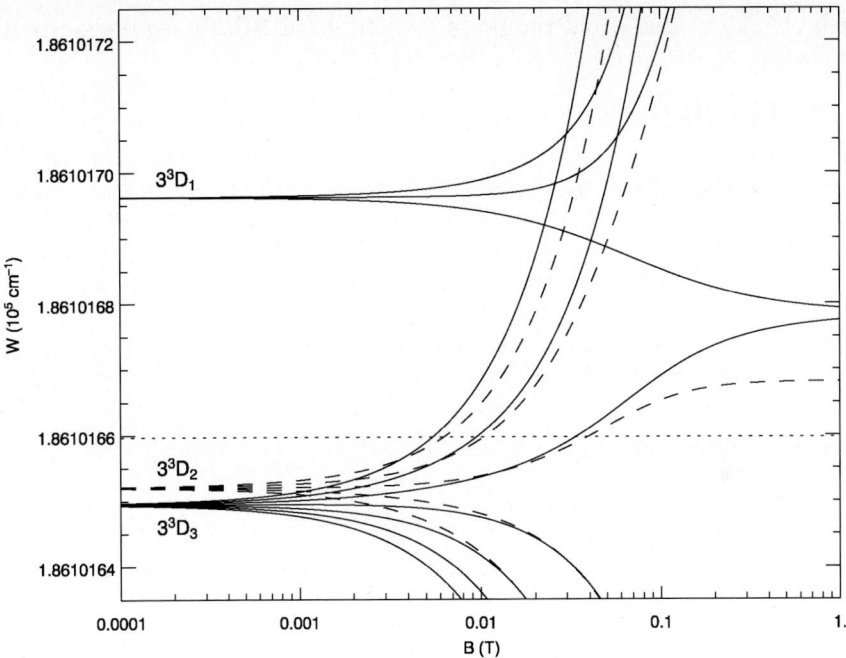

**Fig. 12.18.** Energy diagram for the term $3\,^3\mathrm{D}_{2,1,0}$ of the triplet species of HeI, as a function of the magnetic field. We notice the various crossings between the levels $3\,^3\mathrm{D}_2$ and $3\,^3\mathrm{D}_3$ above $B = 0.003\,\mathrm{T}$, which are responsible for important contributions to the net circular polarization of the HeI $\mathrm{D}_3$ line at $\lambda 587.6\,\mathrm{nm}$

In general, the magnetic Hamiltonian mixes levels with different $J$, and so the Zeeman effect provides a good representation of the energy structure of an atom with fine structure only when $|H_B| \ll |H_{\mathrm{FS}}|$, i.e., when the Zeeman splitting is much smaller than the typical energy separation between the $J$ levels of an $LS$-term. On the contrary, when the magnetic field is large enough so that $|H_B| \sim |H_{\mathrm{FS}}|$, the magnetic-field-induced $J$-mixing is responsible for significant modifications of the energy structure typical of the Zeeman effect, and level-crossing effects become observable in the polarization of scattered radiation (see Fig. 12.18). This is the regime of the so-called *incomplete Paschen-Back effect*. Because of the importance of $J$-mixing in the Paschen-Back regime, the atomic system must be described by density–matrix elements of the form $^{\beta LS}\rho_Q^K(J, J')$, to account for quantum interferences between distinct $J$ levels within the atomic term ($\beta LS$).

To understand how magnetic-induced $J$-mixing modifies the polarization of the scattered radiation, we must consider in some details the form of the SEEs for an atom with fine structure in a magnetic field. The subject is exhaustively treated in [6], so here we focus only on the contribution of the atomic Hamiltonian to the SEEs. This contribution is the so-called *magnetic kernel*, which arises from the $i\omega_{pq}\rho_{pq}$ term of (12.10). Its expression is the following,

$$
iN(\beta LS; JJ'KQ; J''J'''K'Q')
$$

$$
= \delta_{JJ''}\,\delta_{J'J'''}\,\delta_{KK'}\,\delta_{QQ'}\,i\omega_{JJ'}^{\beta LS} + \delta_{QQ'}\,i\omega_B\,(-1)^{J+J'-Q}\,\Pi_{KK'}\begin{pmatrix} K & K' & 1 \\ -Q & Q & 0 \end{pmatrix}
$$

$$
\times\left[\delta_{J'J'''}\,\Gamma_{LS}(J,J'')\begin{Bmatrix} K & K' & 1 \\ J'' & J & J' \end{Bmatrix} + \delta_{JJ''}\,(-1)^{K+K'} \right.
$$

$$
\left.\times\,\Gamma_{LS}(J''',J')\begin{Bmatrix} K & K' & 1 \\ J''' & J' & J \end{Bmatrix}\right], \qquad (12.27)
$$

where we defined $\Pi_{ab\cdots n} \equiv \sqrt{(2a+1)(2b+1)\cdots(2n+1)}$, and

$$
\Gamma_{LS}(J,J') = \delta_{JJ'}\,\Pi_J\sqrt{J(J+1)}
$$

$$
-(-1)^{L+S+J}\,\Pi_{JJ'S}\sqrt{(S+1)S}\begin{Bmatrix} J & J' & 1 \\ S & S & L \end{Bmatrix}. \qquad (12.28)
$$

The magnetic kernel weighs in the contributions from all the density–matrix elements $^{\beta LS}\rho_{Q'}^{K'}(J'',J''')$ to the SEE of the specific element $^{\beta LS}\rho_Q^K(J,J')$.

When quantum interferences between distinct $J$ levels are negligible, we can verify, using basic Racah algebra, that the magnetic kernel reduces to the term $i\omega_B Q$ of (12.16) or (12.24). In particular, we see that it can only be $K' = K$, so the magnetic kernel is strictly diagonal, and its role is simply that of depolarizing the element $^{\beta LS}\rho_Q^K(J,J)$ through the ordinary Hanle effect. This condition is typical of atomic structures where the fine-structure separation of the $J$ levels, $\omega_{JJ'}^{\beta LS}$, is much larger than both the Larmor frequency, $\omega_B$, and the energy widths of the levels, $A(LJ \to L_1J_1)$. In such case, the first depolarizing term of (12.27) dominates the SEE for the element $^{\beta LS}\rho_Q^K(J,J')$, when $J' \neq J$, with the result that such element vanishes at SE. For such atomic structures, it is then completely appropriate to adopt the formalism of the multilevel atom, described through the density–matrix elements $\rho_Q^K(J)$.

On the contrary, in more general cases, quantum interferences between distinct $J$ levels cannot be neglected. We see then that the magnetic kernel brings a contribution to the SEEs also when $|K' - K| = 1$. In particular, starting with an atom that is only aligned ($K = 2$), due to the effects of anisotropic illumination, it is possible to generate orientation ($K = 1$), when a magnetic field is also present [24, 25]. This *alignment-to-orientation conversion mechanism* (hereafter, A–O mechanism) is effective when $\omega_B \sim \omega_{JJ'}^{\beta LS}$. Figure 12.19 shows the atomic alignment and atomic orientation for the three atomic terms of HeI involved in the formation of the lines at $\lambda 1083.0\,\mathrm{nm}$ and $\lambda 587.6\,\mathrm{nm}$. Because in the example of that figure the atom is illuminated by

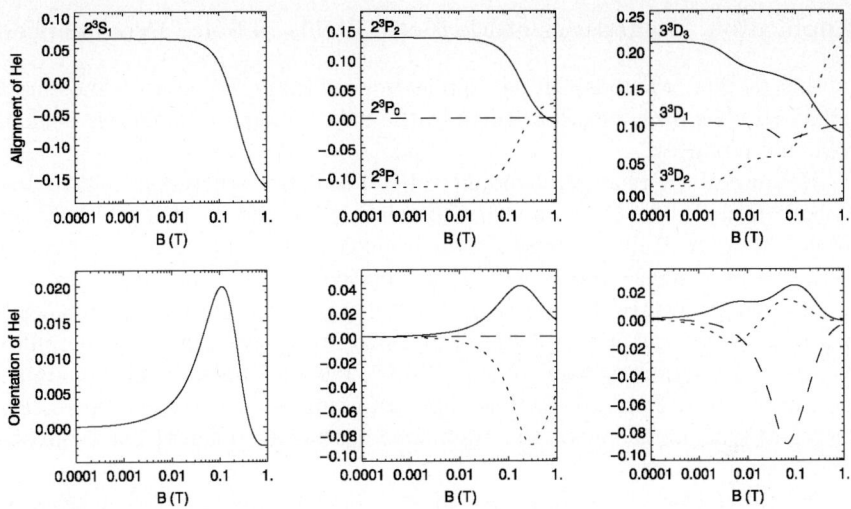

**Fig. 12.19.** Atomic alignment and orientation of the three atomic terms of the triplet species of HeI, which are involved in the formation of the HeI lines at $\lambda1083.0$ nm and $\lambda587.6$ nm. For this example, we considered a magnetic field directed along the vertical, and a scattering height $h = 0.05R$ (see Fig. 12.10.)

a cone of unpolarized radiation, the atomic orientation is totally due to the A–O mechanism.

In our discussion of the two-level atom, we had concluded that the atomic system could only be aligned, but not oriented, unless the incident radiation field was circularly polarized. For the case of a complex atom with fine structure, we just concluded that even if the illumination of the atom is unpolarized, though anisotropic, the presence of a magnetic field, and the ensuing A–O mechanism, is sufficient to create atomic orientation that modifies the polarization of the scattered radiation. In particular, one should expect the appearance of *net circular polarization* (NCP), instead of the perfectly anti-symmetric Stokes $V$ signal characteristic of the Zeeman effect.

The signature of the A–O mechanism in the circular polarization of several spectral lines of solar interest is of great diagnostic interest. Unlike the case of the Zeeman effect, where Stokes $V$ only carries information on the longitudinal component of the magnetic field vector, the NCP resulting from the A–O mechanism is sensitive to both the strength and direction of the magnetic field. Unfortunately, the reluctance of many solar physicists to get acquainted with the subject of non-LTE atomic physics has hindered the application of the A–O mechanism (and, more generally, of the Hanle effect) to the investigation of solar magnetism. In the worst cases, the detection of NCP in these lines has been dismissed as instrumental polarization (and, hence, "cleverly" removed), or otherwise interpreted as an indication of the presence of velocity gradients in the emitting plasma.

## Application: Diagnostics of Magnetic Fields in Solar Prominences

In this section, we consider the application of Hanle-Zeeman diagnostics of polarized lines from complex atoms to the measurement of magnetic fields in solar prominences.

Because of the relative abundance of HeI in these structures, the two lines at $\lambda 1083.0$ nm and $\lambda 587.6$ nm are of particular interest for this problem. Hydrogen lines (typically, H$\alpha$ and H$\beta$) have also traditionally been used, because of their strong signal. However, the large optical thickness of these lines, even in tenuous structures like prominences, generally complicates the diagnostic process (see the general problem of polarized radiative transfer, exemplified by the self-consistency loop of Fig. 12.7).[6] The HeI lines, in particular the $D_3$ line at $\lambda 587.6$ nm, are instead optically thin in most prominences, and therefore, the modeling of the resonance scattering polarization is greatly simplified [6, 27, 28].

Figure 12.20 shows an example of Stokes vector (normalized to peak intensity) of the HeI $D_3$ line, computed under typical prominence conditions.

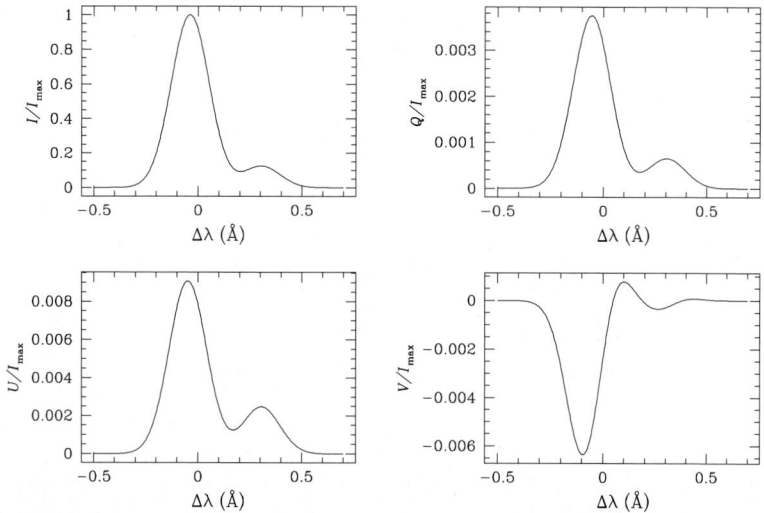

**Fig. 12.20.** Stokes profiles of the HeI $D_3$ line at $\lambda 587.6$ nm, for $90°$ scattering at a height of $h = 0.05R$ (see Fig. 12.10), in the presence of a magnetic field $B = 0.003$ T directed towards the observer. The plasma is supposed to be optically thin, and with a temperature $T = 10,000$ K. We note that the NCP induced by the A–O mechanism completely dominates Stokes $V$ at these field strengths, despite the fact that the contribution of the longitudinal Zeeman effect is maximum for the particular field geometry used for this example

---

[6] We should add that the sensitivity of hydrogen-line polarization to the presence of electric fields in the emitting plasma (e.g., the Holtsmark field from randomly distributed protons [26]) introduces a further element of complication to the modeling of these lines. This subject will be briefly discussed in the following section.

This line corresponds to the transition between the upper term $3\,^3D_{3,2,1}$ and the lower term $2\,^3P_{2,1,0}$ of the triplet species of HeI (see previous section). Because of the important departure of HeI from $LS$-coupling, the component $(J_u = 1) - (J_l = 0)$ is significantly separated in frequency from the other five components of the multiplet. The smaller peak in the profiles of Fig. 12.20 is totally due to this isolated component, whereas all the other components of the multiplet contribute to the larger peak.

The departure from the ordinary Zeeman effect for field strengths typical of solar prominences is immediately noticeable from the strongly asymmetric shape of the Stokes $V$ profile. This is the clear indication of the A–O mechanism, whereas the longitudinal Zeeman effect would result in perfectly antisymmetric shapes for both visible components of the line. As we anticipated in the previous section, the diagnostic advantage brought by the A–O mechanism is that, like the linear polarization signal, Stokes $V$ also becomes sensitive to both strength and direction of the magnetic field in the emitting plasma. The importance of the A–O mechanism as a diagnostic of magnetic fields in solar prominences was originally pointed out in [28].

Figure 12.21 clearly illustrates the dependence of the NCP of the HeI lines on the strength of the magnetic field in the emitting plasma, as well as on

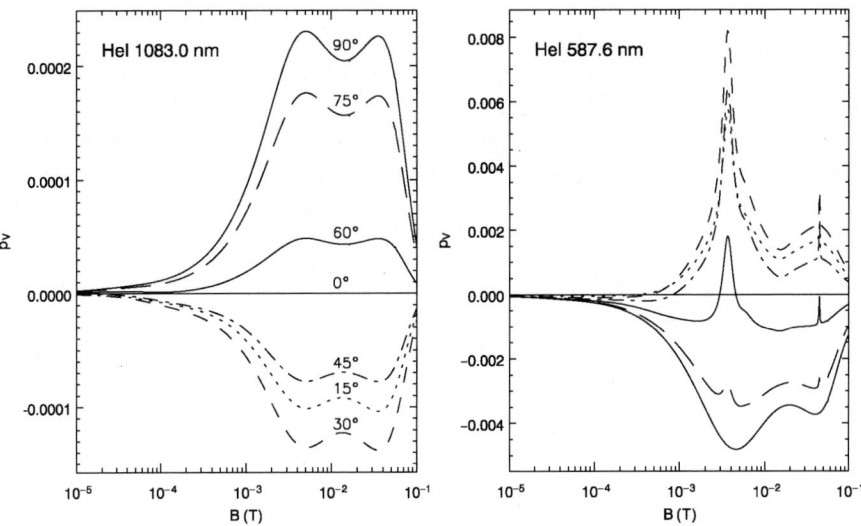

**Fig. 12.21.** NCP of the HeI lines at $\lambda 1083.0$ nm and $\lambda 587.6$ nm as a function of the strength of a magnetic field of different inclinations $\vartheta_B$ from the local vertical, and zero azimuth (see Fig. 12.10). We notice the strong resonance in the NCP of HeI D$_3$ ($\lambda 587.6$ nm) around $B \approx 0.0036$ T, for inclinations between $0°$ and $90°$, which is due to the first $\Delta M = 1$ crossing between the levels $3\,^3D_2$ and $3\,^3D_3$ (see Fig. 12.18). The second notable resonance around $B \approx 0.044$ T is associated instead with the first $\Delta M = 1$ crossing between the levels $3\,^3D_1$ and $3\,^3D_2$. The scattering geometry adopted for this figure is the same as for Fig. 12.20

its direction. We notice that the NCP is always very small for the HeI line at $\lambda 1083.0$ nm, indicating that Stokes $V$ is essentially dominated by the antisymmetric signal from the longitudinal Zeeman effect (see, e.g., Fig. 12.15). The level of NCP that can be attained in the HeI $D_3$ line for particular magnetic-field strengths and directions is instead significantly larger, and it can produce important departures from the antisymmetric shape of Stokes $V$, as demonstrated by Fig. 12.20. This is the reason why the HeI $D_3$ line is a much more powerful diagnostic of prominence magnetic fields than the HeI line at $\lambda 1083.0$ nm. Ideally, both lines should be used in order to increase the reliability of Stokes inversion in solar prominences, because the two lines also have significantly different magnetic regimes for the Hanle depolarization. In fact, the depolarization of HeI $\lambda 1083.0$ nm is already completely saturated for $B \approx 0.002$ T, whereas the depolarization for the $D_3$ multiplet does not reach saturation up to nearly 0.1 T.

Figure 12.22 shows a magnetic map of a solar prominence that was observed on May 25, 2002, using the Dunn Solar Telescope (DST) of the National Solar Observatory at Sacramento Peak (Sunspot, New Mexico, USA). The polarimetric data set of the HeI $D_3$ line used for this inversion was acquired with the High Altitude Observatory/Advanced Stokes Polarimeter deployed at the DST. The inversion was performed by fitting the full Stokes vector of HeI $D_3$ at each point in the map [29]. The spatial resolution is about 2 arcsec (corresponding to the units of the horizontal axis in each panel).

In the map, we report the magnetic field vector in the reference frame of the observer (see Fig. 12.10). For the intensity and inclination (left column, first two top panels), it is possible to give a proper estimate of the inversion errors (right column, first two top panels). The direction of the field projected on the plane of the sky (left column, third panel from the top) is instead affected by the azimuthal ambiguities discussed earlier in this chapter,[7] and for this reason, we omit the inversion error on this quantity. The analysis of the intensity profiles allows a rather direct estimate of the thermal broadening and of the wavelength displacement due to line-of-sight velocities. The two panels in the last row show the corresponding thermodynamic quantities derived from the Stokes inversion.

Finally, we adopted a simple model of polarized radiative transfer for the inversion of this map, assuming that the magnetic and thermodynamic properties of the prominence were constant along the line-of-sight. Although this simplified model does not allow a tomography of the prominence, it permits to include radiative transfer effects that improve significantly the model fit of the observed Stokes profiles. In fact, from the map of the optical depth, $\tau$, we conclude that in some regions of the prominence (with $\tau > 1$) the assumption

---

[7] In this map, the error in $\Phi_B$ is dominated by the 180° ambiguity. However, polarimetric noise affecting the observations creates the conditions for the 90° ambiguity to also be present [23], although it affects only a small fraction of the inverted points in the map.

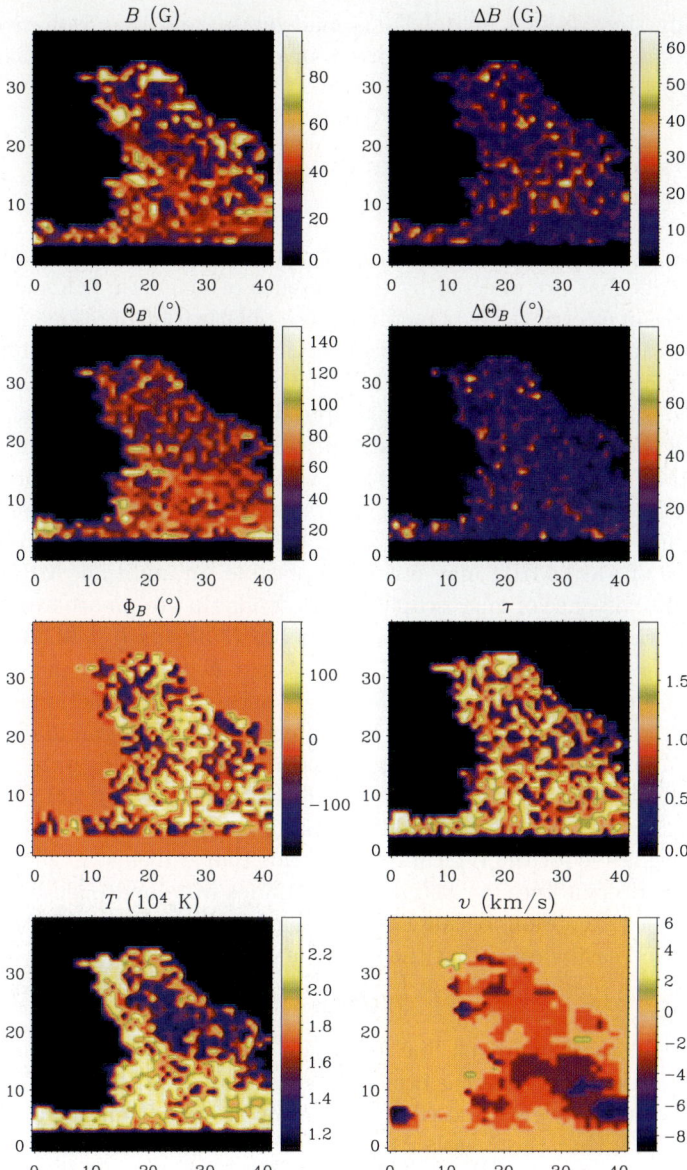

**Fig. 12.22.** Magnetic map of a quiescent prominence observed with the Dunn Solar Telescope (DST) of the Sacramento Peak Observatory, on May 25, 2002. For each point of the map, the full Stokes vector of the scattered radiation in the HeI $D_3$ line was recorded, using the Advanced Stokes Polarimeter deployed at the DST. By model fitting the spectropolarimetric data, it was possible to measure the magnetic field vector (see Fig. 12.10 for the definition of the angular parameters) – with an estimate of the inversion errors in the field strength and inclination with respect to the line-of-sight – as well as to measure the temperature and the line-of-sight velocity of the prominence, and also its optical depth in different regions

of optically thin plasma in HeI D$_3$ is inadequate, and that radiative transfer effects must be included for a correct inversion of the spectropolarimetric observations [30].

## 12.5.2 Hydrogen Polarization in the Presence of Magnetic and Electric Fields

Up to now, we have considered examples of atomic polarization in scattering environments that are affected by the presence of magnetic fields alone. The possible role of electric fields in modifying the polarization of resonance line radiation has so far been neglected for mainly two reasons: (1) According to the theory of magneto-hydrodynamics (MHD), the relaxation time of macroscopic distributions of electric charges in astrophysical plasmas must be very small, because of the characteristic low resistivity of a plasma. Therefore, no macroscopic, stationary electric fields should be expected, even in the presence of electric currents. Exceptions to this general rule can occur in nonequilibrium, fast-evolving plasma processes (e.g., in solar flares), where the fundamental hypotheses of ideal MHD may not be applicable. In addition, MHD cannot rule out the existence of microscopic electric fields that originate from charged perturbers randomly distributed in the plasma (*Holtsmark field* [26]), or the presence of *motional electric fields*, of the form $\boldsymbol{E} = \boldsymbol{v} \times \boldsymbol{B}$, in the atomic reference frame, when the plasma is moving across the magnetic field lines.[8] (2) The large majority of spectral lines of interest for magnetic diagnostics comes from atomic species that are sensitive to electric fields only to second order of stationary perturbation theory (*quadratic Stark effect*). Therefore, one can assume that electric modifications of the polarization of these lines must be negligible for the typical electric fields that can exist in astrophysical plasmas. Hydrogen lines, as well as spectral lines from hydrogen-like species, are an obvious exception, because of their sensitivity to the *linear Stark effect*, and one could easily argue that the importance of this effect cannot be underplayed, on the simple basis that hydrogen is by far the principal constituent of our universe.

The problem of hydrogen line formation in the simultaneous presence of magnetic and electric fields has been of interest to plasma physicists for several decades, mostly in view of possible applications to fusion energy studies [31–33]. It appears that these earlier papers focused exclusively on the problem of polarized line formation under LTE conditions, because they intended to provide a framework to understand the polarized self-emission of hot hydrogen plasmas embedded in stationary electromagnetic fields. To our knowledge, the more general problem of polarized scattering of hydrogen line radiation in the presence of simultaneous magnetic and electric fields was not considered until very recently [34].

---

[8] Because ideal MHD plasmas are frozen-in with the magnetic field lines, bulk velocities perpendicular to $\boldsymbol{B}$ are typically assumed to be very small.

A comprehensive investigation of the polarization properties of hydrogen lines formed under LTE conditions, in the presence of both magnetic and electric fields, was pursued by us nearly a decade ago [35–37]. These properties are very nicely summarized by the 1st- and 2nd-order moments of the Stokes profiles of hydrogen lines, which can be derived algebraically from the hydrogen Hamiltonian with external fields. The $n$th-order frequency moments of the Stokes profile $S$ $(S = I, Q, U, V)$ are defined as

$$\langle \omega^n(S) \rangle \equiv \frac{\sum_k s_k(S)(\omega_k - \bar{\omega})^n}{\sum_k s_k(I)} , \qquad (S = I, Q, U, V) \tag{12.29}$$

where $s_k(S)$ are the polarized intensities of the line components (e.g., for a transition between Bohr's levels $n$ and $m$, there are $2n^2 m^2$ fine-structure components of the line), and $\bar{\omega}$ is the center of gravity of the intensity profile in the absence of the external fields. Given the field vectors $\boldsymbol{B} \equiv (B, \vartheta_{\boldsymbol{B}}, \varphi_{\boldsymbol{B}})$ and $\boldsymbol{E} \equiv (E, \vartheta_{\boldsymbol{E}}, \varphi_{\boldsymbol{E}})$ in a reference frame of choice (e.g., the frame $(O, x', y', z')$ of Fig. 12.10), the algebraic expressions of the 1st- and 2nd-order moments of the Stokes profiles of a transition between Bohr's level $n$ and $m$ (neglecting the hyperfine structure and the diamagnetic term) are [36]

$$\langle \omega(I, Q, U) \rangle = 0 , \tag{12.30a}$$
$$\langle \omega(V) \rangle = -\omega_B \cos \vartheta_{\boldsymbol{B}} , \tag{12.30b}$$

and [37]

$$\langle \omega^2(I) \rangle = \langle \omega^2(I) \rangle_{\mathrm{FS}} + \frac{1}{2} \omega_B^2 (1 + \cos^2 \vartheta_{\boldsymbol{B}})$$
$$+ \omega_E^2 \left[ A_0(n, m) - \frac{1}{2} A_2(n, m)(3 \cos^2 \vartheta_E - 1) \right] , \tag{12.31a}$$

$$\langle \omega^2(Q) \rangle = -\frac{1}{2} \omega_B^2 \sin^2 \vartheta_{\boldsymbol{B}} \cos 2\varphi_{\boldsymbol{B}}$$
$$+ \frac{3}{2} \omega_E^2 A_2(n, m) \sin^2 \vartheta_E \cos 2\varphi_E , \tag{12.31b}$$

$$\langle \omega^2(U) \rangle = \langle \omega^2(Q) \rangle \left\{ \cos 2\varphi_{B,E} \to \sin 2\varphi_{B,E} \right\} , \tag{12.31c}$$

$$\langle \omega^2(V) \rangle = 0 . \tag{12.31d}$$

The fine-structure contribution to the broadening of the intensity profile, $\langle \omega^2(I) \rangle_{\mathrm{FS}}$, and the dimensionless coefficients, $A_0(n, m)$ and $A_2(n, m)$, are conveniently tabulated for all hydrogen transitions up to $n = 50$ [38]. In writing (12.31b) and (12.31c), we assume that the reference direction of positive $Q$ is along the meridional plane of the reference frame (e.g., the $x$-axis of Fig. 12.10).

From the above expressions, we can draw immediately some basic conclusions about the effects of magnetic and electric fields on the polarization of hydrogen lines formed under LTE conditions: (1) The electric fields do not contribute to the 1st-order moments of the Stokes profiles. In fact, only the center of gravity of Stokes $V$ is modified by the presence of a longitudinal magnetic field, through a contribution which is identical for all hydrogen lines.[9] (2) Both magnetic and electric fields contribute to the 2nd-order moments of the Stokes profiles, and their respective contributions are independent (i.e., there are no mixed terms of the form $\omega_B \omega_E$). While the magnetic field again brings a contribution to line broadening, which is identical for all hydrogen lines, the electric contribution depends instead on the particular atomic transition, through the coefficients $A_0(n, m)$ and $A_2(n, m)$. This different behavior is fundamental for any LTE diagnostic method aimed at the simultaneous measurement of magnetic and electric fields in a plasma, since it is possible in principle to distinguish between the magnetic and electric contributions by analyzing the full Stokes vector of two or more hydrogen lines.

Diagnostic methods for estimating electric fields in the solar atmosphere, based on the measurement of the 1st- and 2nd-order moments of polarized hydrogen lines, were applied nearly a decade ago [39, 40]. However, the need for high Stark sensitivity has limited the applicability of these methods to large-$n$, infrared transitions (e.g., 18–3 at $\lambda843.8$ nm, or 15–9 at $\lambda11.54\,\mu$m), which are typically very hard to detect even in the brightest solar features. For this reason, successful applications of these LTE diagnostic methods are scarce at best [39], and one should ask ourselves whether there could be significant advantages in looking at the resonance scattering polarization in more accessible hydrogen lines instead.

While magnetic effects on the resonance scattering polarization in hydrogen lines are rather well known [41], the possible role of electric fields has generally been ignored. The first comprehensive study of the effect of electric fields on the scattering polarization of the Lyman $\alpha$ line can be found in [42]. In that work, the model atom was limited to the lowest two Bohr levels, and the ground level of hydrogen was assumed to be unpolarized. With this simplified model, the authors were able to show that the presence of electric fields in a hydrogen plasma that is illuminated anisotropically can also induce A–O, when the electric Larmor frequency is of the order of the Lamb shift separation between the $2\,^2S_{1/2}$ and the $2\,^2P_{1/2}$ levels ($E \approx 3 \times 10^4\,$V m$^{-1}$). As a consequence, a significant level of NCP can be produced in the Lyman $\alpha$ radiation in the presence of an electric field, even when magnetic fields are absent. This is clearly illustrated in Fig. 12.23, which shows the NCP of the Lyman $\alpha$ and Balmer $\alpha$ lines for a 90°-scattering event, as a function of the electric-field strength. For the calculations in the figure, the field inclination and azimuth are $\vartheta_E = 45°$ and $\varphi_E = 90°$, respectively (see Fig. 12.10 for the definition of the field geometry), while the incident radiation correspond to a Planckian intensity at $T = 20{,}000$ K and maximum anisotropy.

More recently the general problem of resonance scattering from a hydrogen atom in the presence of simultaneous magnetic and electric fields of arbitrary

---

[9] In other words, the *effective Landé factor* is equal to 1 for all hydrogen lines [36].

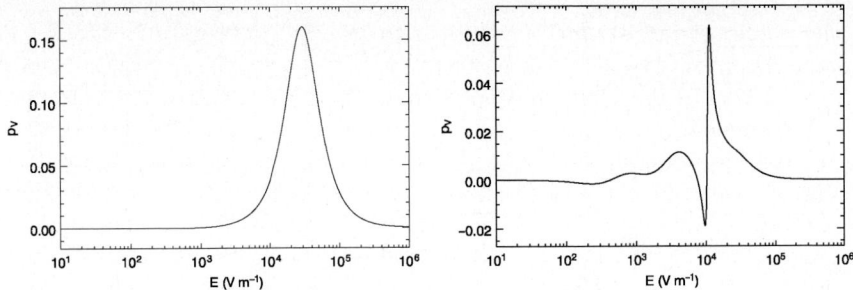

**Fig. 12.23.** NCP of the Lyman $\alpha$ (left panel) and Balmer $\alpha$ (right panel) lines of hydrogen for a $90°$-scattering event, as a function of the strength of an electric field, with $\vartheta_E = 45°$ and $\varphi_E = 90°$. The incident radiation corresponds to a Planckian intensity at $T = 20{,}000\,\mathrm{K}$ and maximum anisotropy. We note the broad NCP resonances due to the various interferences between Lamb-shifted levels. In the case of Balmer $\alpha$, a crossing between the $3\,^2\mathrm{P}_{1/2}$ and the $3\,^2\mathrm{D}_{1/2}$ levels is responsible for the sharp feature around $E = 10^4\,\mathrm{V\,m}^{-1}$

orientations has been the subject of extensive study [34]. A set of general equations describing the SE of such an atom, as well as its emissivity and absorptivity, were derived, allowing for the extension of the model atom to an arbitrary number of Bohr levels (only limited by computational resources), and for the modeling of several interesting lines of the lowest hydrogen series. The NCP curves of Fig. 12.23 were calculated adopting such formalism, with an atomic model of hydrogen including the first three Bohr levels.

One of the major results of the investigation in [34] is that the effects of the magnetic and electric fields on the resonance scattering polarization do not simply combine linearly (unlike the LTE case examined above), but instead they give rise to complicated patterns of resonances in the atomic polarization, which are of high diagnostic interest. Figure 12.24 illustrates the atomic orientation, $\rho_0^1(LJ, LJ)$, for the first four Bohr levels of the hydrogen atom, as a function of the strength of a vertical magnetic field (i.e., aligned with the direction of the incident radiation). In the first column, we show the case of the magnetic field acting alone, whereas in the two other columns, an electric field with $E = 100\,\mathrm{V\,m}^{-1}$ is also present, directed parallel to the magnetic field (second column) and perpendicular to it (third column). We notice the significant changes in atomic orientation introduced by the addition of an electric field, and also their dependence on the field geometry. In particular, we see that the magnetic regime for which the atom reaches a maximum of atomic orientation – e.g., in the level $n = 3$ – is shifted in general towards significantly lower strengths.

This effect has a direct physical explanation. The magnetic Hamiltonian is diagonal in $L$, and therefore the regime at which the A–O mechanism becomes efficient, when only a magnetic field is present, is determined by the fine-structure separation between the two $J$ levels within a term. For the

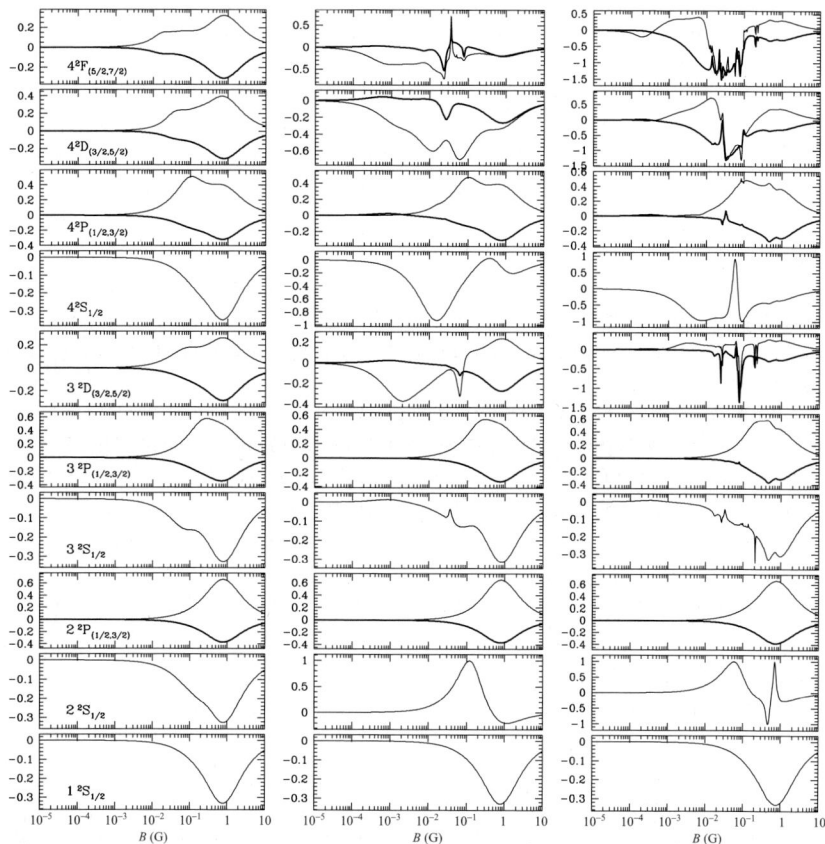

**Fig. 12.24.** Atomic orientation of hydrogen up to level $n = 4$ as a function of the strength of a vertical magnetic field ($\vartheta_B = 0°$): (left) without electric field, (center) with $E = 100\,\mathrm{V\,m^{-1}}$ parallel to $\boldsymbol{B}$ ($\vartheta_E = 0°$), and (right), with $E = 100\,\mathrm{V\,m^{-1}}$ perpendicular to $\boldsymbol{B}$ ($\vartheta_E = 90°, \varphi_E = 0°$). In the panels corresponding to two-level terms, the thicker curves indicate the levels with higher $J$ value

lowest terms of the hydrogen atom, this condition requires magnetic strengths of the order of 0.1–1 T (see left column in Fig. 12.24). When an electric field is also present, instead, $L$ is no longer a good quantum number, because the electric Hamiltonian mixes hydrogen terms with $\Delta L = 1$, due to its dipolar character. Therefore, levels belonging to adjacent terms can also interfere significantly, for appropriate magnetic- and electric-field strengths, and in particular, the A–O mechanism can also involve Lamb-shifted levels. Because the Lamb-shift separations within a given Bohr level are much smaller than the typical fine-structure separation, the magnetic strengths at which the A–O mechanism becomes efficient, in the additional presence of an electric field, are correspondingly smaller. In the case illustrated in Fig. 12.24, for instance, the orientation of the level $3\,^2D_{3/2}$ for parallel fields (central column) reaches an important negative peak already for $B \sim 10^{-3}\,\mathrm{T}$ .

This "electric enhancement" of the atomic orientation of hydrogen has important consequences for the polarization diagnostics of hydrogen lines. Because of the large magnetic-field strengths at which the A–O mechanism becomes efficient when the magnetic field is acting alone, the circular-polarization signal is typically dominated by the characteristic antisymmetric signature of the Zeeman effect. In the additional presence of an electric field, instead, the A–O mechanism can occur at significantly smaller magnetic strengths, so the circular polarization signal from the Zeeman effect can be completely dominated by the symmetric signature of the atomic orientation.

In conclusion, one must expect important modifications of both NCP and the shape of Stokes $V$, when electric fields of various geometries are also present in the emitting plasma. In particular, one should expect that this modification of the NCP of hydrogen lines be manifest also in the presence of microturbulent electric fields, like the Holtsmark field. Figure 12.25 shows that this is actually the case. Recent observations of anomalous levels of NCP in the Balmer $\alpha$ radiation from solar prominences [43] suggest that small electric fields – whether of the motional kind, or simply microturbulent – might be responsible for those signals, acting along with the prominence magnetic field $(B \lesssim 0.01\,\mathrm{T})$.

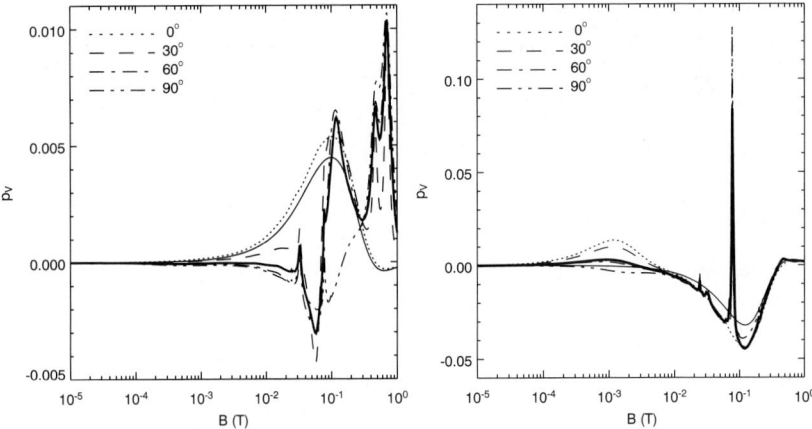

**Fig. 12.25.** NCP of the Lyman $\alpha$ (left) and Balmer $\alpha$ (right) radiation in a forward scattering event, as a function of the strength of a magnetic field directed along the line-of-sight, and in the additional presence of a randomly oriented electric field with $E = 100\,\mathrm{V\,m^{-1}}$. The *thin solid line* corresponds to the solution without electric fields. The *thick solid line* corresponds to the case of isotropically distributed electric fields, while the discontinuous lines show the contributions of azimuth-averaged electric fields of various inclinations $\vartheta_E$. Because of the symmetry properties of the scattering process, the curves for $\vartheta_E$ and $180° - \vartheta_E$ are identical

The sensitivity of hydrogen resonance radiation to the presence of microturbulent electric fields certainly complicates the magnetic diagnostics of astrophysical plasmas based on hydrogen lines, since the A–O mechanism changes its range of efficiency, ultimately depending on the electron density of the plasma. In turn, the atomic alignment is also modified in the presence of microturbulent fields, and therefore one must expect significant modifications of the linear polarization as well. Despite these difficulties, one can envisage a diagnostic procedure that takes advantage of both hydrogen lines and other resonance lines from atomic species that are instead insensitive to the linear Stark effect, like the HeI lines at $\lambda 1083.0$ nm and $\lambda 587.6$ nm. Such an approach, which opens to the possibility of a simultaneous diagnostics of magnetic fields and plasma densities, is the subject of current investigation.

# References

1. J.O. Stenflo, C.U. Keller, Nature **321**, 927 (1997)
2. M. Born, *Atomic Physics* (Blackie, London, 1958)
3. U. Fano, Rev. Mod. Phys. **29**, 73 (1957)
4. F. Mandl, G. Shaw, *Quantum Field Theory* (Wiley, New York, 1984)
5. E. Landi Degl'Innocenti, Solar Phys. **85**, 3 (1983)
6. E. Landi Degl'Innocenti, M. Landolfi, *Polarization in Spectral Lines* (Kluwer, Dordrecht, 2004)
7. J. Trujillo Bueno, R. Manso Sainz, Astrophys. J. **516**, 436 (1999)
8. L.D. Landau, E.V. Lifshitz, *Classical Theory of Fields* (Pergamon, New York, 1975)
9. W. Hanle, Z. Phys. **30**, 93 (1924)
10. W. Happer, Rev. Mod. Phys. **44**, 169 (1972)
11. R. Loudon, *The Quantum Theory of Light* (Clarendon, Oxford, 1983)
12. C. Cohen-Tannoudji, J. Dupont-Roc, G. Grynberg, *Atom-Photon Interactions* (Wiley, New York, 1992)
13. R. Manso Sainz, E. Landi Degl'Innocenti, Astron. Astrophys. **394**, 1093 (2002)
14. R. Manso Sainz, J. Trujillo Bueno, Phys. Rev. Lett. **91**, 1102 (2003)
15. D.M. Brink, G.R. Satchler, *Angular Momentum* (Clarendon, Oxford, 1993)
16. R. Becker, *Electromagnetic Fields and Interactions* (Dover, New York, 1982)
17. E. Landi Degl'Innocenti, Solar Phys. **91**, 1 (1984)
18. R. Casini, P.G. Judge, Astrophys. J. **522**, 524 (1999)
19. S. Tomcyzk, *Multi-Channel Polarimeter for Coronal Magnetic Field Measurements*, American Geophysical Union, Fall Meeting 2003 (2003)
20. J. Trujillo Bueno, E. Landi Degl'Innocenti, M. Collados, L. Merenda, R. Manso Sainz, Nature **415**, 403 (2002)
21. H. Lin, M. Penn, J. Kuhn, Astrophys. J. **493**, 978 (1998)
22. J. Trujillo Bueno, L. Merenda, R. Centeno, M. Collados, E. Landi Degl'Innocenti, Astrophys. J. **619**, 191 (2005)
23. R. Casini, R. Bevilacqua, A. López Ariste, Astrophys. J. **622**, 1265 (2005)
24. J.C. Lehmann, J. Physique **25**, 809 (1964)
25. J.C. Kemp, J.H. Macek, F.W. Nehring, Astrophys. J. **278**, 863 (1984)
26. H.R. Griem, *Spectral Line Broadening by Plasmas* (Academic, New York, 1974)

27. V. Bommier, S. Sahal-Bréchot, Astron. Astrophys. **69**, 57 (1978)
28. E. Landi Degl'Innocenti, Solar Phys. **79**, 291 (1982)
29. R. Casini, A. López Ariste, S. Tomczyk, B.W. Lites, Astrophys. J. **598**, L67 (2003)
30. A. López Ariste, R. Casini, Astrophys. J. **575**, 529 (2002)
31. Nguyen-Hoe, H.-W. Drawin, L. Herman, J. Quant. Spectrosc. Radiat. Transfer **7**, 429 (1967)
32. C.A. Moore, G.P. Davis, R.A. Gottscho, Phys. Rev. Lett. **52**, 538 (1984)
33. F.M. Levinton, G.M. Gammel, R. Kaita, H.W. Kugel, D.W. Roberts, Rev. Sci. Instr. **61**, 2914 (1990)
34. R. Casini, Phys. Rev. A **71**, 062505 (2005)
35. R. Casini, E. Landi Degl'Innocenti, Astron. Astrophys. **276**, 289 (1993)
36. R. Casini, E. Landi Degl'Innocenti, Astron. Astrophys. **291**, 668 (1994)
37. R. Casini, E. Landi Degl'Innocenti, Astron. Astrophys. **300**, 309 (1995)
38. R. Casini, Astron. Astrophys. Suppl. Ser. **114**, 363 (1995)
39. P.V. Foukal, B.B. Behr, Solar Phys. **156**, 293 (1995)
40. R. Casini, P.V. Foukal, Solar Phys. **163**, 65 (1996)
41. V. Bommier, S. Sahal-Bréchot, Solar Phys. **78**, 157 (1982)
42. B. Favati, E. Landi Degl'Innocenti, M. Landolfi, Astron. Astrophys. **179**, 329 (1987)
43. A. López Ariste, R. Casini, F. Paletou, S. Tomczyk, B.W. Lites, M. Semel, E. Landi Degl'Innocenti, J. Trujillo Bueno, K.S. Balasubramaniam, Astrophys. J. **621**, 145L (2005)

# 13

# Electromagnetic Waves

R.M. More, T. Kato, Y.S. Kim, and M.G. Baik

Real plasmas are often heated with high-power electromagnetic waves with frequencies closely related to the Zeeman splitting. In this chapter, we ask whether high-power microwaves can alter the polarization properties of light emitted from a plasma.

## 13.1 Introduction

The chapter began with a seminar discussion of a mysterious polarization spectrum from a Tokamak fusion experiment [1]. The spectrum seemed to imply that the main toroidal magnetic field in the machine had been rotated through a significant angle. In searching for a better explanation, the idea emerged that maybe high-power microwaves used for electron cyclotron-resonance heating could cause an atomic polarization to rotate.

In magnetic resonance imaging technology, nuclear spins in a magnetic field are controlled by an applied microwave field. Special microwave pulses are designed to invert the nuclear-spin polarizations or rotate them through a specified angle [2].

This raises a question: Can time-dependent electric and magnetic fields $E(t)$ and $B(t)$ in a plasma alter the polarization of emitted light?

To study this question, we wrote a small computer code to solve the time-dependent Schroedinger equation for an atom in time-dependent electric and magnetic fields. It is a pleasant exercise to write such a code. This chapter describes the code and shows some calculations performed with it.

The code is a tool, which can be used to answer several questions. If the electric and magnetic fields are constant, the code will describe Stark, Zeeman, and/or mixed Stark-Zeeman effects, as occur in the motional-Stark effect [3]. If the fields are constant plus an oscillating field, the code can be used to investigate the effect of microwave heating in a fusion machine. If the fields are pulsed, the code gives an approximate theory for excitation by a fast charged particle, and perhaps can give an estimate of the cross-section for polarized excitation of a ground-state atom (i.e., excitation to specific magnetic sublevels).

There are ideas for a new generation of high-intensity X-ray sources, and it would be interesting to predict the possible nonlinear optical processes involving high-intensity X-rays. Therefore we used the code to study two-photon absorption by hydrogenlike high-$Z$ ions [4]. We also compared the code described here to other methods of calculation [5].

This chapter can be thought of as a homework exercise for the preceding chapters: What happens to the atomic polarization when the ambient electric and/or magnetic fields are time-dependent? As with a homework question, we want to get an answer but do not need to survey the literature. The calculations seem correct as far as they go, but because they are so easy and straightforward, we cannot advance any claims of priority or originality. The calculations could be improved at the price of more formalism and computation.

## 13.2 Effect of Environment on Atomic Dynamics

Can there be a dynamical perturbation of hydrogen line emission ($n = 2$–$3$ transitions) by high-power microwaves? As much as 1–5 MW of 100 GHz ECR heating are pumped into typical magnetic fusion machines. These microwaves are approximately resonant with the Zeeman splitting in the main toroidal magnetic field. The rough numbers motivate this study, but when the calculation is finished, we do not have strong reason to believe the effects considered here are observed on existing machines. It would require tuning the microwaves to the Zeeman splitting and/or more intense microwaves. Of course both the microwave intensity and the main magnetic field vary in a fusion machine and there might be special locations where the resonance is strongest. As there are several Zeeman components, there are several possibilities for resonance coupling of the microwaves. In any case, the purpose of this chapter is to ask whether such a coupling could lead to any observable change of the emission.

### 13.2.1 An Atomic Computer Code

The computer code solves the time-dependent Schroedinger equation for the amplitudes of a set of states of a hydrogen-like atom in electric and magnetic fields. With a finite basis set, the time-dependent Schroedinger equation is a matrix differential equation. The computer code works with complex (Hermitian) matrices.

The one-electron basis states are hydrogenic wave-functions, $\phi_{n,l,m,\sigma}$. We work with the amplitudes of these basis states but never directly use the functions $\phi_{n,l,m,\sigma}(r)$. If the matrices are too large, the calculation is slow and so it is necessary to decide which states to include. Our attention will be on the hydrogen $3 \rightarrow 2$ transitions. We define large and small sets of states:

*Large set (60 states):*
1s, 2s,2p, 3s,3p,3d, 4s,4p,4d,4f, including spin states
*Small set (13 states):*
2s,2p, 3s,3p,3d, spin is ignored

All the results shown below are calculated with the "large" set. The computer code is written so that the states included can easily be altered.

## 13.2.2 Matrices for Quantum Operators

The most important operators are the Hamiltonian $H(t)$ and density–matrix $\rho(t)$, which describes the atomic state. The density–matrix is a matrix of products of amplitudes for the different basis states. The expectation value of an operator $O$ is expressed in terms of $\rho$ as

$$\langle O \rangle = \mathrm{Tr}[\rho O] . \tag{13.1}$$

We require matrices for the coordinate vector $\boldsymbol{r} = (X, Y, Z)$ and the orbital and spin angular momenta, $\boldsymbol{L}$ and $\boldsymbol{S}$:

$$\boldsymbol{r} = (X, Y, Z); \quad \boldsymbol{L} = (L_x, L_y, L_z); \quad \boldsymbol{S} = (S_x, S_y, S_z) .$$

The Hamiltonian $H(t)$ is the sum of three terms,

$$H(t) = H_0 + H_\mathrm{S} + H_\mathrm{d}(t) . \tag{13.2}$$

Here $H_0$ is the usual isolated-atom Hamiltonian for hydrogen. $H_0$ includes the Bohr energy $(= 13.6\,\mathrm{eV}\,n^{-2})$ and also includes spin–orbit interaction and first-order relativistic mass-shift. In our basis-set, the spin–orbit interaction is not diagonal. The Lamb shift is omitted.

$H_\mathrm{S}$ represents the perturbation produced by static electric and magnetic fields $\boldsymbol{E}_0$ and $\boldsymbol{B}_0$:

$$H_\mathrm{S} = -\boldsymbol{\mu} \cdot \boldsymbol{B}_0 - e\boldsymbol{r} \cdot \boldsymbol{E}_0 . \tag{13.3}$$

The fields $\boldsymbol{E}_0$ and $\boldsymbol{B}_0$ are vectors, which can point in any directions. The electron magnetic moment is taken to be

$$\boldsymbol{\mu} = \mu_0 \boldsymbol{L} + 2\mu_0 \boldsymbol{S} . \tag{13.4}$$

Here $\mu_0 = 5.78838 \times 10^{-5}\,\mathrm{eV\,T^{-1}}$ is the Bohr magneton. While (13.4) is a good approximation, it omits the "anomalous $g$-factor" of the electron, a small (0.1%) correction to the factor 2 in the spin moment. This omission is consistent with leaving out the Lamb shift.

In addition the total Hamiltonian includes the effect of time-dependent or dynamical electromagnetic fields:

$$H_\mathrm{d}(t) = -\boldsymbol{\mu} \cdot \boldsymbol{B}(t) - e\boldsymbol{r} \cdot \boldsymbol{E}(t) \ . \tag{13.5}$$

If we were only interested in constant fields, it would be most efficient to simply diagonalize the Hamiltonian and find the spectrum. If we were only interested in sinusoidal electromagnetic fields it would be natural to use the Floquet method, which exploits the periodicity of the Hamiltonian by solving only for the time-dependence in one cycle of the field. In (13.5), the radiation field is represented by a time-dependent classical electromagnetic field $\boldsymbol{E}(t)$ and $\boldsymbol{B}(t)$. In the Floquet method, it is possible to distinguish states of a quantum electromagnetic field containing different numbers of photons, but the calculations described here neglect such subtle quantum effects [6–8].

There are no transitions except those caused by radiation, i.e. there are no collisional processes. A single collision could be imitated by introducing a pulsed electromagnetic field.

The finite set of excited states is a severe limitation. This limitation means the code does not describe ionization. The code would make an inaccurate prediction of the static electric dipole polarizability, because there is a contribution from coupling to higher excited states (and even from the continuum). For weak perturbations that mainly mix the nearby levels, we can expect reasonable results. More accurate results could be obtained by introducing "pseudo-states" to imitate the effect of the omitted levels.

### 13.2.3 Density–Matrix Equation of Motion and Line Profile

We solve the time-dependent Schroedinger equation for the evolution operator $U(t)$, which generates a time-dependent wave-function

$$\psi(t) = U(t)\psi_0 \ . \tag{13.6}$$

If the total Hamiltonian $H(t)$ were independent of time, the evolution operator would be

$$U(t) = \exp(-\mathrm{i}Ht/\hbar) \ . \tag{13.7}$$

When $H(t)$ contains time-dependent electric or magnetic fields $U(t)$ is found by solving the matrix equation:

$$\mathrm{i}\hbar\frac{\mathrm{d}}{\mathrm{d}t}U(t) = H(t) \circ U(t) \ . \tag{13.8}$$

From the solution, we form the auto-correlation function of the emitting dipole:

$$\Phi_x(t) = \mathrm{Tr}\left[\hat{x}\hat{U}^+(t)\hat{x}\hat{U}(t)\hat{\rho}_0\right] = \langle\hat{x}(0)\hat{x}(t)\rangle \ , \tag{13.9}$$

where $\hat{x}$ is the position operator (and likewise for $\hat{y}$, $\hat{z}$ coordinates). The choices for the initial density matrix $\rho_0$ are discussed below. For an observer looking from the $y$ direction, the line profile has the form

$$P(\omega) = \frac{2e^2\omega^4}{3c^3} \mathrm{Re} \int_0^\infty e^{i\omega t} e^{-\lambda t} \left[ \Phi_x(t) + \Phi_z(t) \right] dt \ . \tag{13.10}$$

Since it is not yet decided what are the directions of the electric and magnetic fields, we hesitate to apply labels $\sigma, \pi$ to the two linear-polarized components; they are most precisely labeled as above by the directions of their electric field vectors. $\lambda$ is an artificial damping that makes a Lorentz line profile. Equations (13.9) and (13.10) are the standard theory of line-profiles due to Anderson, Kubo, and Baranger [9–11].

Nuclear magnetic resonance is often described by the more intuitive Bloch equations [12–14]. For a simple one-spin system, the Bloch equations and the quantum equations are exactly equivalent but would not be for the present case. The quantum equations are also only approximate because the set of excited states is limited to keep the matrices small ($60 \times 60$).

### 13.2.4 Computer Time

Depending on the size of the magnetic fields, we must follow the evolution for 40 ps to 1 ns to resolve the Zeeman splittings. However we want a small time-step $dt \sim 0.4 \times 10^{-17}$ s to resolve the characteristic time of the 1s electron. That means the code must take something like 10 million time-steps.

We solve (13.8) by a second-order difference scheme patterned on Simpson's rule. Unitarity of $U(t)$ is used as a test of the calculation. With double-precision FORTRAN, $U(t)$ remains unitary to better than a part per million even after $10^7$ time-steps; with ordinary (single-precision) compilation the calculation is faster but $U$ has errors ~0.03% at the end of a typical calculation. To obtain second-order accuracy, several matrix-multiplications are needed for each time-step. For a ($60 \times 60$) Hamiltonian, each matrix multiplication requires 216,000 complex arithmetic operations.

For these reasons, it is best to organize the calculation. We use home-made subroutines for complex-number matrix-algebra. We wrote simple subroutines to (1) set a matrix to zero, (2) make a linear combination of 2 or 3 matrices, (3) find the Hermitian adjoint of a matrix, and (4) multiply two matrices. To speed up the matrix-multiplications, we use pointers to the nonzero elements. When only the diagonal part of a matrix is needed, that is all that is calculated. For test-runs, the small set of states ($n = 2, 3$ only) can be used.

### 13.2.5 Atomic Data for Hydrogen

The code requires atomic data for the position $X, Y, Z$ and angular momentum operators $\boldsymbol{L}, \boldsymbol{S}$. For polarization phenomena *we even care about the signs,* and it is not completely satisfactory to take the formulas from books. Bethe-Salpeter [15] use special signs for spherical harmonics $Y_{lm}$ (for $m < 0$) and seem to have a typographical error for the sign of the radial matrix-element $R_{n,l}^{n,l+1}$ (see their (63.5)). The textbook of Landau-Lifshitz [16] does not give all the needed formulas. We use the Gordon formula [15] for radial matrix-elements of $r$, correct the sign for $n = n'$, adopt Landau-Lifshitz signs for $\langle n, l, m | R | n', l', m' \rangle$, and modify the Landau-Lifshitz formula for $\langle n, l || R || n'l' \rangle$; for safety everything must be *checked for consistency*.

We use matrix multiplication to check the algebraic properties of our matrices (here $L$ is in units of Planck's constant $\hbar$):

$$L_x^2 + L_y^2 + L_z^2 = L(L+1) \quad [L_x, L_y] = iL_z$$
$$[L_y, L_z] = iL_x \qquad [L_z, L_x] = iL_y$$
$$[X, L_z] = -iY \qquad [X, L_y] = iZ$$
$$[Y, L_x] = -iZ \qquad [Y, L_z] = iX$$
$$[Z, L_y] = -iX \qquad [Z, L_x] = iY$$
$$[Z, L_z] = 0$$

Our matrices are Hermitian and obey the equations listed.

The matrix-elements neglect any relativistic effects on the wave-functions. For hydrogen-like ions these nonrelativistic matrices are obtained from textbook formulas; to make a similar code for other ion species, there would be a need for atomic data.

## 13.2.6 Calculations

Figure 13.1 shows the calculated line profiles in a static magnetic field $B_z = 6\,\text{T}$. The initial density-matrix equally populates $n = 3$ states. The components of the emitted spectrum are labeled by the polarization vector of the emitted light. It is assumed the emitting atom is at $r = 0$ and is observed from the $y$ direction. In this case, the component $I_z$ is the $\pi$ component and the component $I_x$ is the $\sigma$ component, but in other cases these labels ($\sigma, \pi$) might be confusing. Figure 13.1 is the expected Zeeman pattern.

Figure 13.2a shows the components in a static electric field, $E_z = 1.5 \times 10^6\,\text{V m}^{-1}$; this is the usual Stark effect. Comparison cases are given in Schroedinger's original paper [17]. The spectrum has 9 main lines and 6 weaker lines that are visible only on a logarithmic plot. The line positions agree with the second-order formula for the Stark splitting [16]. A strong-field experimental spectrum given as Fig. 25c in Bethe-Salpeter [15] is not quite symmetrical and our calculation has a similar asymmetry.

The reason for this lack of symmetry is an interesting question. We could test two conjectures. The first idea is that $n = 4$ levels make a stronger perturbation of the $n = 3$ states than do $n = 2$ levels. To test this we repeat the calculation omitting $n = 4$ levels. The result is not visibly changed, i.e., is still asymmetric. The second idea is that the relativistic effects cause the

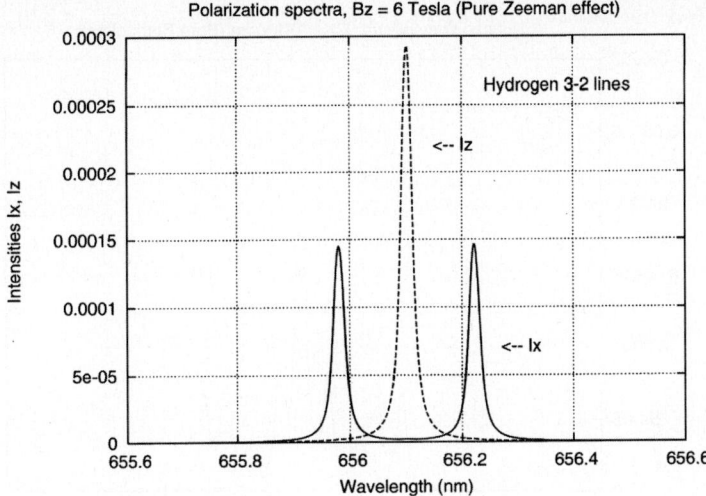

**Fig. 13.1.** Calculated profiles for $n = 3$ to $n = 2$ lines of hydrogen in a static magnetic field $B_z = 6\,\mathrm{T}$. The initial state is unpolarized. Linearly polarized components of the emitted light are labeled by the polarization vector of the emitted light. The emitting atom is at the origin and is observed from the $+y$ direction. $I_z$ is the $\pi$ component and $I_x$ is the $\sigma$ component, but in other cases these labels ($\sigma,\pi$) might be confusing. The intensities $I_{x,z}(\omega)$ are given in arbitrary units

asymmetry. To test this we repeat the calculation with spin-orbit interaction and relativistic mass-shift omitted from the zero-order Hamiltonian. The profile resulting from this calculation is symmetric to graphical accuracy as shown in Fig. 13.2b. We conclude the asymmetry is a real effect caused by relativity.

While it is possible to diagonalize the Hamiltonian for an atom in mixed static $E$, $B$ fields, we could not find convenient charts of the mixed Stark–Zeeman spectra in the literature. We give two examples here. The first case, Fig. 13.3, has $B_z = 6\,\mathrm{T}$, $E_z = 1.5 \times 10^6\,\mathrm{V\,m^{-1}}$. The second case, Fig. 13.4, has $B_z = 6\,\mathrm{T}$, $E_x = 1.5 \times 10^6\,\mathrm{V\,m^{-1}}$. These two spectra are surprisingly different.

For the "motional Stark" spectrum produced by a high-energy neutral-beam hydrogen atoms crossing a magnetic field, we find that both electric and magnetic fields are important.

(Note added by the Editors: The code to calculate the Stark–Zeeman profile in Chap. 2 reproduces the spectra of Figs. 13.1–13.4. So both calculations give identical results for constant or static fields.)

All the results presented here assume unpolarized populations of the $n = 3$ shell. The code shows that alignment or polarization changes the ratios of Zeeman components – just as expected. Alignment and polarization are imposed by choice of the initial density matrix. For normal (unpolarized) spectra, each $n = 3$ state has equal amplitude. For the aligned case, we assume

$$\rho_0 = 1 + \frac{1}{2}m^2$$

(a)

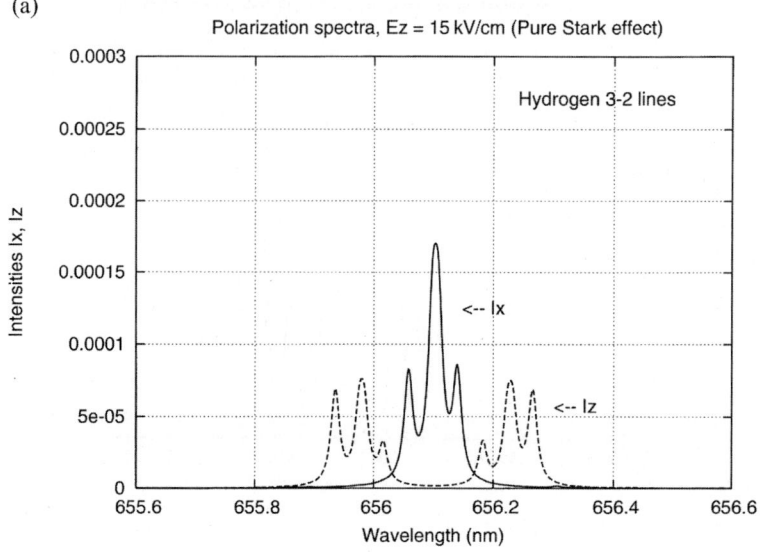

(b)

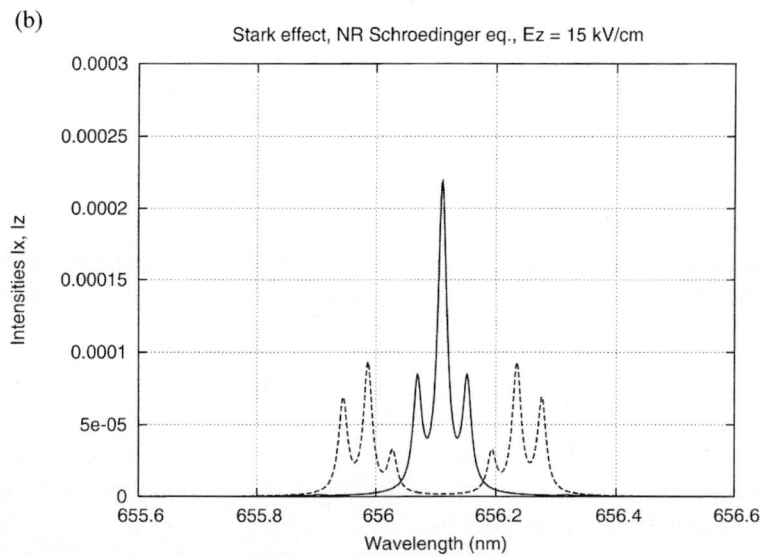

**Fig. 13.2.** Spectrum of hydrogen in a static electric field, $E_z = 1.5 \times 10^6 \,\mathrm{V\,m^{-1}}$. The spectrum has nine main lines and six weaker lines that are visible on a logarithmic plot. Comparison calculations show the asymmetry is caused by the fine-structure. **(a)** WITH fine-structure and **(b)** WITHOUT fine-structure

as the initial density matrix for $n = 3$ states (subsequently normalized). For a polarized initial state, we assume

$$\rho_0 = 1 + \frac{1}{2}m$$

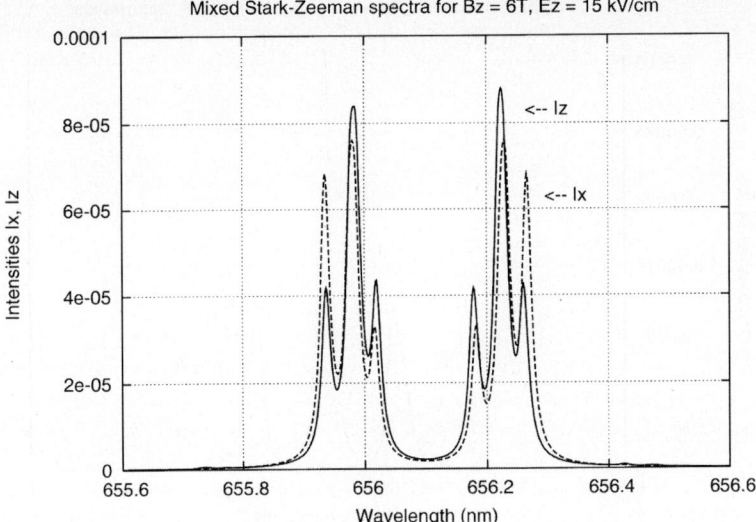

**Fig. 13.3.** Emission from hydrogen in mixed static electric and magnetic fields. Here $B_z = 6\,\mathrm{T}$, $E_z = 1.5 \times 10^6\,\mathrm{V\,m}^{-1}$

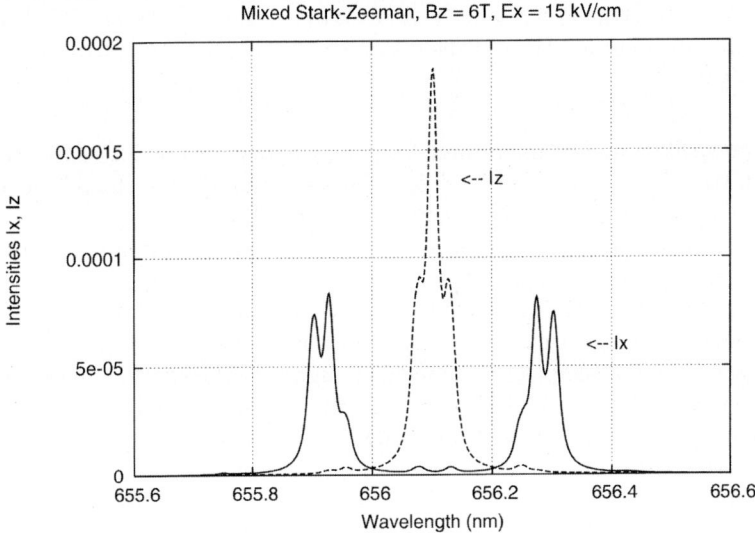

**Fig. 13.4.** Emission from hydrogen in mixed static electric and magnetic fields. Here $B_z = 6\,\mathrm{T}$, $E_x = 1.5 \times 10^6\,\mathrm{V\,m}^{-1}$

(again, subsequently normalized). Here, $m\hbar$ is the orbital $z$-component ($l_z$) and the spin-states are equally populated.

Finally, we come to the original question. We find that a resonant micro-wave field of high power $\sim 30\,\mathrm{kW\,cm}^{-2}$ has a $\sim 15\%$ effect on the hydrogen

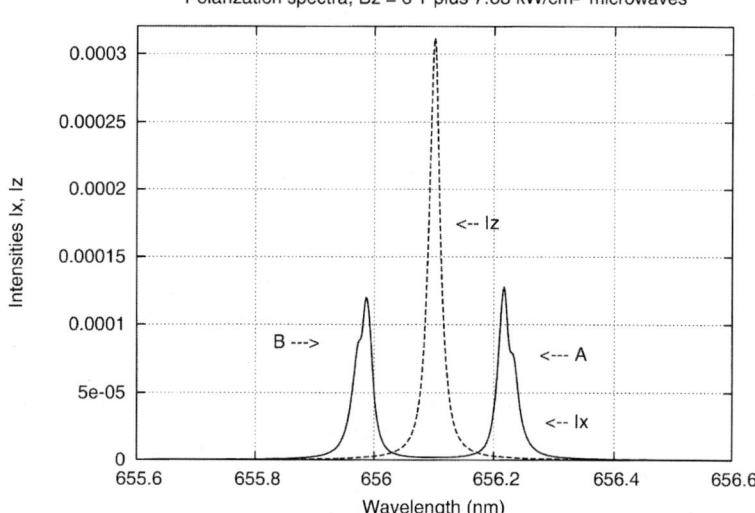

**Fig. 13.5.** Calculated spectrum for hydrogen in a 6 T static $B$ field (along $z$) irradiated by $7.63\,\mathrm{kW\,cm^{-2}}$ of microwaves. The microwave frequency is taken to be $5.17 \times 10^{11}\,\mathrm{rad\,s^{-1}}$ (82.3 GHz). The frequency chosen here is close to a resonance. In comparison with Fig. 13.1, we see the central ($\pi$) component has a higher peak intensity and the shapes of the $\sigma$ components are distorted (marked A and B in the figure)

Zeeman spectrum. The microwave frequency is taken to be $5.17\times10^{11}\,\mathrm{rad\,s^{-1}}$ (82.3 GHz). The change in the line-shapes is large and should be easily detected, but $30\,\mathrm{kW\,cm^{-2}}$ is a very high microwave intensity.

Figure 13.5 shows the calculated spectra for hydrogen in a 6 T static field (along $z$) irradiated by $7.63\,\mathrm{kW\,cm^{-2}}$ of microwaves. The electric field vector of the microwave field is assumed to be oriented at $\pi/4$ to the $x$-axis in the $x$–$y$ plane (the magnetic field is perpendicular to that also in the $x$–$y$ plane). Here again the line-shapes are changed enough that the effect should probably be observable. There are changes in the line shape (marked) and the central component has a higher peak intensity, in comparison to the pure-Zeeman spectra (Fig. 13.1).

Surprisingly in both cases, while the line-shapes are changed, the integrated intensities of the two components is not changed by the microwaves. However, because of background emission, the integrated intensity probably would be difficult to observe.

Figure 13.6 shows the effect of an microwave AC field frequency on the $H\alpha$ line profiles. The 3D picture shows the calculated spectra as functions of wavelength and AC frequency. This quantity plotted is the sum of the two polarization intensities. The microwave intensity and the polarization vector are the same as those of Fig. 13.5. It is clear that the microwave causes dramatic effect on line profiles.

Intensity (AU)

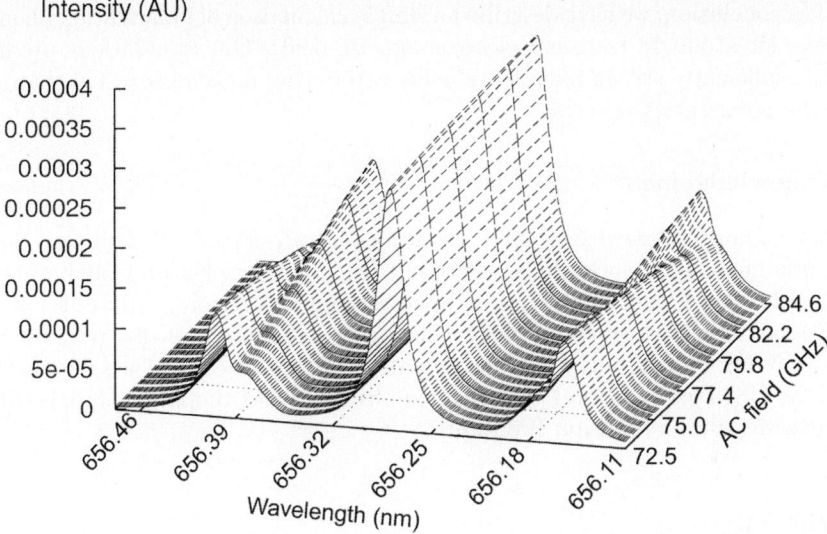

**Fig. 13.6.** The effect of an AC field frequency on the Hα line profiles. The magnetic field vector and the AC field are the same as that of Fig. 13.5. The spectra depend strongly on the microwave frequency

### 13.2.7 Limitations of the Calculations

The calculations use an accurate Hamiltonian and accurate numerical methods, but of course do not get an exact answer. It is important to understand the simplifications in the calculation that have an effect on the answers.

First, the set of atomic states is truncated. The truncation omits all continuum states and also highly-excited states. The truncation has an effect, for example, on the electric-dipole polarizability. The truncation error will be more or less serious for different atomic properties. For normal Stark/Zeeman spectra, it may not be a severe approximation. (Calculations with and without $n = 4$ levels for the Stark profiles gave indistinguishable results.) For Stark ionization or other strong-field effects, the truncation would have serious effects.

Second, we treat the electromagnetic field as a classical field. The full quantum theory is more complicated. In the quantum theory the EM field should be represented as a sum of amplitudes to have various numbers of photons, and the atomic equations will have matrix-elements that change the numbers of photons. A virtue of the Floquet method is that it can follow this physics for a limited class of radiation fields. In our case, the classical EM field is assumed, and it is usually thought that this approximation is good enough for "low frequencies" and high amplitudes of a coherent EM field. Specialists still study the precise conditions for applicability of the classical EM field in this domain.

In conclusion, we have described a simple calculation of polarization phenomena for atoms in transient electromagnetic fields. Our calculations predict that sufficiently strong microwave fields can indeed make observable changes in the polarization spectra.

## Acknowledgement

The authors acknowledge kind assistance of Professors T. Fujimoto and H. Summers. This work was presented at the US–Japan Fusion Collaboration Workshop on Plasma Polarization Spectroscopy, Livermore, California, June 2001, and the Korea–Japan Core Universities collaboration program workshop, NIFS, Japan, July 2001. Participation in those meetings was supported by the US–Japan Fusion Collaboration Program and the Korea-Japan Core Universities Collaboration Program.

## References

1. T. Fujimoto et al., private communication.
2. H. Lee, G. Wade (eds.), *Imaging Technology* (IEEE Press, New York, 1986)
3. D. Den Hartog, D. Craig, G. Fiksel, J. Reardon, V. Davydenko, A. Ivanov, in *Advanced Diagnostics for Magnetic and Inertial Fusion*, ed. by Stott et al. (Kluwer, Academic/Plenum Publishers, New York, 2002), pp. 237–240
4. R.M. More, H. Yoneda, in *Science of Superstrong Field Interactions*, ed. by K. Nakajima, M. Deguchi (AIP Conference Proceedings #634, Melville, New York, 2002), p.139
5. Y.S. Kim, J.S. Choi, B.I. Nam, M.G. Baik, T. Kato, I. Murakami, R.M. More, *Key Eng. Mater.* 277–279, 1049 (2005)
6. C. Joachain, in *Laser Interaction with Atoms, Solids and Plasmas,* ed. by R. More (NATO ASI series B 327, Plenum Press, New York, 1994)
7. C. Cohen-Tannoudji, J. Dupont-Roc, G. Grynberg, *Photons and Atoms* (John Wiley, NY, 1989); C. Cohen-Tannoudji, J. Dupont-Roc, G. Grynberg, *Processus d'interaction entre photons et atomes* (CNRS, Paris, 1988)
8. N. Kroll, in *Quantum Optics and Electronics*, Les Houches 1964 ed. by C. deWitt, A. Blandin, C. Cohen-Tannoudji (Gordon and Breach, New York, 1965)
9. R. Kubo, in *Many-Body Theory* (Tokyo Summer Lectures in Theoretical Physics) Part 1 (Syokabo, Tokyo and Benjamin, New York, 1966); see also [12]
10. P.W. Anderson, Phys. Rev. **76**, 647 (1949)
11. M. Baranger, Phys. Rev. **112**, 855 (1958); A.C. Kolb and H.R. Griem, Phys. Rev. **111**, 514 (1958); H.R. Griem, *Principles of Plasma Spectroscopy* (Cambridge University Press, Cambridge, 1997); T. Fujimoto, *Plasma Spectroscopy* (Clarendon Press, Oxford, 2004)
12. R. Kubo, T. Nagamiya (eds.), *Solid State Physics* (McGraw-Hill, New York, 1969)
13. C. Kittel, *Introduction to Solid-State Physics*, 7th ed. (Wiley, New York, 1996)
14. A. Abragam, *Principles of Nuclear Magnetism* (Clarendon Press, Oxford, 1961)

15. H.A. Bethe, E.E. Salpeter, *Quantum Mechanics of One- and Two-Electron Atoms* (Academic Press, New York, 1957)

16. L. Landau, E. Lifchitz, *Physique Theorique, t. 3, Mecanique Quantique*, 3rd ed. (Editions Mir, Moscou, 1988)

17. E. Schroedinger, *Collected Papers on Wave Mechanics*, 3rd ed. (Chelsea Publishing Company, New York, 1982)

# 14

## Instrumentation I

A. Iwamae

In PPS experiments, in contrast to conventional intensity spectroscopy, the light should be separated into two orthogonal linearly polarized components, and the intensities of each component should be determined with high accuracy. Since, in many cases, polarization degree is a couple of percent, uncertainty in the intensity determination should be low. Therefore, instrumentation is a very important element in PPS. In the visible-uv wavelength region, PPS observation can be practiced relatively easily, because various polarization selection optical components and detectors with high efficiency are available.

In this chapter, we review instrumentations in this wavelength region with examples of PPS observations. As mentioned above, experimental uncertainty is an essential issue in PPS observations, and we discuss this problem, too.

## 14.1 PPS Instrumentation in the UV–Visible Region

### 14.1.1 Sheet Polarizer and Narrow Bandpass Filters: Polarization Map

The most widely used linear polarizer is the so called *H-Sheet*. A sheet of polyvinyl alcohol impregnated with iodine is heated and stretched in a direction. The hydrocarbon molecules align in a straight long chain. The conduction electrons of the iodine atoms can move along the chain; this is analogous to the wire grid polarizer. The component of the electric field in the incident light that is parallel to the molecular chains is strongly absorbed, while the perpendicular component is transmitted. The transmission axis of the polarizer is therefore perpendicular to the direction in which the film is stretched. Figure 14.1 shows the spectral transmission of an H-sheet, the HN42 polarizer, having a neutral color. It can be used over the visible wavelength region, but it is less effective at the blue end.

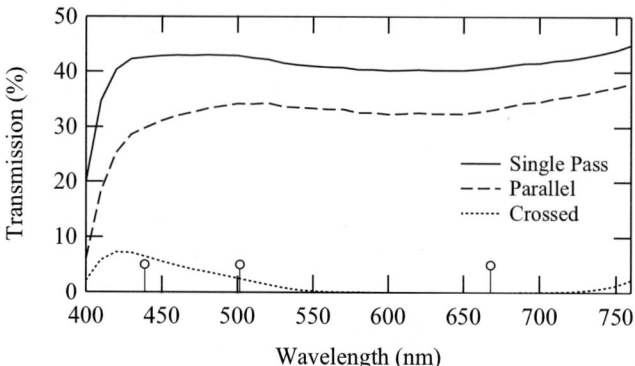

**Fig. 14.1.** Typical spectral transmission of a sheet polarizer HN42. *Solid curve:* Single path. *Dashed curve:* Transmission axes of the two polarizers are parallel. *Dotted curve:* Transmission axes of the two polarizers are perpendicular to each other. The wavelength positions of the three He I lines ($\lambda$438.8 nm, $\lambda$501.6 nm, and $\lambda$667.8 nm) are indicated with the *open circles* with bars

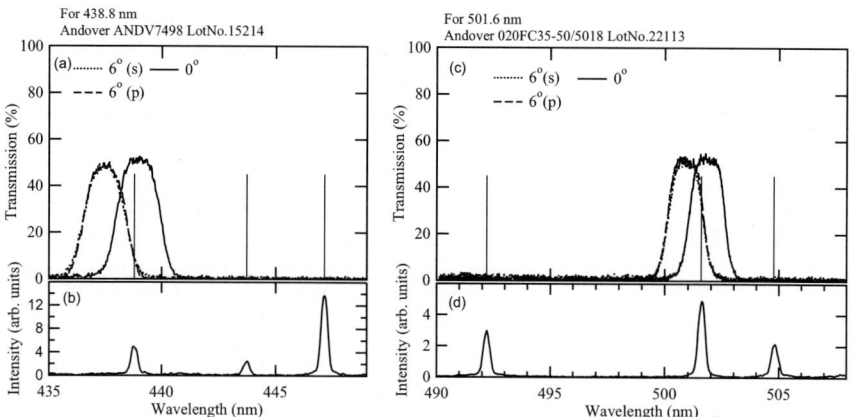

**Fig. 14.2.** Spectral transmission of the bandpass filters for emission lines $\lambda$438.8 nm and $\lambda$501.6 nm are shown in the upper panels (**a**) and (**c**), respectively. He I emission lines from a cusp plasma are shown in the lower panels (**b**) and (**d**). The filter is set at the tilted angle of $0°$ and $6°$ with respect to the line of sight. *Solid curves:* $0°$. *Dotted curves:* $6°$ with s-polarization. *Dashed curves:* $6°$ with p-polarization. The wavelength positions of emission lines $\lambda$438.8 nm ($2^1P - 5^1D$), $\lambda$443.8 nm ($2^1P - 5^1S$), and $\lambda$447.1 nm ($2^3P - 4^3D$) are indicated with the vertical lines in (**a**). The wavelength positions of emission lines $\lambda$492.2 nm ($2^1P - 4^1D$), $\lambda$501.6 nm ($2^1S - 3^1P$), and $\lambda$504.8 nm ($2^1P - 4^1S$) are indicated with the vertical lines in (**c**)

We discuss below the experimental procedure by which the polarization map of Fig. 1.1 is obtained. A narrow bandpass filter and the sheet polarizer are placed in front of the camera lens. Figure 14.2a,b shows the transmission curves of the dielectric bandpass filters used for HeI emission lines $\lambda$438.8 nm

$(2^1P - 5^1D)$ and $\lambda 501.6\,\text{nm}$ $(2^1S - 3^1P)$, respectively. With an increase in the angle of incidence the center wavelength shifts to shorter wavelength. The shift is expressed in terms of the angle of incidence $\theta$ and the effective refractive index, $N_{\text{eff}}$, of the filter:

$$\lambda_\theta = \lambda_0 \left\{ 1 - \left( \frac{N_{\text{ext}}}{N_{\text{eff}}} \right)^2 \sin^2 \theta \right\}^{\frac{1}{2}} , \qquad (14.1)$$

where $\lambda_\theta$ is the shifted wavelength from $\lambda_0$ which is the wavelength at the normal incidence and $N_{\text{ext}}$ is the refractive index of the external medium, which is virtually unity. From the shifts in Fig. 14.2, $N_{\text{eff}}$ is calculated to be 1.42 and 2.10 for $\lambda 438.8\,\text{nm}$ and $\lambda 501.6\,\text{nm}$, respectively.

Since the filter is used with a digital camera (Canon PowerShot G5), the effective field angle is rather wide. A diffuse reflectance standard plate (Labsphere: Spectralon, SRT-99-050) is illuminated with a helium spectrum lamp (Edmund: Spectrum Analysis Tube). Images of the diffuse plate recorded by the digital camera equipped with the H-sheet and one of the filters are shown in Fig. 14.3. The zoom of the camera lens is fixed at the wide-angle side 7.2 mm, ($f$35 mm for 35 mm film equivalent). The signal intensity is nearly uniform in the central region, and it decreases rapidly toward the four corners of the images, where the incident angle is larger than $\sim 7°$. Faint interference fringe circles appear, which is probably due to the bandpass filter.

The transmission efficiency of the narrow bandpass filter depends on the polarization of the incident light as shown in Fig. 14.2. The transmission efficiency of the s-polarized component is slightly lower than that of the p-component. Figure 14.4 shows the ratio of the observed signal intensity with the transmission axis set at $0°$ and $90°$. Within the field at which the angle of incidence is less than $7°$, the ratio is slightly larger than unity and almost constant. In the following we use the signal within this region.

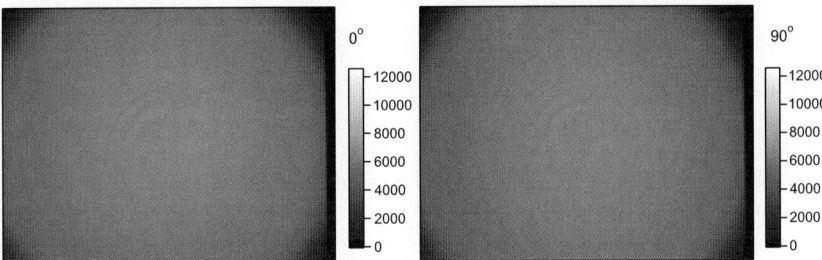

**Fig. 14.3.** The intensity images of the illuminated reflectance standard recorded with a linear polarizer and a narrow bandpass filter for $\lambda 501.6\,\text{nm}$. *Left*: the angle of the transmission axis is $0°$. *Right*: $90°$. The signal intensity decreases rapidly toward the four corners of the image. The black vertical strip on the right-hand side is the edge of the diffuse plate

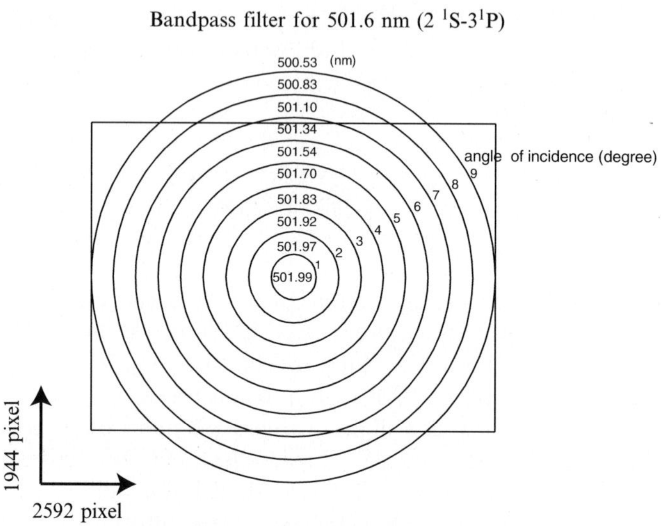

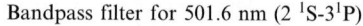

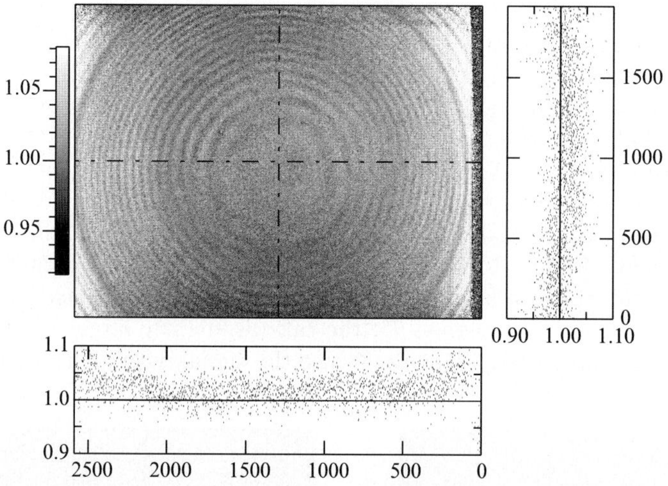

Bandpass filter for 501.6 nm (2 $^1$S-3$^1$P)

Fig. 14.4. (a) Angle of incidence of the light in the field view and the corresponding center wavelength of the transmission curve of the narrow bandpass filter for the $\lambda$501.6 nm line. (b) The signal intensity ratio $I_{0°}/I_{90°}$ is shown in the image. The ratios along $x = 1{,}296$ and $y = 972$, which are indicated with the single *dash-dotted lines*, are plotted at the bottom and on the right-hand side

The digital camera uses the Bayer grid of 3-color red–green–blue (RGB) filters. An AD converter of 12-bit digitizes the signal intensity for each pixel. An image with 2592×1944 pixels is recorded in a raw data format. The program based on *dcraw.c* [1] is used to process the raw data images. Figure 8.16a shows an image of a plasma produced in a cusp magnetic field taken with the

$\lambda 501.6\,\mathrm{nm}$ $(2^1\mathrm{S}\text{--}3^1\mathrm{P})$ emission line. Magnetic field lines on the plane perpendicular to the line of sight (LOS) including the plasma axis are also shown along with the ECR surface. Four images are taken with the transmission axis of the polarizer at $0°$, $90°$, $45°$, and $135°$. The signal digits over a square cell of $108{\times}108$ pixels are summed to yield signal intensities of this cell. The whole image consists of $24{\times}18$ cells. The four intensity data are processed and the Stokes parameters (See Appendix A) are calculated for each cell, where circular polarization is assumed absent. The polarization map thus obtained is shown in Figs. 1.1 and 8.16b.

## 14.1.2 Birefringent Polarizers

### A Calcite Plate Behind the Entrance Slit

Although the H-sheet is a handy polarizer, it has several shortcomings, the most serious of which is that it transmits only one of the two polarized components. The wavelength region is limited, too, as mentioned earlier.

Calcite, a calcium carbonate ($\mathrm{CaCO_3}$) crystal, has high birefringence and wide spectral transmission as shown in Fig. 14.5. The transmission curve, however, strongly depends on the quality of the crystal. Reasonable size rhombs are available for manufacturing a polarizer. It has also an advantage of being nonhygroscopic, so that protection from the humidity in the atmosphere is unnecessary, although it is fairly soft and easily scratched. When a narrow beam of light is sent into a calcite crystal normal to a cleavage plane, it splits into two components and they emerge as two parallel beams (Fig. 14.6). They are known as the ordinary (o-)ray, and the extraordinary (e-)ray.

The o-ray has its electric field vector perpendicular to the principal plane, which contains the optical axis, and the field vector of the e-ray is parallel to the principle plane. The refractive indices for the o-ray and the e-ray, $n_\mathrm{o}$ and $n_\mathrm{e}$, respectively, are shown in Fig. 14.7.

When the PPS experiment was performed on the WT-3 tokamak [2] (see Sect. 8.4.1), a calcite plate of 5.4 mm thickness was used and the crystal

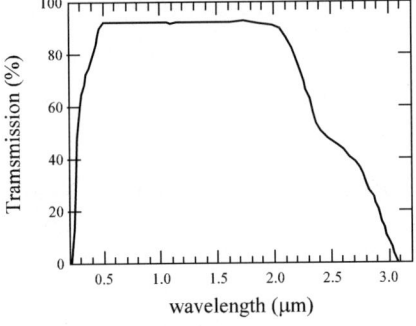

**Fig. 14.5.** Transmission curve of calcite

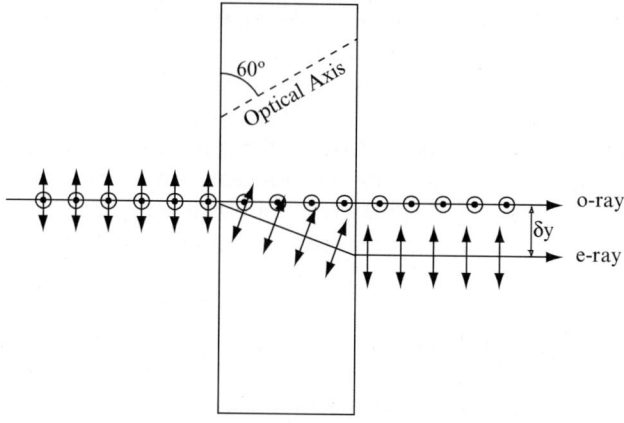

**Fig. 14.6.** A calcite plate used in the WT-3 experiment. The crystal optical axis is at 60° with respect to the surface, and the normal incident light is separated into the o-ray and e-ray [5]

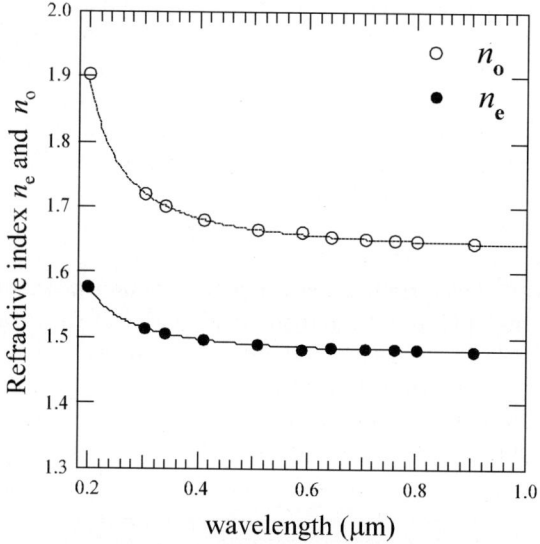

**Fig. 14.7.** Refractive indices of calcite; $n_o$ is for the ordinary ray and $n_e$ is for the extraordinary ray

optic axis was chosen at 60° with respect to the surface [3]. It was placed just behind the entrance slit of a spectrometer (Nihonbunko CT-100) with the crystal optical axis in the horizontal plane. The o-ray and the e-ray were separated in the horizontal direction. The spectrometer had 1 m focal length and $f/10$. The grating was holographic with 3,600 grooves/mm. The impurity emission lines from the plasma, CV $\lambda$227.792, 227.725, and 227.091 nm ($2s^3S_1 - 2p^3P_{1,0,2}$) and OV $\lambda$278.985, 278.699, and 278.101 nm

$(3s^3S_1 - 3p^3P_{0,1,2})$, were observed. An example of the results is shown in Figs. 8.13 and 8.14.

Although the above polarization separation scheme is direct and has high throughput in ultraviolet wavelength region, it has several disadvantages. The spectrometer should be located close to the plasma and the light should be transmitted or reflected with minimum influence on its polarization characteristics. The slit itself may even influence the polarization of the incident light. A calcite plate inserted into the optical path of the spectrometer affects its focusing characteristics, and the shift of the focusing positions at the exit of the spectrometer is slightly different for the e-ray and the o-ray. The different efficiencies of the grating for the two polarized components are also directly reflected. Absolute sensitivity calibration has been performed for the two polarized components for CT-100 with the imaging optics and an intensified-CCD camera. An external horizontal slit limited the height of the vertical slit. The calcite plate was placed with the crystal optic axis in the vertical plane. A spectral irradiance standard deuterium lamp (ORIEL 68840, 63945 Deuterium Calibrated Lamp) and a tungsten-halogen lamp (USHIO JPD100V-500WCS) were used with a diffuse reflectance standard (Labsphere Spectralon SRS-99-020). Spectral continuum light was observed in two horizontal images. The result is shown in Fig. 14.8. With an increase in wavelength, the efficiency of the horizontally polarized light (the p-light) increases whereas that of the vertically polarized (the s-light) decreases. Here p and s refers to the grating surface.

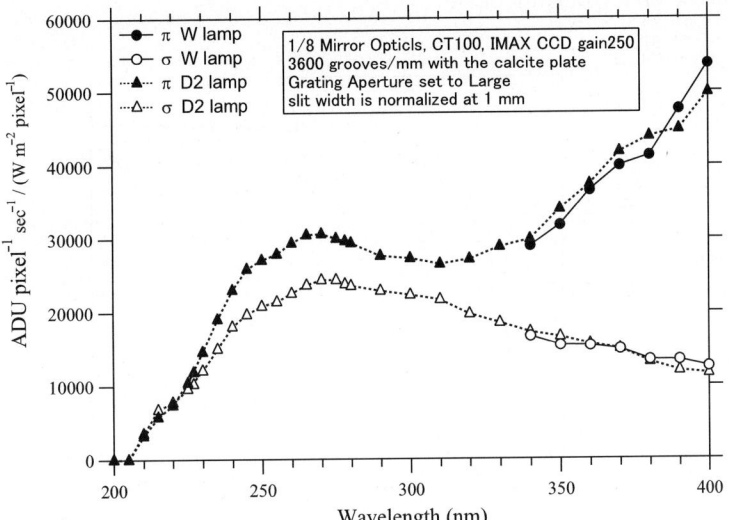

**Fig. 14.8.** The sensitivity calibration curves for p and s light. Relative efficiency from 200 to 400 nm is calibrated with a deuterium lamp. Absolute sensitvity is calibrated from 340 to 400 nm with a calibrated tungsten-halogen lamp and a white reflectance standard plate

## B Beam-Splitting Polarizer with Optical Fibers

Optical fibers are often used to transmit the radiation emitted from a plasma to a spectrometer. Linearly polarized light through a Glan–Taylor prism of $\lambda 253.7$ nm emission line from an Hg pen-ray lamp is transmitted through an 8-m 400 μm core multimode optical fiber (Mitsubishi cable STU-400-ES: UV grade). The output light from the fiber is almost unpolarized: the polarization degree of the output light is found to be 0.06. The intensity transmission also decreases with an increase in the length of the fiber. This decrease is pronounced in the ultraviolet wavelength region. The transmission of 200 nm light through an 8-m fiber of the UV grade is 28%, and a 25-m fiber transmits 2%.

Polarization of the light should be separated before entering optical fibers. A Glan–Thompson polarizer is made of two cemented prisms of calcite. The $n$-butyl methacrylate index matching cement ($n = 1.48$) results in transmission from 300 nm to beyond 2,500 nm (see Fig. 14.9). A beam-splitting Glan–Thompson polarizer shown in Fig. 14.10 is designed to transmit the e-ray beam undeviated and to reflect the o-ray beam to emerge from the side exit surface. The two orthogonally polarized output beams exit at normal incidence to their respective output faces. The beam separation angle of 45° is independent of wavelength. A Glan–Thompson polarizer with 2.5:1 length to aperture ratio has a full acceptance cone angle of more than 15°, symmetric about the input axis (see Table 14.1.). There is a slight refractive index mismatch of the cement for the extraordinary beam. Owing to this mismatch, a small portion ($<1\%$) of the extraordinary component is reflected and admixed into the ordinary output beam.

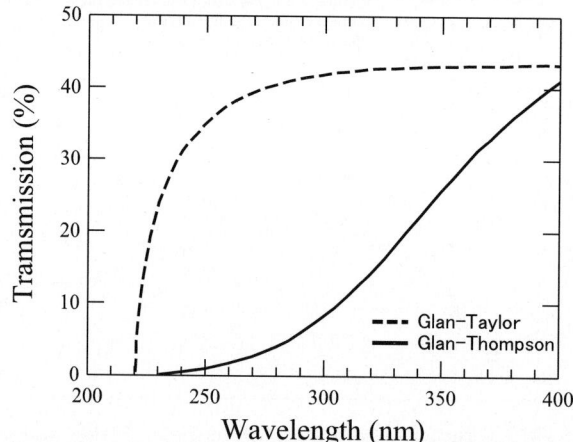

**Fig. 14.9.** Typical transmission curves of the Glan–Thompson (*solid*) and the Glan–Taylor (*dashed*) polarizers in the ultra-violet region. For the former, the absorption below 400 nm is caused by the index matching cement, which results in transmission above 300 nm

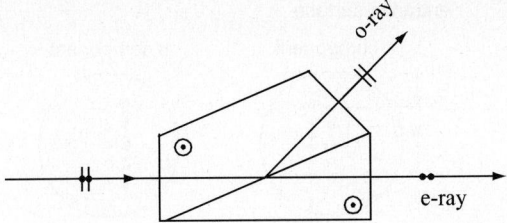

**Fig. 14.10.** A beam-splitting Glan–Thompson polarizer is designed to transmit the e-ray component beam undeviated and to reflect the o-ray component beam to the side exit surface. The optical axis of calcite is perpendicular to the plane

**Table 14.1.** Specification of Glan–Thompson polarizer and Glan–Taylor polarizer

| Design | Material | Interface | Extinction ratio | Field angle | Wavelength (nm) |
|--------|----------|-----------|------------------|-------------|-----------------|
| Glan–Thompson | Calcite | Cement | $10^{-5}$ | 15° | 300 − 3000 |
| Glan–Taylor | Calcite | Air Gap | $10^{-5}$ | 8° | 220 − 3000 |
| Glan–Taylor | $\alpha$ BBO | Air Gap | $<10^{-5}$ | 6° | 200 − 300 |

For the PPS measurements on the cusp plasma device (see Sect. 8.4.2), a beam-splitting Glan–Thompson polarizer is used for polarization separation of emission lines from helium atoms in the plasma. Almost all emission lines from the upper levels with $n \leqslant 6$ to lower $n = 2$ levels lie in the wavelength region from 300 to 730 nm, suitable for the Glan–Thompson polarizer. The exceptions are those of $2^3$S – $5^3$P and $2^3$S – $6^3$P lines. The large acceptance angle of the Glan–Thompson prism allows us to couple the light beam by a lens to the prism and further to optical fibers of numerical aperture NA = 0.20.

Figure 8.15 shows the experimental setup for the PPS experiment on the cusp ECR plasma [4]. The polarization separation optics system consists of a condenser lens ($f = 50$ mm), the Glan–Thompson prism and an optical-fiber bundle. Figure 14.11 shows the fiber arrangements on the entrance surface and the exit surface of the optical-fiber bundle. On each output side of the beam splitting Glan–Thompson prism, five optical fibers of 400 μm core diameter (Mitsubishi cable, STU-400: NA = 0.20) are aligned 1 mm apart each other in a column. Two branches of the optical fiber bundle for the o-ray and e-ray on the entrance side merge to form a single fiber bundle with a length of 8 m, which consists of 10 individual optical fibers. The polarization separation optics system is designed so that each corresponding pair of the o-ray ($\pi$ component) and e-ray ($\sigma$ component) falls on the same line of sight. For fine adjustment a beam of helium–neon laser is fed into the optical fibers from the exit surface side. An image of each 400 μm core at the entrance surface is

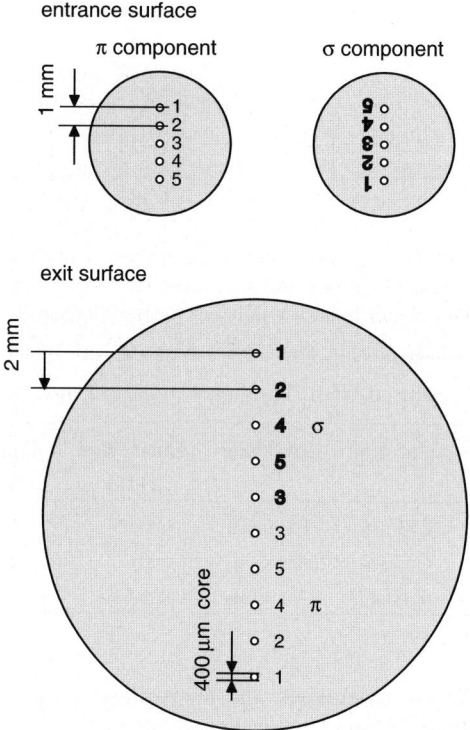

**Fig. 14.11.** Optical fiber bundle: entrance surface and exit surface

focused on a screen at the distance of 550 mm from the prism along the line-of-sight through the prism and the lens. The shape of the image of the e-ray component is a circle with a diameter of 4 mm. The shape of the reflected o-ray component is an ellipse with minor and major axes of 4 and 6 mm, respectively. The birefringence of the calcite causes the difference in the effective numerical apertures for different directions of the o-ray component. As shown in Fig. 14.11 at the exit end of the fiber bundle 10 optical fibers are aligned 2 mm apart each other in a column. The exit surface of the fiber bundle is placed at a distance of 2 mm from the entrance slit of a spectrograph (Nikon G500: $f = 500$ mm, $F/8.5$); this is for the purpose of focusing the fiber image on the spectrometer focal plane. A 1,200 grooves/mm grating provides reciprocal linear dispersion of 1.52 mm/nm near 500 nm in the first order. Figure 14.12 shows an example of ICCD images of the polarization resolved spectra. Three emission lines, HeI $\lambda$492.2 nm ($2^1$P-$4^1$D), $\lambda$501.6 nm ($2^1$S-$3^1$P), and $\lambda$504.8 nm ($2^1$P-$4^1$S), appear at five line-of-sight (LOS)s. Fifty vertical pixels corresponding to the spectral image of each optical fiber are binned and five pairs of the polarization separated spectra are obtained simultaneously. An example of the observed spectra is shown in Fig. 8.17a.

ICCD image

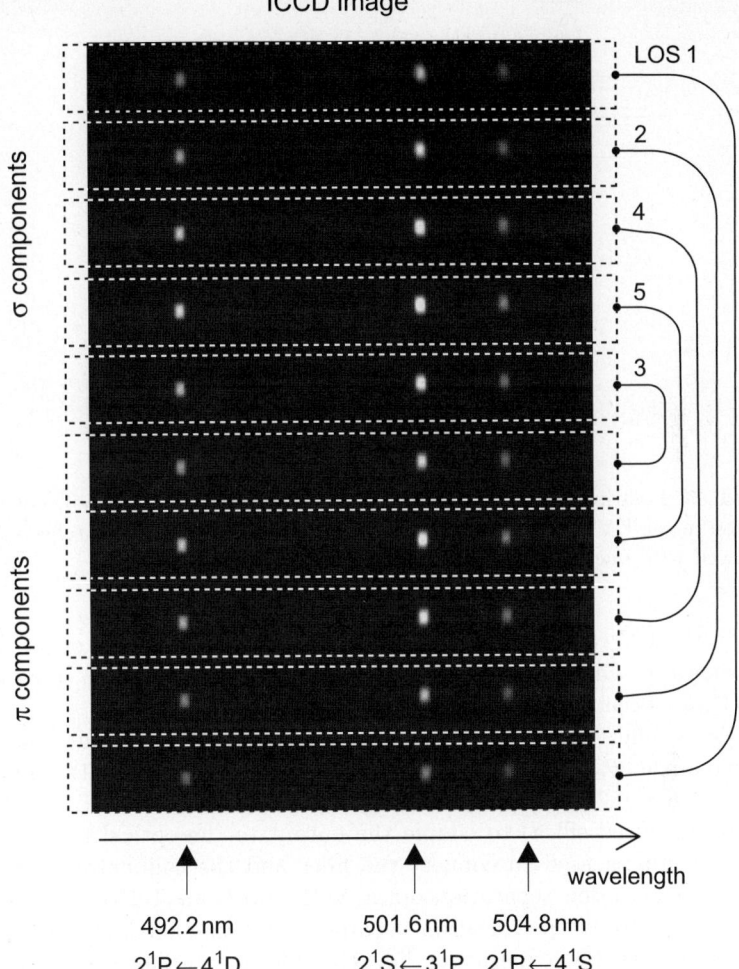

**Fig. 14.12.** The image of the ICCD (Princeton Instruments ICCD-578: $578 \times 384$ pixels, 1.5:1 fiber coupled): The polarization separated spectra of emission lines, HeI $\lambda492.2$, $\lambda501.6$, and $\lambda504.8$ nm, are observed

## UV Observation with a Beam-Splitting Polarizer with Optical Fibers

For observation of polarization of emission lines in the uv region, e.g., BIV $\lambda282$ nm, OV $\lambda278$ nm, and CV $\lambda227$ nm triplet lines, we adopt Glan–Taylor prisms that are air spaced polarizers. The calcite Glan–Taylor polarizer transmits the light in the wavelength region from 220–2,500 nm. Its transmission

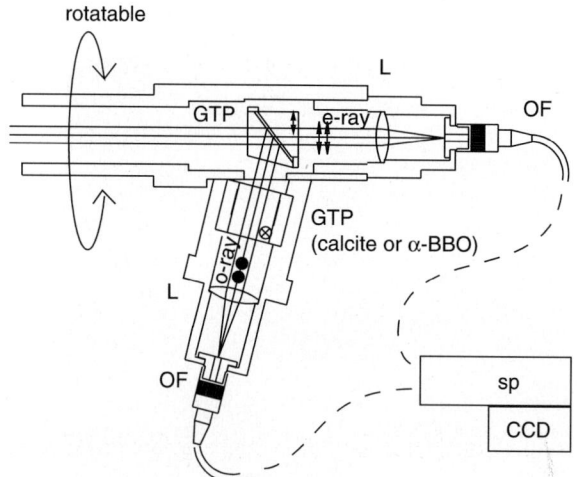

**Fig. 14.13.** Polarization separation optics system (Glan–Taylor type). For the observation in the UV region, the additional Glan–Taylor prism in the reflected beam is replaced with one made of αBBO crystal for higher transmission

curve in the UV region is plotted in Fig. 14.9 with the dashed curve. Figure 14.13 shows the Glan–Taylor beam-splitting polarizer to which a conventional Glan–Taylor prism is added.

Since the effective field angle 8° of the Glan–Taylor prism is narrower than the cone angle $2\theta_{max} = 23°$ of the optical fiber determined by the numerical aperture NA = 0.2, we use collimators (Mitsubishi cable B95BHL, $\phi = 19$ mm, $f = 30$ mm, fused silica) to couple the output to the optical fiber. An FC-type connector is used to connect the fiber and the collimator. For visible light the polarization separation optics with two Glan–Taylor prisms works well and is used for polarization separation of emission lines in the large helical device (LHD) experiment [5]. The additional Glan–Taylor prism in the reflected beam eliminates the spurious e-ray polarized component. This addition of calcite causes a substantial decrease in the signal intensity of the o-ray component of UV light, e.g., the CV λ227 nm emission line.

An αBBO crystal ($BaB_2O_4$) is also an excellent birefringent crystal. It is characterized by a large birefringent coefficient and a wide transmission window ranging from 189 to 3,500 nm. A Glan–Taylor prism made of αBBO is designed to separate linearly polarized components in the UV wavelength region (200–300 nm) with the extinction ratio $5 \times 10^{-6}$. In the wavelength region above 300 nm, the extinction ratio of the αBBO Glan–Taylor prism increases. The polarization separation optics system for the UV wavelength region (220–300 nm) consists of a beam splitting Glan–Taylor prism made of calcite and an additional Glan–Taylor prism made of αBBO with two lens couplers and optical fibers. Ultraviolet λ266 nm light pulses of the forth harmonic of Q-switched Nd:YAG laser are used to align the two lens couplers and optical fibers of the

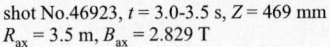

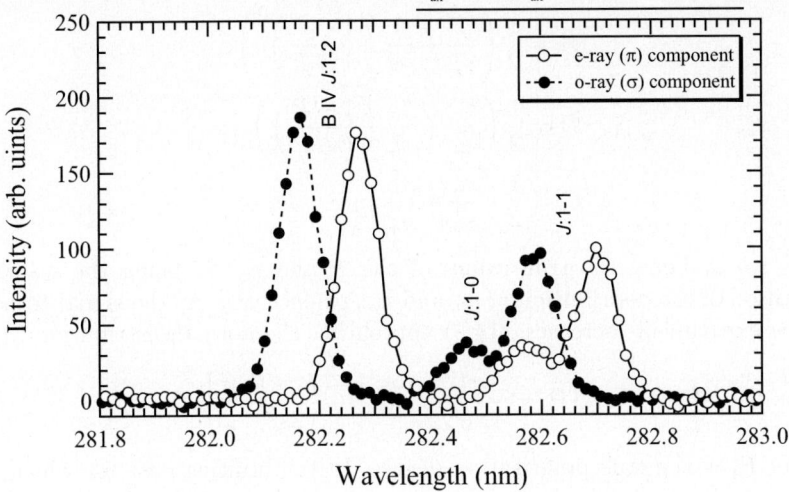

**Fig. 14.14.** An example of polarization resolved spectra BIV 1s2s $2^3S_1$ – 1s2p $2^3P^o_{1,0,2}$, by use of the $\alpha$BBO prism. The e-ray component is displaced by 0.1 nm

polarization separation optics. The image by the 266 nm light is converted to visible light with fluorescence paper.

Figure 14.14 shows an example of polarization resolved spectra of BIV 1s2s $2^3S_1$−1s2p $2^3P_{1,0,2}$ taken on the neutral-beam-injection heating plasma in LHD. The e-ray component is displaced by 0.1 nm for easy comparison.

## 14.2 Polarization Degree

### 14.2.1 Uncertainty of Polarization Degree for Low Signal Intensity

The polarization degree $P$ is written as

$$P = \frac{n_\pi - n_\sigma}{n_\pi + n_\sigma}, \tag{14.2}$$

where $n_\pi$ and $n_\sigma$ are the number of photons recorded for the $\pi$ and $\sigma$ light, respectively. In the determination of the uncertainties it may be assumed that for a set of repeated measurements the distribution of $P$ is normal. However, it has been shown in [6,7] that the general probability distribution of $P$ takes the form

$$f(P) = \frac{B \exp\left(\frac{B^2}{A} - C\right)}{(\pi A)^{1/2} A s_\pi s_\sigma (1 + P)^2}, \tag{14.3}$$

where

$$A = \frac{1}{2} \left( \frac{1}{s_\pi^2} + \frac{1}{s_\sigma^2} \left( \frac{1-P}{1+P} \right)^2 \right),$$

$$B = \frac{1}{2} \left( \frac{z_\pi}{s_\pi^2} + \frac{z_\sigma}{s_\sigma^2} \left( \frac{1-P}{1+P} \right) \right),$$

$$C = \frac{1}{2} \left( \frac{z_\pi^2}{s_\pi^2} + \frac{z_\sigma^2}{s_\sigma^2} \right),$$

with $z_\pi$, $z_\sigma$ being the true values of the $P$ and $s_\pi$, $s_\sigma$ being the standard deviation of the distribution of $n_\pi$ and $n_\sigma$, respectively. As the signal-to-noise of the experiment increases, (14.3) approaches the normal distribution

$$f_n(P) = \frac{1}{s_n \sqrt{2\pi}} \exp \left[ \frac{-(P - P_0)^2}{2s_n^2} \right], \tag{14.4}$$

where $P_0$ is the true polarization degree. In the limiting case for which the uncertainties are solely due to the noise associated with the counting of photoelectrons, the standard deviation is shown to be

$$\sigma_n = \left( \frac{1 - P_0^2}{z} \right)^{1/2}, \tag{14.5}$$

where $z = z_\pi + z_\sigma$. Figure 14.15 shows the distribution functions $f(P)$ and $f_n(P)$ for various $(z_\pi, z_\sigma)$ with constant $z = 10$. Deviations from the normal distribution are barely seen. The $f(P)$ distribution is slightly peakier than the normal distribution. The far wings of the distribution are broader than the normal distribution. The ratio $R(P)$ of (14.3) to (14.4) has been investigated in [6,7]. Figure 14.16 shows $R(P)$ at $P_0 = 0$ for values corresponding to $1s_n (68.2\%)$, $2s_n$ $(95.4\%)$, and $3s_n$ $(99.7\%)$. It can be seen that for $1s_n$, the curve lies below unity, while for $2s_n$ and $3s_n$ it lies above. As is suggested by Fig. 14.15 the central part of $f(P)$ is quite close to that of $f_n(P)$. The effect of the wings of the distribution is substantial for low $z$, but for $z$ higher than 1,000, for example, $R(P)$ value corresponding to $3s_n$ is 1.012.

If the factors associated with the noises for the two photon counts are different so that the two uncertainties are expressed in the form $s_\pi = \sqrt{K_\pi n_\pi}$ and $s_\sigma = \sqrt{K_\sigma n_\sigma}$ with $K_\pi \neq K_\sigma$, then $f(P)$ is not symmetrical about $P_0$ and exhibits a bias; repeated measurements for this case would provide a mean value, which would not be equal to $P_0$.

Effects of background subtraction is discussed in [7]. The procedure for removing the background is to record the photon counts over a certain wavelength width in the absence of the emission line and to take means of the repeated ($M$ times) measurements;

$$\bar{n}_{\pi B} = \frac{1}{M} \sum_{i=1}^M n_{\pi B i} \quad \text{and} \quad \bar{n}_{\sigma B} = \frac{1}{M} \sum_{i=1}^M n_{\sigma B i}. \tag{14.6}$$

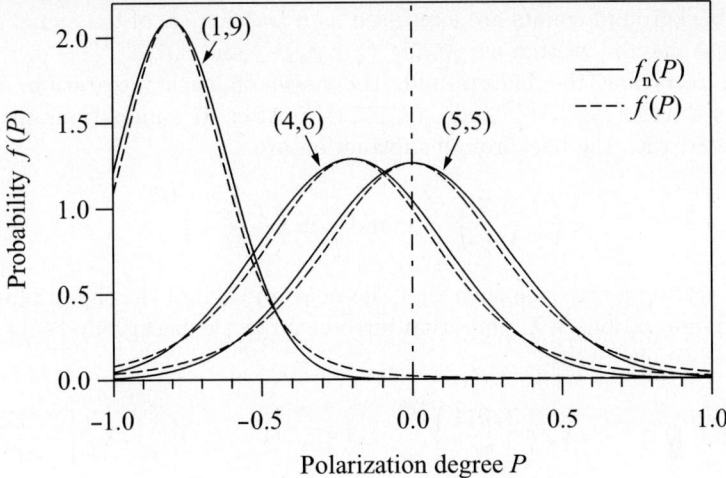

**Fig. 14.15.** The general probability distribution of polarization degree $f(P)$ and the normal distribution $f_n(P)$ for various $(z_1, z_2)$ with constant $z = 10$

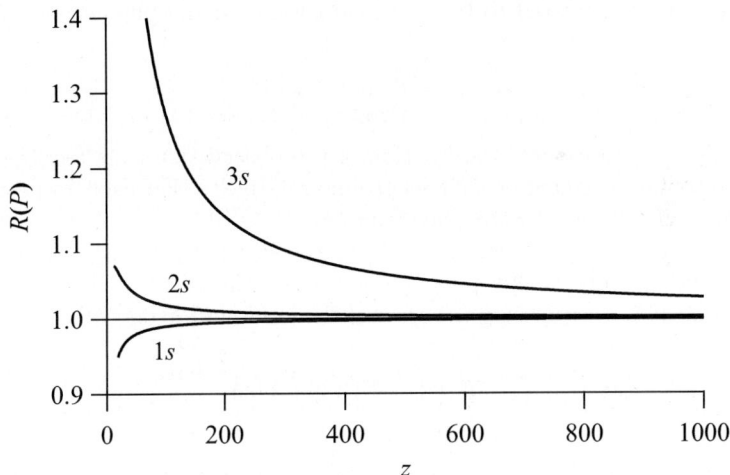

**Fig. 14.16.** The far wings of the distribution are broader than the normal distribution. The effect of positive kurtosis of the $f(P)$ distribution of $P_0 = 0$ is indicated as a function of the total photon count $z$ by displaying the behavior of the distribution normalized to the normal curve for values $(x)$ corresponding to $1s_n$ (68.2%), $2s_n$ (95.4%), and $3s_n$ (99.7%)

Subsequent measurements of the signal including the intensity of the emission line are subtracted by these mean background counts to yield $n_{\pi*}$ and $n_{\sigma*}$.

By taking into account the photon shot noise, the noises of the signal plus background summation are $\pm(\overline{n_{\pi*} + n_{\pi\mathrm{B}}})^{1/2}$ and $\pm(\overline{n_{\sigma*} + n_{\sigma\mathrm{B}}})^{1/2}$. If the mean background counts are expressed as a fraction, $f$, of the signal counts, the noise may be written as $\pm \overline{n_{\pi *}}^{1/2}(1 + f_\pi)^{1/2}$ and $\pm \overline{n_{\sigma *}}^{1/2}(1 + f_\sigma)^{1/2}$.

On recording the background, the noise on each integration may be expressed as $\pm(f_\pi \overline{n_{\pi *}})^{1/2}$ and $\pm(f_\sigma \overline{n_{\sigma *}})^{1/2}$. After $M$ summations the noises associated with the background subtraction are

$$\pm \left( \frac{f_\pi \overline{n_{\pi *}}}{M} \right)^{1/2} \quad \text{and} \quad \pm \left( \frac{f_\sigma \overline{n_{\sigma *}}}{M} \right)^{1/2}. \tag{14.7}$$

By setting a criterion such that the noise in each of the signal plus background integrations is $X$ times that introduced by the background subtraction process, we may write

$$\{\overline{n_{\pi *}}(1 + f_\pi)\}^{1/2} = X_\pi \left\{ \frac{f_\pi \overline{n_{\pi *}}}{M} \right\}^{1/2} \quad \text{and} \quad \{\overline{n_{\sigma *}}(1 + f_\sigma)\}^{1/2} = X_\sigma \left\{ \frac{f_\sigma \overline{n_{\sigma *}}}{M} \right\}^{1/2}.$$

Hence

$$M = \frac{X_\pi^2 f_\pi}{(1 + f_\pi)} = \frac{X_\sigma^2 f_\sigma}{(1 + f_\sigma)}.$$

If the criterion is related to the $N$ integration of signal plus background, we may write

$$\frac{M}{N} = \frac{X_\pi^2 f_\pi}{(1 + f_\pi)} = \frac{X_\sigma^2 f_\sigma}{(1 + f_\sigma)}.$$

The ratio $\left( \frac{M}{N} \right)$ represents the fraction of the observation time required to be devoted to measurement of the background relative to that used for measuring the signal in order to achieve the criterion.

For an unpolarized background (e.g., the noise is primarily from the detector) and an only weakly polarized source so that $f_\pi \approx f_\sigma = f$, it may be shown that (14.5) expressing the error of the polarization degree is modified to

$$s_P = s_n \left( 1 + f \left( 1 + \frac{M}{N} \right) \right)^{1/2}. \tag{14.8}$$

## 14.2.2 Signal Intensity and Photoelectron Number in CCD Detector

In the history of plasma spectroscopy the absolute intensity of emission lines has been a quantity that is difficult to determine experimentally. In PPS experiments the light intensity is the essential quantity, and the amplification characteristics of the device are particularly important in this respect. Developments in instrumentation techniques have enabled us to perform quantitative spectroscopy more easily. Multichannel detectors with digital signal processing techniques such as a charge-coupled device (CCD) [8] have been widely used in the area of spectral measurements. In CCD imaging, the gain refers to the magnitude of amplification which a given system produces. The gain is given in terms of electrons/ADU (analog-to-digital units). Methods to calculate the system gain are examined in [9,10]. The procedure used in [9] is based on the use of an equation

$$V_{\mathrm{m}}(S) = C_{\mathrm{s}}^2 F^2 + C_{\mathrm{s}}\overline{S}, \tag{14.9}$$

where $V_{\mathrm{m}}(S)$ is the variance of the measured signal in terms of ADU, $C_{\mathrm{s}}$ is the system sensitivity in ADU per electron, $F$ is the r.m.s. system noise expressed in terms of equivalent electrons, and $S$ is the measured unbiased signal. The gain is given by $1/C_{\mathrm{s}}$. The validity of this equation is examined in [9] and actual value of gain is given in the data sheet provided by the manufacturer. In this approach we implicitly assume that there is statistical independence between pixels, i.e., the signal intensities measured in each pixels are mutually unrelated.

### 14.2.3 Experiments on the Uncertainty in Polarization Degree

We examine the uncertainty of polarization degree with a CCD detector (Andor DV435-BV). The gain of the CCD detector is measured by the manufacturer with the procedure described in the previous subsection and is 2 electrons/ADU at the A/D conversion rate of 1 MHz with 16 bit resolution. The CCD is Peltier-cooled at $-20°$C. At this temperature the dark current is $0.3$ electrons $(\text{pixel s}^{-1})$.

Emission lines from a Th-Ar hollow cathode lamp are used. Figure 14.17 shows the schematic of the experimental setup. A Glan–Taylor prism makes the transmitted light linearly polarized in the $45°$ direction. The Glan–Thompson polarization-separation optics is used to separate the horizontally $(0°)$ and vertically $(90°)$ polarized components. Figure 14.18 shows an example of the polarization separated spectra. The apparent intensity difference

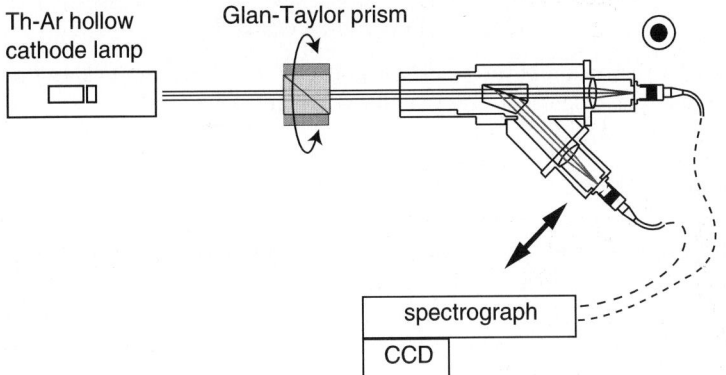

**Fig. 14.17.** Experimental setup to examine the uncertainty of the polarization degree

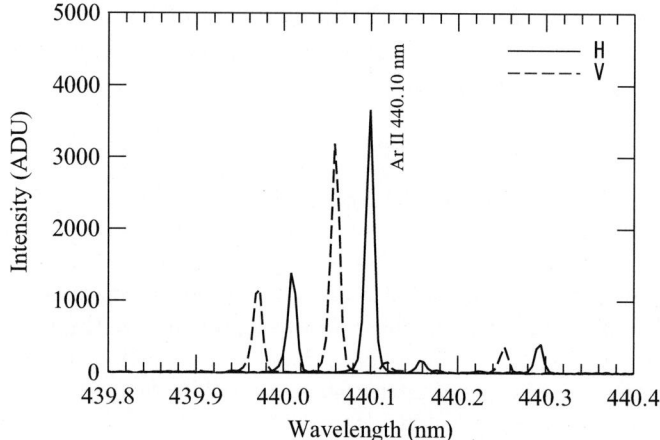

**Fig. 14.18.** Polarization resolved spectra of ArII observed from Thorium–Argon hollow cathode lamp for uncertainty evaluation of polarization degree

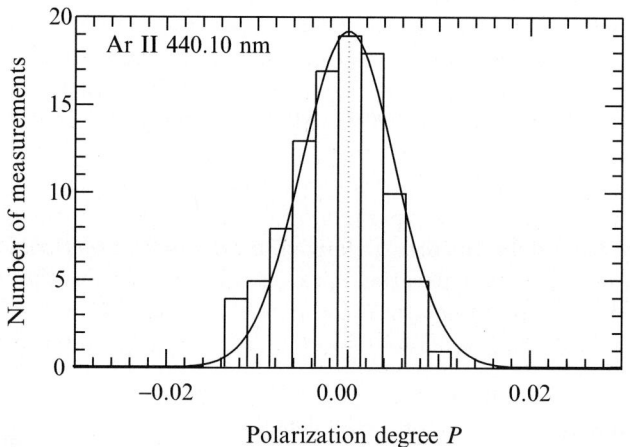

**Fig. 14.19.** Histogram of the polarization degree $P$. Histogram follows the normal distribution. The dispersion depends on the line intensity

is instrumental. For the ArII 440.1 nm line, the observed intensity ratio of the vertically polarized component and the horizontally polarized component $I_{\mathrm{V}}^{\mathrm{obs}}/I_{\mathrm{H}}^{\mathrm{obs}}$ is determined to be $R_{\mathrm{s}} = 0.9035$. The intensity fluctuates owing to the photoelectron statistics, for example. The observed polarization degree $P = (I_{\mathrm{V}}^{\mathrm{obs}} - R_{\mathrm{s}}I_{\mathrm{H}}^{\mathrm{obs}})/(I_{\mathrm{V}}^{\mathrm{obs}} + R_{\mathrm{s}}I_{\mathrm{H}}^{\mathrm{obs}})$ is distributed around the mean value of 0.0. The histogram of the polarization degree from 100 measurements is shown in Fig. 14.19. The histogram follows the normal statistical distribution, and the dispersion depends on the line intensity. Since the observed intensity is sufficiently high, the histogram has been fitted with the normal distribution

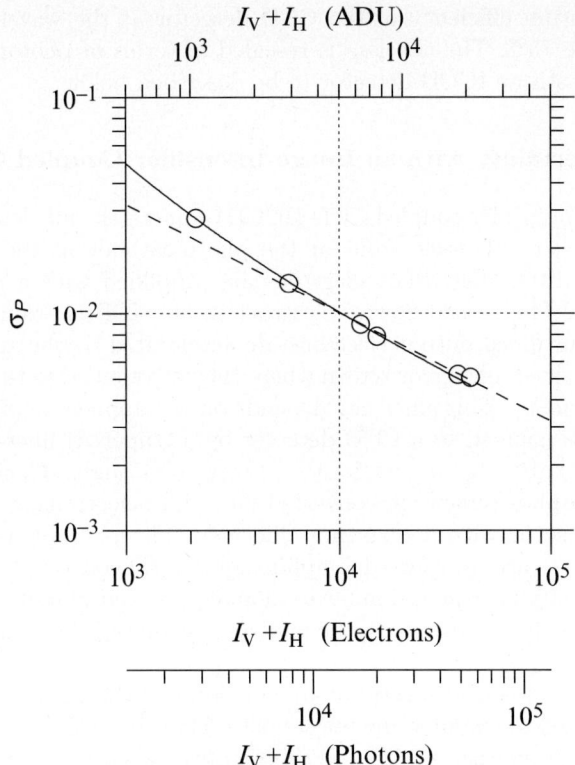

**Fig. 14.20.** The standard deviations $\sigma_P$ against the measured spectral line intensity in ADU or photoelectron numbers (*open circle*). The value of $\sigma_P$ is deduced from fitting the histograms to a Gaussian distribution. The gain of the CCD detector is $2\,\mathrm{pe/ADU}$. The theoretical $\sigma_P$ derived from photoelectron numbers is represented by the *solid line*. $\sigma_P$ decreases with an increase in the intensity with the dependence of $1/\sqrt{n_{pe}}$. The quantum efficiency of the CCD detector is 75% at the wavelength 440 nm

given by (14.4). The fitted standard deviation $\sigma_P$ is $(0.0052 \pm 0.0002)$. The $\sigma_P$ values of the distribution are plotted against the sum of the observed intensity $I_V^{obs} + I_H^{obs}$, as shown in Fig. 14.20. The measured dependence of $\sigma_P$ (open circles) on the electron numbers is consistent with (14.5), i.e., the inverse square root of photoelectron numbers $1/\sqrt{n_{pe}}$, which is plotted with the dashed line. The average intensity of the spectra corresponding to the histogram is $z \sim 38{,}000$ electrons.

This distribution analysis on the polarization degree is a gain evaluation at the same time; the slope of the log-log plot of $\sigma_P$ with signal intensity in terms of ADU gives the gain coefficient. The measured $\sigma_P$ is fitted with (14.8) to yield the solid curve. This fitting gives gain of $2.3 \pm 0.1$ electrons/ADU. This value is slightly larger than the value provided by the manufacturer.

The quantum efficiency of the CCD detector at the wavelength 440 nm is given to be 75%. The abscissa is rescaled in terms of photon numbers for comparison with an ICCD detector to be described below.

### 14.2.4 Uncertainty with an Image Intensifier Coupled CCD

In an image intensifier coupled CCD (ICCD) camera the incident photons are first converted to photoelectrons at the photo-cathode at the front surface of the image intensifier. Photoelectrons are amplified with a micro-channel plate (MCP). The cascad flow of electrons in an MCP is examined in detail in [11,12]. Multiplied output electrons are accelerated by the applied electric field. They strike a phosphor screen where they are converted to photons with a certain efficiency. This efficiency depends on the applied voltage. The phosphor screen is coupled to a CCD detector by a (tapered) fiber-optics, which transmit the light with a certain acceptance solid angle. Then the photons from the phosphor screen are converted into photoelectrons at each pixel on the CCD detector with a quantum efficiency. The photoelectrons are read out, i.e., charges are transferred, amplified, and A/D converted. The observed spectral intensity is displayed in terms of analog-to-digital units (ADU). Since the process involves many steps, the uncertainty analysis becomes accordingly complicated.

Polarization resolved spectra of an emission line HgI 253.7 nm from a penlay lamp is used to examine the uncertainty. Light from the lamp is polarized by a Glan–Taylor prism at 45° and illuminates the entrance slit of the spectrometer (CT-100, $f = 1$ m, with a grating of 3,600 grooves/mm). A calcite plate is placed behind the entrance slit of the spectrometer. Its optical axis is in the horizontal direction. The horizontally polarized component and the vertically polarized component are separated in the wavelength dispersion direction. Since the incident light is polarized at 45°, the horizontally and vertically polarized components of the incident light should have equal intensities. The signal intensities are measured with an ICCD camera (Princeton Instruments IMAX-512: $19 \times 19 \, \mu m^2$ pixel with 1.2:1 fiber optics coupling). Figure 14.21 shows two examples of polarization separated spectra. The width of a spectral line is 90 μm FWHM at the slit width of 20 μm. The line profiles recorded with the ICCD are almost the same for different slit widths from 5 to 20 μm. This is understood from the fact that the observed spectral profile is determined by the spread of the secondary photoelectrons hitting the phosphor screen. The MCP used in the ICCD has 70 μm FWHM spot size, which corresponds to three pixels. The accumulated signal intensities in neighboring pixels on the CCD are not independent. The signal intensities of the immediate four or five pixels are mutually related. Therefore the procedure described in Sect. 14.2.2 may result in an incorrect gain value.

A calibrated photodiode (Hamamatsu Photonics S1226-18BQ) is placed at the exit surface of the spectrometer. The dispersed light from a deuterium

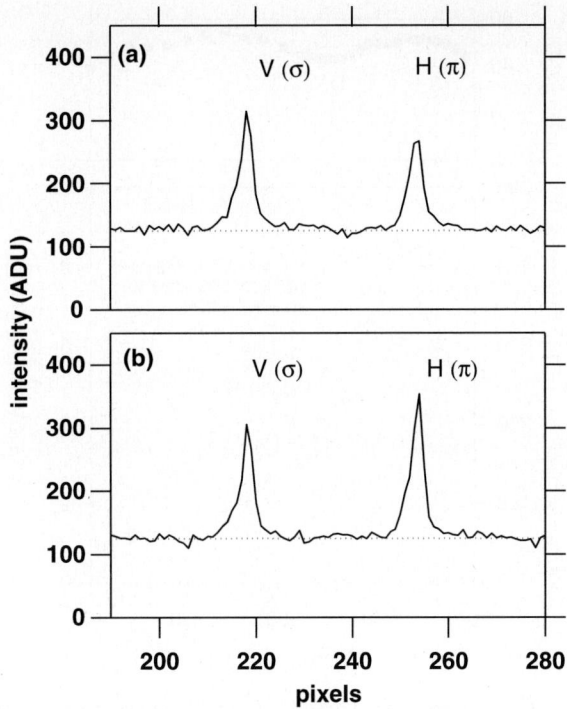

**Fig. 14.21.** Example of polarization separated spectra of an emission line HgI 253.7 nm from a pen-lay lamp recorded with an ICCD camera (Princeton Instruments IMAX-512). The incident light polarized at 45° was separated into the linear horizontal (H) and vertical (V) polarized components. The apparent intensity difference on the whole spectral profile is due to the statistical variation. The MCP used in the ICCD has 70 μm spot size, which corresponds to three pixels on the CCD detector. The accumulated signal intensities in neighboring pixels are not independent

lamp is measured. The current from the photodiode is recorded with a pico-ammeter. In Fig. 14.22b, the photon numbers obtained from the photodiode current are shown by the open circles. The photodiode is replaced with the ICCD camera. The area of $43 \times 43$ pixels of the ICCD corresponds to the effective area of the photodiode $(1.1 \times 1.1 \text{ mm}^2)$. The wavelength dependence of the conversion coefficient from photon-numbers to the signal intensity is shown by Fig. 14.22a. For example, it is 49 photons/ADU at 255 nm. The quantum efficiency of the photocathode in the intensifier is 32.7% at the wavelength $\lambda 270 \text{ nm}$; this means that 16 electrons/ADU are created at the photocathode.

The histogram of the polarization degree of the emission line HgI 253.7 nm is fitted with a Gaussian distribution and the standard deviation $\sigma_P$ of the distribution is plotted against the sum of the observed intensity $I_V + I_H$, as shown in Fig. 14.23. Measured $\sigma_P$ decreases with an increase in intensity.

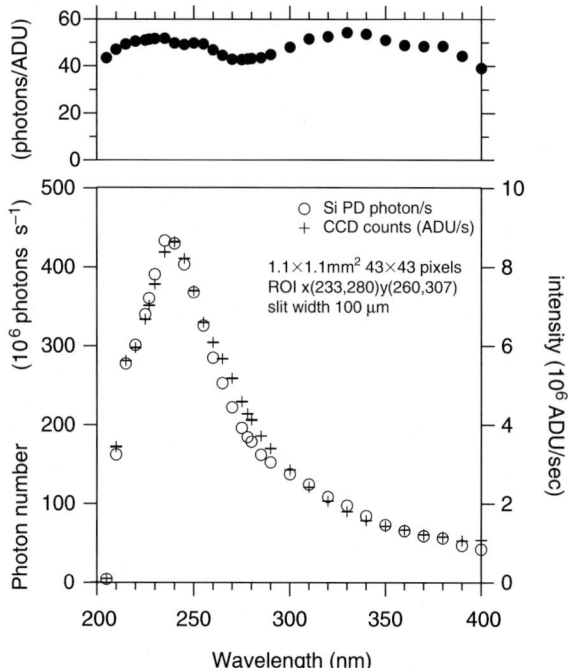

**Fig. 14.22.** (a) The wavelength dependence of the conversion coefficient between the signal intensity in ADU and photon-numbers. (b) The wavelength dependence of the signal intensity

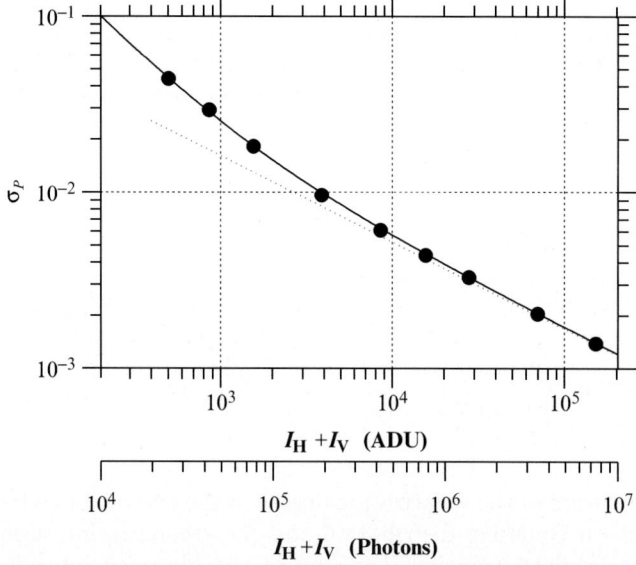

**Fig. 14.23.** The standard deviations $\sigma_P$ against the measured spectral line intensity in photon numbers (*closed circle*). The $\sigma_P$ values are deduced from fitting of the histograms to a Gaussian distribution. The solid line is (14.8)

The dependence is proportional to the inverse square root of the intensity at high intensities. The deviation increases with the decrease in the intensity. The reader may tend to identify this deviation with the increase in Fig. 14.16 in the low signal region. However, this is not the case; the present deviation is due to the increase in $f$ of the background noise. In fact, the measured $\sigma_P$ is fitted with (14.8) to yield the solid curve. This fitting gives the "effective" gain coefficient of $3.44 \pm 0.16$ electrons/ADU. The post amplification and photon conversion processes in the image intensifier increase the variance of the signal intensity.

We now compare the present result with that of the CCD detector as shown in Fig. 14.19. For a same number of photons, $\sigma_P$ for the ICCD detector is at least three times as high as that for the CCD detector. For low intensity the background noise in ICCD further increases $\sigma_P$. In order to achieve the same uncertainty with the ICCD detector, almost ten times photon number is required.

# References

1. D. Coffin: *Decoding raw digital photos in Linux*
   http://www.cybercom.net/~dcoffin/dcraw
2. T. Fujimoto, H. Sahara, T. Kawachi, T. Kallstenius, M. Goto, H. Kawase, T. Furukubo, T. Maekawa, Y. Terumichi: Phys. Rev. E **54**, R2240 (1996)
3. T. Kallstenius: Thesis *"Plasma Polarization Spectroscopy: Impurity Ion Emission Lines From the WT-3 Tokamak"*, (Royal Institute of Technology, Stockholm, Sweden, 1994)
4. A. Iwamae, T. Sato, Y. Horimoto, K. Inoue, T. Fujimoto, M. Uchida, T. Maekawa: Plasma Phys. Control. Fusion **47**, L41 (2005)
5. A. Iwamae, M. Hayakawa, M. Atake, T. Fujimoto, M. Goto, S. Morita: Phys. Plasmas **12**, 042501 (2005)
6. D. Clark, B. G. Stewart, H. E. Schwarz, A. Brooks: Astron. Astrophys. **126**, 260 (1983)
7. D. Clark, B. G. Stewart: Vistas in Astronomy **29**, 27 (1986)
8. J. Janesick: *Scientific Charge Coupled Devices*, (SPIE Press Monograph Vol. PM83, 2001)
9. L. Mortara, A. Fowler: Solid State Imagers for Astronomy SPIE **290**, 28 (1981)
10. J. R. Janesick, K. P. Klaasen, T. Elliot: Opt. Eng. **26**, 972 (1987)
11. R. L. Bell: IEEE Transactions on Electron Devices, ED-**22**, 821 (1975)
12. S. E. Moran, B. L. Ulich, W. P. Elkins, R. L. Strittmatter, M. J. DeWeert: SPIE Proceedings 3173-43 (1997).

# 15

# Instrumentation II

E.O. Baronova, M.M. Stepanenko, and L. Jakubowski

Following the preceding chapter, we discuss in this chapter PPS instrumentations in the X-ray region. Since X-ray spectroscopy based on the reflection by a crystal surface is intrinsically polarization selective, X-ray spectroscopy is, in principle, polarization spectroscopy in its nature. Furthermore, PPS diagnostics is highly demanded in various plasma researches. However, enormous difficulties accompany practical X-ray PPS experiments. In this chapter we discuss various facets of the X-ray PPS instrumentation.

## 15.1 X-ray Polarization Measurements

In the visible and ultraviolet regions, studies of polarization of radiation have contributed to investigating the anisotropic characteristics of the matter which emits, transmits, or reflects the radiation [1]. This is because various types of polarizers are available in these wavelength regions, as discussed in the previous section. This is in strong contrast with the X-ray regions where optical materials are quite limited. X-rays are reflected by the periodic parallel atomic planes according to Bragg's law: $2d \sin \theta_B = k\lambda$, where $d$ is the spacing between the atomic planes, $k$ the order number, and $\lambda$ the wavelength. The integrated reflectivity $P_D$ for Bragg reflection from an ideal crystal is expressed by

$$P_D = \left( \frac{16}{3\pi} \right) \left( \frac{e^2}{4\pi\varepsilon_0 mc^2} \right) \left( \frac{d^2}{V} \right) \tan \theta_B K |F_{hkl}| , \qquad (15.1)$$

where $e$ and $m$ are the electron charge and mass, $F_{hkl}$ is the structure factor for the $hkl$ reflection, which can be found in [2], $V$ is the volume of a unit cell (if we take quartz as an example it is $112 \, \text{Å}^3$), and $K$ is the polarization factor and is equal to $(1 + |\cos 2\theta_B|)/2$ for natural rays: $K = 1/2$ for the s-component and $K = |\cos 2\theta_B|/2$ for the p-component. Here the s-component has the electric field vector in the direction perpendicular to the dispersion plane and the p-component parallel to the plane. These values are for an ideal crystal, and if the crystal is ideally mosaic, the above $|\cos 2\theta_B|$ is replaced by $\cos^2 2\theta_B$. The temperature dependence and the absorption and extinction effects inside the crystal have been ignored. It is obvious that at the Bragg angle of $45°$ the polarization separation efficiency is 1.

Even if the Bragg angle is not exactly $45°$, the reflection efficiency of the crystal is substantially different for the s- and p-components. This is the idea on which the X-ray polarization measurement is based. Therefore, a reflecting crystal works as a polarizer as well as a dispersing element at the same time. In this sense, an X-ray spectrometer is a polarizer-spectrometer. An X-ray graphite polarizer was first suggested by Barkla [3] and then implemented in [4]. In this scheme, one polarized component is first measured, and the reflecting crystal is rotated by $90°$ to measure another polarized component. This approach is valid for stationary or reproducible light sources like magnetic or resonant scattering experiments, ion traps [5], synchrotron radiations [6]. In the following, we discuss an alternative arrangement which is capable of handling irreproducible light sources. As discussed already in Chap. 8, $z$-pinch plasmas emit polarized line radiations, but they are quite irreproducible. We thus take these plasmas as an example of the objects of polarization measurements, and discuss problems accompanying the X-ray polarization measurements by this scheme.

A natural way to analyze polarization of X-ray emission from irreproducible light sources is to use a scheme involving two identical crystal-polarizers [7], i.e., the two-crystal scheme. The dispersion planes of the two crystals have to be oriented in mutually perpendicular directions, and the Bragg angle be as close to $\theta_B = 45°$ as possible. Each polarizer detects radiation emitted from a particular plasma region; this region is called the field of view. The two-crystal scheme works on the assumptions that (1) the light source is axially symmetric, (2) the reflection characteristics of both the polarizers are identical, and (3) both the polarizers have the same field of view; i.e., they detect radiation from the same plasma region.

Flat, convex, and concave crystals serve as a polarizer. Flat and convex ones have a relatively wide spectral range, and spectra can be registered in several orders of reflection; this latter point provides an additional advantage to these schemes in the polarization analysis. However, in these schemes the width of spectral lines depends on the size of the source, so that they are suitable only to a point-like X-ray source. Concave (say, cylindrical) crystals are the principal unit of Johann/Johansson spectrometers. Although the spectral range is relatively small, the width of lines depends only slightly on the source size with the Johann spectrometers and is independent of the size with the Johansson spectrometers. As discussed later, the width of a recorded line is the result of the sum of components due to (1) aberrations of the geometry, (2) diffraction, (3) quality of the detector, (4) accuracy of the crystal and detector alignment, (5) geometry of the experiment, and (6) intrinsic width of the line emitted by the source. The more detailed discussion on resolution of

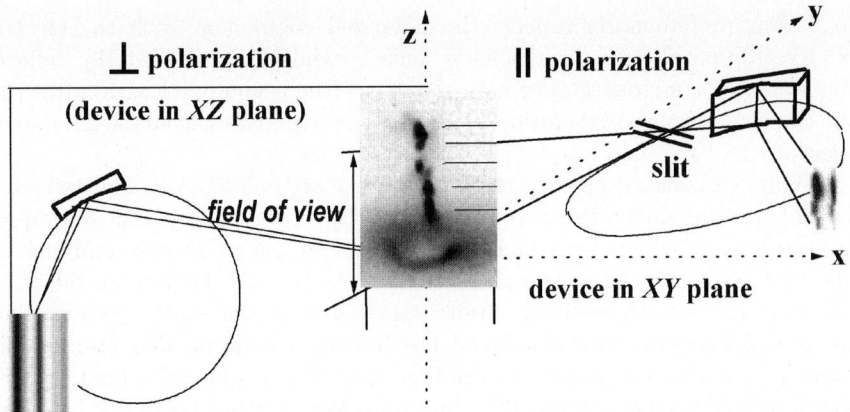

**Fig. 15.1.** Two-crystal scheme

various concave spectrometers is given in [8–10]. Obviously the spectral resolution and the detection efficiency of the concave spectrometers are much better than those achieved with flat or convex crystals. We discuss below the focusing Johann spectrometer, which is commonly used in X-ray spectroscopy. Among the assumptions mentioned above, assumption (3) is difficult to be fulfilled. We examine this point below. It will turn out that polarization measurements of the *point* X-ray source is easier to interpret, while the interpretation of those of *extended* sources is difficult or even impossible.

We take a pulsed X-ray source, i.e., the plasma focus of the Mather type. The working gas is argon. Spectra of Ar XVII are emitted by the plasma ($I = 500\,\text{kA}$, $U = 35\,\text{kV}$). The experimental setup and the diagnostics are shown in Fig. 15.1. A pinhole camera gives the image of the plasma as shown in the central part of the figure. The plasma consists of 1–10 small plasma regions, called the hot spots. The hot spots are usually formed around the discharge axis ($z$) near the anode. Electron temperature and density of these plasmas are widely distributed, but the typical values are $T_e = 0.5$–$1\,\text{keV}$, $n_e = 10^{25}$–$10^{27}\,\text{m}^{-3}$ [11]. Those parameters were estimated rather approximately from relative intensities of the satellites to the resonance line and the intercombination line to the resonance line, respectively, under the assumption of a Maxwellian distribution function for electrons. At this preliminary stage we did not take into account the effects of opacity, charge exchange, the presence of an electron beam, the polarization, etc.

Two polarizer-spectrometers of identical characteristics are used; each of the spectrometers is equipped with a quartz crystal, $2d = 6.67$ Å and area $8 \times 20\,\text{mm}^2$. Radius of curvature of the crystal is 500 mm. The source-to-crystal distance is 500 mm. The field of view of the spectrometers is calculated from the length of the crystal to be 12 mm in the dispersion direction. The dispersion plane of the right-hand spectrometer, rhs, is perpendicular to the $z$-axis and the left-hand spectrometer, lhs, has the dispersion plane parallel to $z$. Rhs preferentially reflects the polarized component with the electric field vector parallel to the discharge axis $z$, while lhs preferentially reflects the polarized component perpendicular to $z$. Rhs is equipped with a 200 μm slit to provide spatial resolution along $z$, while lhs gives the space-integrated spectra.

Figure 15.1 shows the spectra of helium-like argon (Ar XVII) emitted from the hot spots as shown by the pinhole image. Spectrum contains the resonance line $(1s^2\,{}^1S_0 - 1s2p\,{}^1P_1)$, which is the strongest one, the intercombination line $(1s^2\,{}^1S_0 - 1s2p\,{}^3P_1)$, and the satellites. As is seen, the spectra taken by rhs correspond to the four hot spots, starting from the anode surface. This corresponds to the field of view of rhs in the $z$ direction. For the lhs the geometry of the experiment does not permit us to confirm the field of view in the same direction. Figure 15.1 illustrates how a polarimeter-spectrometer works on real plasmas.

We further discuss a few important features of the given polarization experiment. The Bragg angle for the resonance line $\lambda 3.946$ Å is $36.27°$. As shown in the beginning of this section, if the crystals are ideal crystal the relative efficiency is $I_s/I_p = \cos 2\theta = 0.3$, where $I_s$ and $I_p$ are the intensities of the s- and p-components, respectively, on the assumption of the unpolarized incident light. If the crystal is ideal mosaic, then the same relative efficiency is proportional to $\cos^2 2\theta = 0.09$. Commercial flat quartz crystals are very close to ideal, but bent crystals may be close to mosaic and have to be investigated in each particular case. Calibration of the bent crystals with a known polarized radiation is desirable, which, however, is very difficult. Thus, the effectiveness of the separation of the s- and p-components $(I_s - I_p)/(I_s + I_p)$ is in the range 54–82%, depending on the degree of mozaicity of the crystal. This is one of the reasons why an X-ray polarization measurement is difficult.

Although the field of view is calculated from the geometry of the experiment, the actual field of view is slightly different for different wavelengths. Figure 15.2 illustrates how the resonance and intercombination lines, for example, have different field of views; the displacement of these lines on the focal plane is 1.9 mm, which is of the order of distance between the hot spots. In this figure, hot spot 1 does not give the yield into the intercombination line, while hot spot 4 does not give the yield into the resonance line.

As an example of confusing spectra which is due to different field of views for different lines, we look at Fig. 15.3; on the left-hand side pinhole images are shown and on the right-hand side the corresponding recorded spectra are shown. Spectrum in SX65-2 shot contains five lines (from left to right): the resonance line, the intercombination line, the weak satellite lines of lithium- and beryllium-like ions. Spectrum SX65-3 mainly contains only the resonance line and SX65-4 contains mostly satellites. This example clearly indicates that, for an interpretation of a spectrum, we should take into account the geometry of the experiment, i.e., field of view of the device, its orientation, the number of hot spots, their locations with respect to the anode surface, etc.

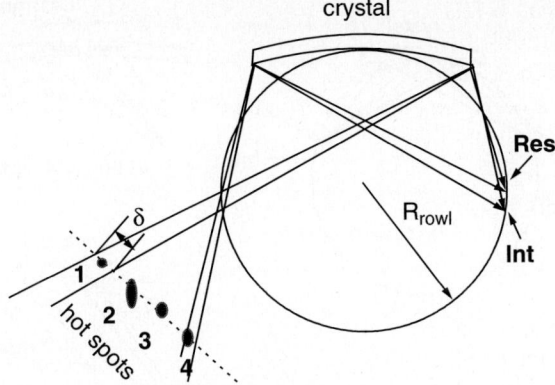

**Fig. 15.2.** Field of view for the resonance and intercombination lines

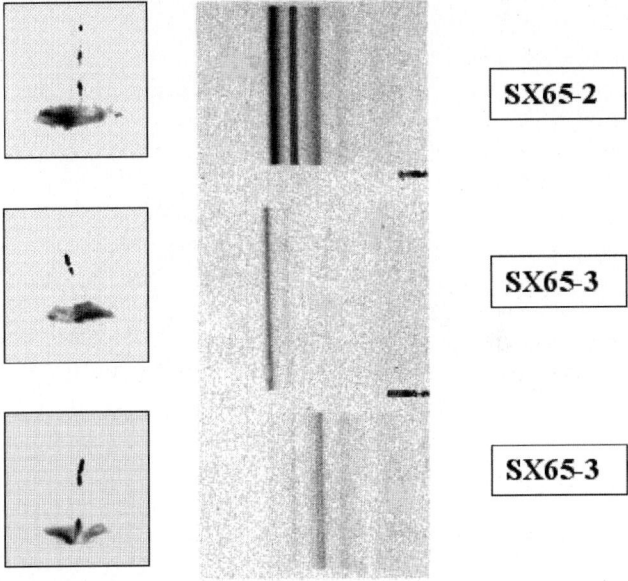

**Fig. 15.3.** Pinhole images of the plasma-focus plasmas in several shots and corresponding spectra of helium-like argon

Figure 15.4 demonstrates the next feature; the resonance and intercombination lines are reflected by different zones of the crystal surface. In this figure a single hot spot is considered for simplicity. The distance between the zones that reflect the resonance and intercombination lines is calculated to be 2 mm for the present geometry of experiment. The length of the crystal zone, reflecting a particular line, is 1 mm for the typical hot spot size 500 μm, and 0.3 mm for the hot spot size of 150 μm. Therefore the intensity of a reflected spectral line depends on the local reflection coefficient of the crystal.

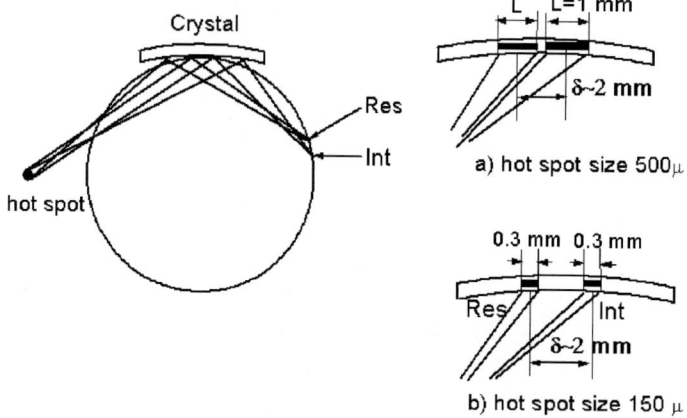

**Fig. 15.4.** Resonance and intercombination lines are reflected by different parts of the crystal

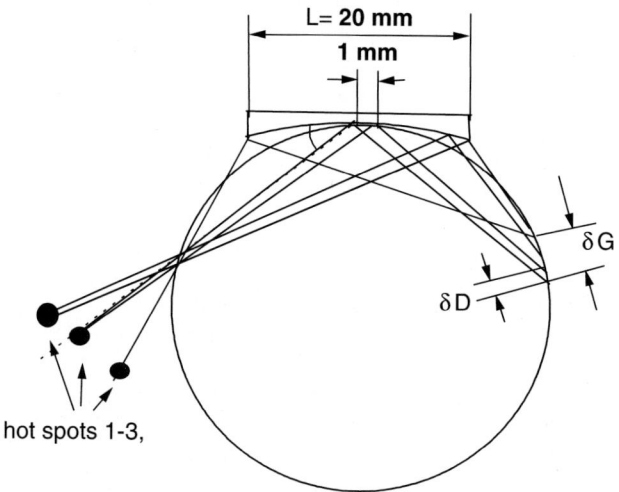

**Fig. 15.5.** For Johann device the shape of a line can depend on the local reflection coefficient and the local brightness of the source

Flat quartz crystals are usually homogeneous in reflection. For bent crystals their reflection coefficient can be different along the crystal surface [12]. It is desirable that the local reflection coefficient of a given crystal is studied before starting polarization experiments.

The profile of a recorded line is also affected by the geometry. We take the example of Fig. 15.5. The distance between hot spots 1 and 3 is taken to be 12 mm, corresponding to the field of view of the present geometry. The width of a particular line is calculated from the diffraction theory to be $\delta D = R\delta\theta_{\mathrm{di}} = 75\,\mu\mathrm{m}$, where $R$ is the crystal curvature radius and $\delta\theta_{\mathrm{di}}$ is the half width of the rocking curve at the given point of the bent crystal. Here the half width of the rocking curve is usually around 30–40 arcsec, which is compared with 25–30 arcsec for best quality bent crystals. The line width due to the astigmatism, or the geometrical defocusing, is $\delta G = L^2 \cot \theta_B / 8R = 140 \,\mu\text{m}$, where $L$ is the length of the crystal. We note here that $140 \,\mu\text{m}$ is of the order of the distance between the reflecting positions of the same line emitted by hot spots 2 and 3, for example, and recorded on the film. Therefore, if the source is extended in the dispersion direction, the intensity distribution within the width of a particular line depends on the local reflection coefficient of the crystal and/or on the distribution of brightness inside the source. This kind of problem is absent with Johansson spectrometers: the crystal surface has the radius of curvature $R$ while the atomic planes have the radius $2R$. Local reflection coefficient in Johansson crystals might be inhomogeneous along the crystal surface, but this does not affect the line profile because the geometrical focusing is ideal, i.e., $\delta G = 0$.

If the light source is a point source, the number of problems discussed earlier is reduced. In the Mather plasma focus machine some rare shots produce only one single hot spot. The pinhole image in Fig. 15.6 shows an example. Spectra of helium-like argon are registered by the spectrometers in the two-crystal scheme, where care has been taken that the near central zones of the crystals reflect the lines in both the spectrometers. The densitograms of the spectra are on the right-hand side. In this figure panel A is taken with the device with the dispersion plane in the $x$-$z$ plane (see Fig. 15.1) and B is

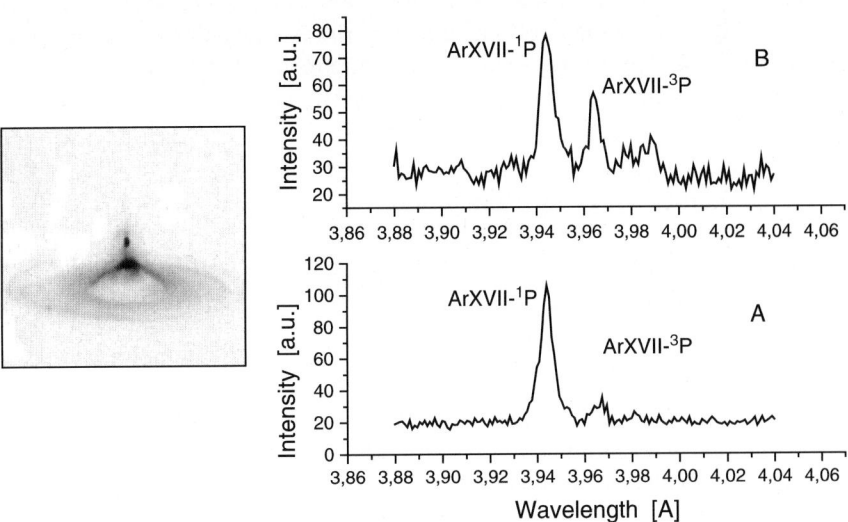

**Fig. 15.6.** A single hot spot created in the plasma focus machine. The pinhole image is shown on the left side and the spectra taken with the two-crystal scheme is on the right side

taken with the dispersion in the $xy$ plane. It is seen that the resonance line in spectrum A is much more intense than the intercombination line, while in spectrum B the two lines have comparable intensities. Such obvious difference cannot be explained by a possible inhomogeneity of the local reflection coefficient, which is usually very small [5], nor the possibility that the resonance and intercombination lines are emitted by different plasma regions, since the hot spot is quite small. The most probable explanation is the presence of polarization of these lines. For the shot analyzed the resonance line may be polarized in the direction perpendicular to the discharge axis.

(E.O. Baronova, L. Jakubowski)

## 15.2 Novel Polarimeter–Spectrometer for X-rays

In the X-ray region polarization study is less developed as compared with the visible–UV regions. This is partly because a polarimeter has been unknown for X-rays. Here we define a *polarimeter* as a device that separates the two perpendicularly polarized components with equal efficiencies whereas a *polarizer* is a device to transmit or reflect only one polarized component. We describe in this section a polarimeter–spectrometer for X-rays [13].

As shown in the previous section X-rays are reflected by the periodic parallel atomic planes according to Bragg's law: $2d \sin \theta_B = k\lambda$. When the Bragg angle $\theta_B$ is close to $45°$, the reflection efficiency is substantially different for two polarized components. A natural consequence is to use two identical spectrometers in mutually perpendicular directions: the two-crystal scheme. As discussed in the previous section, however, the X-ray polarization study by the conventional two-crystal scheme is accompanied by various difficulties.

In the following, we present a novel polarimeter, one-crystal spectropolarimeter, which is free from these difficulties.

### 15.2.1 Principle of X-ray Polarimeter

In Fig. 15.7 suppose A and B are two series of atomic planes that reflect the beam incident from the $z$-direction. These series may be two crystals/polarizers in the conventional two-crystal scheme, or two series of identical atomic planes within the same crystal in the case of the polarimeter [13] with which we are concerned in this section. The incident beam is a superposition of two perpendicularly polarized components 11 $\parallel Ox$ and 22 $\parallel Oy$. Let the s-component be the component having the electric field vector perpendicular to the plane of incidence and the p-component with the electric field vector directed in the plane. In Fig. 15.7a the plane of incidence is parallel to $yOz$, and in Fig. 15.7b the plane is parallel to $xOz$.

The question we have to solve here is: Is it possible to select the two polarized components using one single crystal? Or in other words, Are there inside one crystal two series of identical atomic planes A and B (Fig. 15.8),

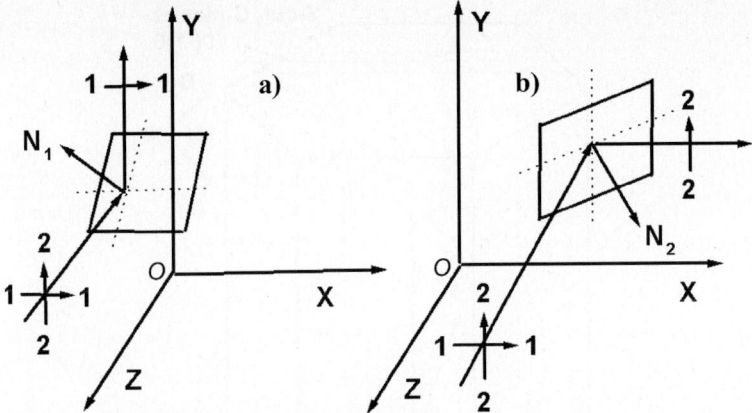

**Fig. 15.7.** Geometry of the reflecting planes of the polarimeter. The light beam is incident along the $z$-direction

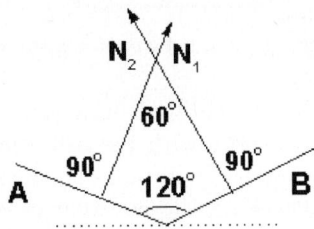

**Fig. 15.8.** Angle between polarizing planes A and B

operating like two crystals A and B in Fig. 15.7? If such A and B series of planes exist, what should be the angle between them? The unit vectors $N_1$ and $N_2$ in Fig. 15.7 are expressed as

$$N_1 = \{0; 1/\sqrt{2}; 1/\sqrt{2}\} , \tag{15.2}$$

$$N_2 = \{1/\sqrt{2}; 0; 1/\sqrt{2}\} , \tag{15.3}$$

so that the scalar product is

$$\cos(N_1 N_2) = 1/2 . \tag{15.4}$$

As shown in Fig. 15.8, the angle between $N_1$ and $N_2$ is 60°. This means that the angle between the series of atomic planes A and B (we will call them the polarizing planes) should be 120°. This situation is realized in any trigonal, or hexagonal, or rhombohedral crystals. It will be shown below that the trigonal structure can also be found in cubic crystals [14, 15].

### 15.2.2 How to Cut a One-Crystal Polarimeter from a Crystal

Figure 15.9 shows the simplified geometry of a boule of a quartz crystal. The two possible series of identical atomic planes, or the polarizing planes, are

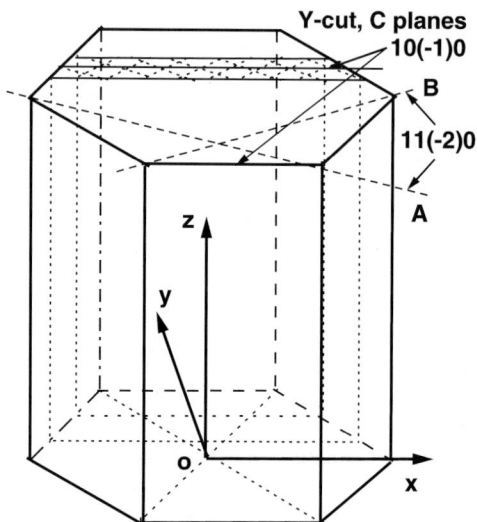

**Fig. 15.9.** How to make a polarimeter from a boule of quartz

A and B marked with the thin dotted lines, and C planes, which we call the mechanical planes, are drawn with the solid lines. The $xyz$-coordinates are used for simplicity, while the notation of planes are given in hexagonal coordinate system. In Sect. 15.2.4 the notation of planes in trigonal crystal are given within trigonal and hexagonal coordinate systems for comparison.

The so-called Y cut of this crystal, or expressed by Bravais indices $10(-1)0$, has atomic planes with $2d = 8.51\,\text{Å}$, which are parallel to the plane $xOz$ in Fig. 15.9. The crystal slab having atomic planes parallel to the mechanical plane has the two polarizing planes inside: planes A ($2d = 4.91\,\text{Å}$) and B ($2d = 4.91\,\text{Å}$), inclined at 120° each other. The Bravais indices of these planes are $11(-2)0$ (the more detailed discussion on the indices are given in Sect. 15.2.4, and strictly speaking, A and B planes have different combination of indices, A has $11(-2)0$, B has $(-2)110$).

Therefore, the slab of the Y cut is suitable for the polarimeter of the wavelength $k\lambda_\mathrm{p} = 4.91\times\sin45° = 3.47\,\text{Å}$ in the first order of reflection. The same cut can be used for $1.74\,\text{Å}$ in the second order and for $1.16\,\text{Å}$ in the third order; for this particular quartz cut the reflection coefficient in the third order is high enough for a practical use.

Corresponding to the above combination, a quartz crystal slab having the mechanical plane $11(-2)0$ has polarizing planes A and B of $10(-1)0$, so that this slab can be used as polarimeter for the wavelength $k\lambda_\mathrm{p} = 8.51\times\sin45° = 6.02\,\text{Å}$ in the first order of reflection, and so on. The mechanical plane was at 30° with respect to the both polarizing planes. The important feature of polarizing and mechanical planes is that they are interchangeable. More detailed discussion about the mechanical and polarizing planes will be given later in this section.

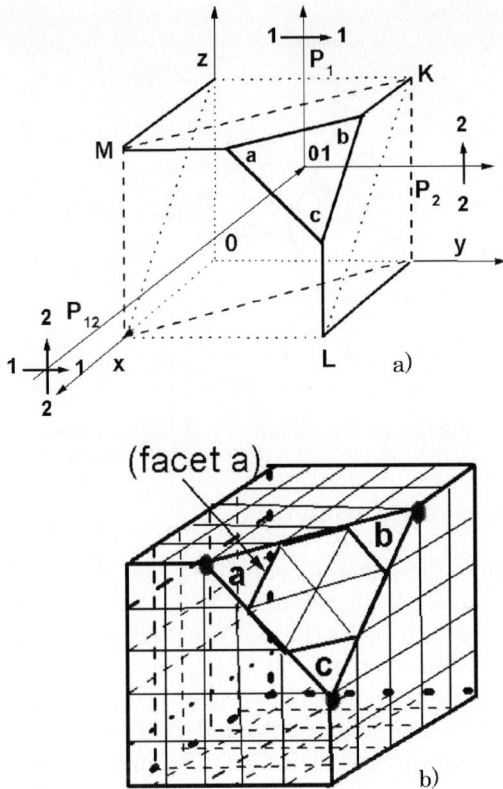

**Fig. 15.10.** (a) First way to make a polarimeter from a cubic crystal. (b) Second way: trigonal structure inside the cubic crystal

We now turn to cubic crystals [14]. Figure 15.10a shows the two polarizing planes A ($xzKL$-plane, dotted) and B ($xMKy$-plane, dashed) inside the crystal. The angle between A and B is 120° as is obvious from Figs. 15.7 and 15.8. As shown in Fig. 15.10a, a prism is cut out from the corner, and $abc$ surface is at the angle 35.3° to the both polarizing planes A, B. From the discussion concerning Fig. 15.7, it is obvious that, for the incident beam $P_{12} \parallel Ox$, the reflected beams $P_1$ and $P_2$ are directed into the $z$- and $y$-directions, respectively, and are fully polarized. One can use the same plane A and $MzyL$ as plane B. Then to get two polarized beams $P_{12}$ should lie in the plane $Oz01L$. As will be shown in Sect. 15.2.3 the angle between $P_{12}$ and $zL$-line (the crossline of the planes A and B) has to be 54.7°.

Figure 15.10b illustrates the presence of a trigonal structure inside the cubic crystal. In this case, the procedure to cut a one-facet polarimeter from this crystal is in 2 steps: (1) to cut out a prism from the cubic crystal, (2) to cut crystal through facet $a$ (or $b$, or $c$) perpendicularly to the plan $abc$. An example of a polarimeter utilizing the crystal thus cut is described later.

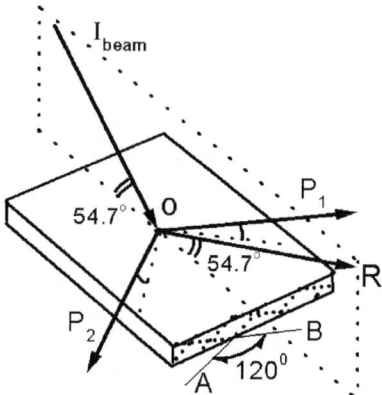

**Fig. 15.11.** Geometry of polarimeter

### 15.2.3 The Optics of Polarimeter

A hexagonal (quartz), trigonal, or cubic (Al, Ge, NaCl, LiF) crystal can be used for the polarimeter. As shown in Fig. 15.11, the polarizing planes A and B making angle $\zeta = 120°$ to each other are oriented symmetrically with respect to the mechanical plane C: the polarizing planes A, B are marked with the dotted lines on the front side of the crystal. First, we specify the direction of the incident beam and those of the reflected beams with respect to the mechanical plane, or the crystal surface. The incident beam $P_{12}$ ($I_{\text{beam}}$) is reflected into the three directions: (1) in the $P_1$ and $P_2$ directions by the series of planes A and B, respectively, (2) in the $R$ direction by the atomic planes C. Let $P_1$, $P_2$, and $R$ denote also the directions of the respective beams, $\theta_I$ the angle of incidence onto the crystal surface (the Bragg angle for plane C), and $\theta_B$ the Bragg angle with respect to the planes A and B. Simple geometry leads to the relationship between $\theta_I$ and $\theta_B$,

$$\sin\theta_B = \sin\theta_I \times \sin(\zeta/2) . \tag{15.5}$$

For a given value of $\lambda$, by remembering that $\theta_B$ is given by Bragg's law ($2d\sin\theta_B = k\lambda$), we can find $\theta_I$ from (15.5). Geometrical considerations give the angle $P_1P_2$ between the directions of $P_1$ and $P_2$,

$$\sin(\angle P_1P_2/2) = \sin\theta_I \times \sin\zeta . \tag{15.6}$$

The angle between the incident $P_{12}$ beam and the reflected $P_1$ or $P_2$ beam is

$$\angle P_{12}P_1 = \angle P_{12}P_2 = 180° - 2\theta_B . \tag{15.7}$$

The angle between projection $OB_1$ and $OB_2$ of beams $P_1$ and $P_2$ to crystal plane is

$$\angle B_1OB_2 = 2\arctan(\tan\theta_I \times \sin\zeta) . \tag{15.8}$$

The ideal polarimeter is to reflect the p- and s-components with $\theta_B = 45°$, one along $P_1$ and the other along $P_2$. Since the s- and p-components have mutually perpendicular electric field vectors, the angle $P_1 P_2$ is $90°$. See Fig. 15.7. Thus the ideal polarimeter has $P_{12} \perp P_1$, $P_{12} \perp P_2$, and $P_1 \perp P_2$. These relations and the incident angle $\theta_I$ can be used to determine the positions of the X-ray source and the detectors with respect to the mechanical crystal surface. Obviously, $\boldsymbol{E}_1 \parallel P_2$, $\boldsymbol{E}_2 \parallel P_1$, $\boldsymbol{E}_1 \perp P_{12}$, and $\boldsymbol{E}_2 \perp P_{12}$, where $\boldsymbol{E}_1$ and $\boldsymbol{E}_2$ are the electric field vectors of the beams $P_1$ and $P_2$, respectively, which are fully polarized.

From (15.5) it is obvious that, for $\theta_B = 45°$, we have $\theta_I = 54.7°$. For a given spacing $d$ (for A and B planes) the wavelength $\lambda_p$ is fixed by Bragg's law: $k\lambda_p = 2d \sin 45° = d\sqrt{2}$. The beam $R$ is reflected according to $k\lambda = 2d_0 \sin 54.7° = 2d_0\sqrt{2/3}$, where $d_0$ is for plane C.

In practice, many X-ray sources radiate lines with various wavelengths, according to the transitions in the ions or atoms in the source. It is difficult to find a crystal having $d$ that exactly matches the above condition. We would then use a crystal with the Bragg angle, the deviation of which from $45°$ should be as small as possible. In this case, the angle $\theta_I$ is again found from (15.5). The relative intensity of the s- to p-components in beams $P_1$ and $P_2$ is proportional to $|\cos 2\theta_B|$ (or $\cos^2 2\theta_B$) for an ideal (or ideal mosaic) crystal, as mentioned earlier. The ideal mosaic crystals provide higher purity for a given Bragg angle $\theta_B \neq 45°$. For instance, unpolarized radiation reflected in $R$ (C planes, $\theta_I = 54.7°$ for $\theta_B = 45°$) would be more polarized if the crystal is ideal mosaic, i.e., the degree of polarization is 0.80 against 0.49 for an ideal crystal.

Even though no any real crystals are perfect, commercial flat quartz is nearly perfect, and its diffraction properties are satisfactorily described by the dynamical theory. For many other crystals, calibration is desirable, which provides reflection coefficients for each polarized component at arbitrary Bragg angles.

### 15.2.4 Relationship between Bravais Indices of Polarizing and Mechanical Planes

An important practical problem is how to define the three planes in a crystal for a given wavelength $\lambda$. The procedure is as follows: (1) from the list of crystals one chooses polarizing planes according to the relation $d\sqrt{2} = k\lambda$; for example, any crystal planes parallel to the $z$-axis in Fig. 15.9 could be polarizing planes A and B, (2) using the Bravais indices of the polarizing planes A, B one determines Bravais indices of the mechanical plane C and cuts the boule parallel to this mechanical plane. Reflectivity of the polarizing planes should be high enough for the photon numbers and the detector sensitivity in the particular experiment.

Step (1): Table 15.1 shows the series of atomic planes, or crystal planes, in a quartz designated with a combination of numbers 1, 2, and 3 and another

**Table 15.1.** Some of quartz cuts, their interplanar spaces as well as angle with 10(−1)0

| cut | −5140 | 41−50 | 21−30 | 12−30 | 14−50 | 1−540 | 1−320 | 2−310 | 4−510 | −5410 | −3210 | −3120 | −5140 |
|---|---|---|---|---|---|---|---|---|---|---|---|---|---|
| $d$ Å | 0.929 | 0.929 | 1.608 | 1.608 | 0.929 | 0.929 | 1.608 | 1.608 | 0.929 | 0.929 | 1.608 | 1.608 | 0.929 |
| $\delta$ ° | −10.89 | 10.89 | 19.11 | 40.89 | 49.11 | 70.89 | 79.11 | −79.11 | −70.89 | −49.11 | −40.89 | −19.11 | −10.89 |

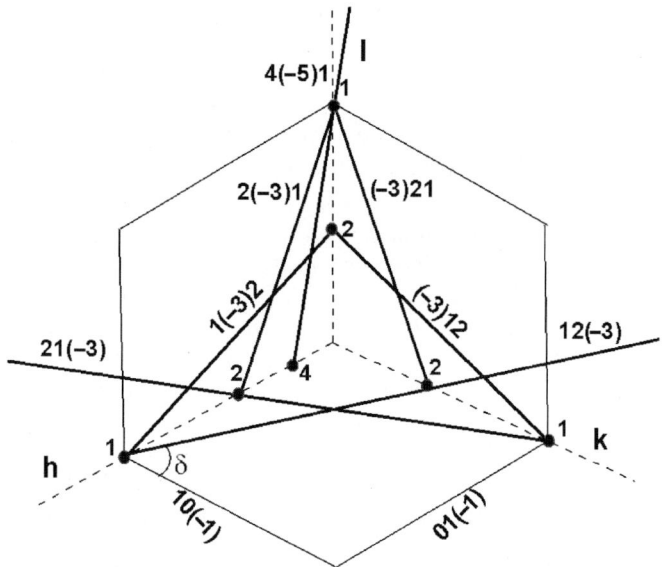

**Fig. 15.12.** Polarizing and mechanical planes in quartz

series of planes, designated with numbers 1, 4, and 5. This table also shows the spacing $d_{hkl}$ of the atomic planes and the corresponding angles $\delta$ between the particular plane and the plane 10(−1)0 (see Fig. 15.12). The last Bravais index (zero for all the cases) is not shown in the figure for simplicity.

For the planes parallel to the $z$-axis in a hexagonal crystal, Fig. 15.12, spacing $d_{hkl}$ and angle $\delta$ can be calculated by

$$d_{hkl} = \frac{1}{2}\sqrt{3}\frac{a}{\sqrt{h^2 + k^2 + h \cdot k}} \ , \tag{15.9}$$

$$\sin\delta = \frac{1}{2}\sqrt{3}\frac{|k| \cdot sign\,(|l| - |h|)}{\sqrt{h^2 + k^2 + h \cdot k}} \ , \tag{15.10}$$

where $h$, $k$, and $l$ are the first three Bravais indices and $a = 4.913\,\text{Å}$ is the lattice length.

For a trigonal (rhombohedral) crystal in Fig. 15.13 the spacing can be calculated in the two ways:

$$d_{hkl} = \frac{a\sin(\alpha/2)}{\sqrt{h^2 + k^2 + h \cdot k}} \ , \tag{15.11}$$

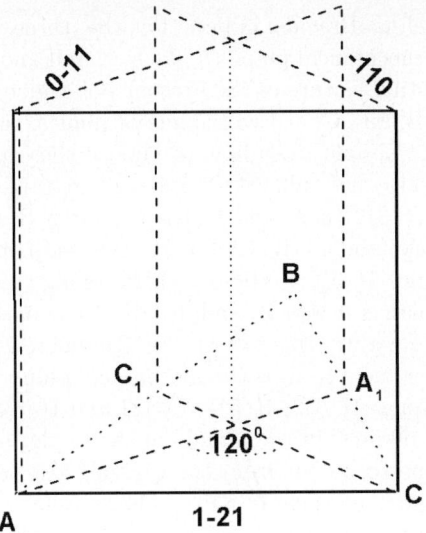

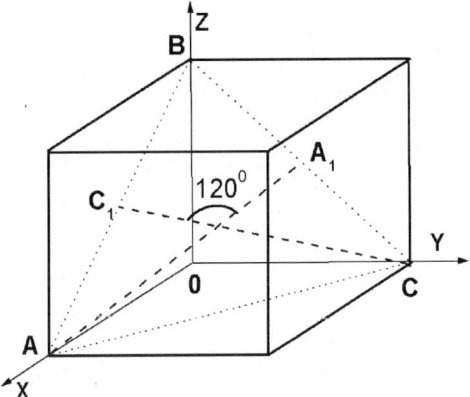

**Fig. 15.13.** (a) How to find the polarizing planes in a cubic crystal. (b) Disposition of mechanical and polarizing planes in cubic crystals

$$d_{hkl} = \frac{a \sin (\alpha/2)}{\sqrt{h^2 - k \cdot l}} \, , \tag{15.12}$$

where $\alpha$ is the angle between ribs. Formula (15.11) is the general formula and (15.12) is valid for the planes that satisfy the relation $h + k + l = 0$ (polarizing planes). A cubic crystal (Fig. 15.13) is a special case of a trigonal crystal in the sense that $\alpha = 90°$, so that $\sin(\alpha/2) = 1/\sqrt{2}$. Equation (15.11) for a trigonal crystal is similarly transformed into (15.9) for a hexagonal crystal with $\alpha = 120°$, so that $\sin(\alpha/2) = \sqrt{3}/2$.

Step (2): We define Bravais indices for the three planes: a polarizing plane $(h_1k_1l_1)$, the mechanical plane $(h_mk_ml_m)$, and another polarizing plane $(h_2k_2l_2)$. An interesting feature of the present polarimeter is we have two sets of three planes, A, B, and C, or two triplets of planes, as shown below. There are two independent simple rules how to choose these planes. The first rule which we call the "skip one" rule (SOR) is based on Table 15.1. As an example we take the plane 1(−3)2 (here and below the forth Bravais indices, zero, is omitted for simplicity) and go right following the "skip one" rule by two steps. We then obtain planes 1(−3)2, 4(−5)1, (−3)21; among them 1(−3)2 and (−3)21 are the polarizing planes A and B, and 4(−5)1 is the mechanical plane C. We also might go left to get 1(−3)2, 14(−5), 21(−3), where 1(−3)2 and 21(−3) are the polarizing planes and 14(−5) is the mechanical plane. Thus, we obtain two sets of triplets of planes: 1(−3)2, 4(−5)1, (−3)21 and 1(−3)2, 14(−5), 21(−3). We put the mechanical plane C in the middle of the triplet to show its important role as the real plane to be cut from the crystal. The second rule, which we call "cyclic rearrangement" rule (CRR), is for a polarizing plane $hk(−l)$ one finds two polarizing planes $h_1k_1l_1$ and $h_2k_2l_2$ by cyclically interchanging the Bravais indices in the way

$$h_1 = (−l);\ k_1 = h;\ (−l_1) = k; \quad \text{and} \quad h_2 = k;\ k_2 = (−l);\ l_2 = h;$$

or more succinctly $(h_1k_1l_1) = ((−l)hk)$ and $(h_2k_2l_2) = (k(−l)h)$. Such cycling means rotation of the initial plane by 60°. The corresponding mechanical planes are

$$(h_{B1}k_{B1}l_{B1}) = (h − h_1;\ k − k_1;\ (−l) − (−l)_1) = (h − (−l);\ k − h;\ (−l) − k)$$

$$(h_{B2}k_{B2}l_{B2}) = (h − h_2;\ k − k_2;\ (−l) − (−l)_2) = (h − k;\ k − (−l);\ (−l) − h)$$

The above procedure means rotation by 30°. It is seen that, in a hexagonal crystal, for the initial plane $hk(−l)$ one can find two polarizing planes, each of which is paired to the initial plane, and two corresponding mechanical planes. For the example of 13(−4) plane we have two sets of polarizing planes: 13(−4), (−4)13 and 13(−4), 3(−4)1. The corresponding mechanical planes are 1−(−4), 3−1, (−4) −3 = 5, 2(−7) and 1−3, 3−(−4), (−4)−1 = 2(−7)5, respectively. As we mentioned earlier, in a hexagonal, trigonal, or cubic crystal, any crystal plane which is parallel to the crystallographic axis can be a polarizing plane. In Fig. 15.14 the equation for the crystallographic axis is $x = y = z$, and the plane $ABC$ with Bravais indices $hkl$ is described as $hx + ky + lz = na$. That is the reason why any polarizing plane satisfies the relation $h + k + l = 0$.

Figure 15.15 shows the notation of some planes in a trigonal crystal, along with the trigonal $h_Tk_Tl_T$ and hexagonal $h_hk_hl_h$ coordinate systems for comparison. The trigonal system rotated by 30° coincides with the hexagonal coordinate system. Remembering that the mechanical plane C have the same 30° angle with polarizing planes A, B, we may conclude that the procedures like SOR, CRR work also for trigonal (rhombohedral) crystals and cubic crystals (as a special case of a trigonal crystal).

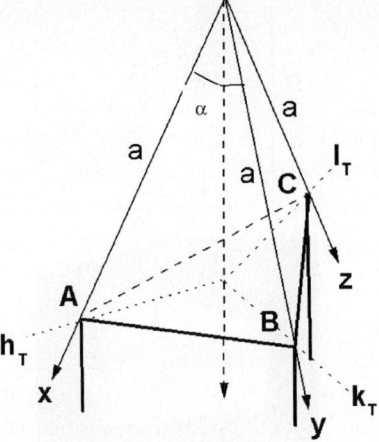

**Fig. 15.14.** Geometry of the boule for a trigonal crystal

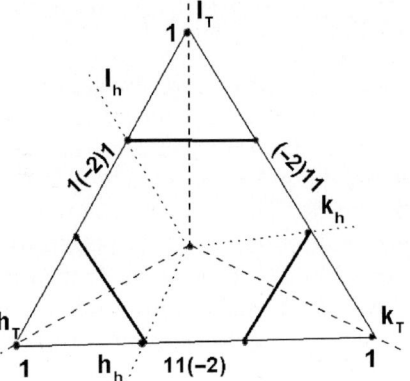

**Fig. 15.15.** Polarizing and mechanical planes in a trigonal crystal, and hexagonal and trigonal coordiantes

Figure 15.13 shows the case of a cubic crystal; the C plane is parallel to 1(−2)1 (solid), and the A and B planes are parallel to (−1)10 and 0(−1)1 (dotted). Other possible C planes are (−2)11 and 11(−2).

## 15.2.5 Characteristics of the Four-Facet Quartz X-ray Polarimeter

We fabricated our first polarimeter from a quartz crystal, because quartz is well known for its high quality (narrow rocking curve, high elasticity, and other mechanical properties). We use the term "polarimeter" to express a properly cut crystal. Figure 15.16 shows the four-facet polarimeter, although it may appear like a simple prism. Each of the four facets is surface C, which is parallel to the crystallographic axis, as shown outside the prism. These planes

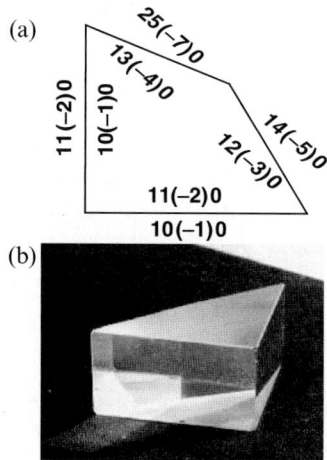

**Fig. 15.16.** (a) Geometry of four-facet polarimeter. (b) Appearance of four-facet polarimeter

are 10(–1)0, 11(–2)0, 25(–7)0, and 14(–5)0. Each prism facet has its own pair of A and B polarizing planes, one of which (A) is given inside the prism; 11(–2)0, 10(–1)0, 13(–4)0, and 12(–3)0, respectively. For the mechanical surface 14(–5)0, for example, the polarizing cuts are 12(–3)0. On the contrary, if the mechanical surface are 10(–1)0, then the polarizing planes are 11(–2)0.

The above polarimeter allows one to study polarization of X-rays at twelve wavelengths: $6\,\text{Å}$, $3.47\,\text{Å}$, $2.27\,\text{Å}$, $1.67\,\text{Å}$ in the first order, $3\,\text{Å}$, $1.74\,\text{Å}$, $1.13\,\text{Å}$ in the second order, $2\,\text{Å}$, $1.16\,\text{Å}$ in the third order, $1.5\,\text{Å}$, $0.87\,\text{Å}$ in the forth order, and $0.69\,\text{Å}$ in the fifth order. At these wavelengths, the integrated reflectivity, (15.1), is enough high, and the polarization purity is 100%. As noted already, a polarimeter works fairly well even at Bragg angles that are close to $45°$. The present polarimeter was tested using Cu K-alpha radiation, $1.5404\,\text{Å}$, emitted from an X-ray tube. The radiation was reflected in the forth reflection order at $\theta_B = 46.4°$. Two signals of equal intensities were registered by Si detectors mounted in two mutually perpendicular directions.

Examples of the transitions that can be studied with this prism polarimeter are listed below. Notations are as follows: 20(–2)0 means the second order of reflection from 10(–1)0, 22(–4)0 the second order from 11(–2)0, and 33(–6)0 the third order from 11(–2)0. Helium-like argon and helium-like potassium are frequently encountered in thermonuclear fusion experiments like laser produced plasmas, z-pinches, and tokamak plasmas. An investigation of the degree of polarization of these lines can provide important information on the anisotropy of the plasma, e.g., the distribution of electromagnetic fields, the distribution of the current flow, and the anisotropic electron velocity distribution function.

| Cut | $2d$ | $\lambda_{45}$ | Element | | Transition | $\lambda$ | $\theta_B$ |
|---|---|---|---|---|---|---|---|
| 10(−1)0 | 8.51 | 6.02 | Al, | H-like | 1s–3p | 6.05 | 45.3 |
| 20(−2)0 | 4.25 | 3.01 | K, | He-like | $1s^2$–1s3p | 3.02 | 45.2 |
| 11(−2)0 | 4.91 | 3.47 | Ar, | He-like | $1s^2$–1s3p | 3.37 | 43.3 |
| 22(−4)0 | 2.46 | 1.74 | Mn, | He-like | $1s^2$–1s3p | 1.72 | 44.5 |
| 33(−6)0 | 1.64 | 1.16 | Zn, | He-like | $1s^2$–1s3p | 1.18 | 46.1 |
| 12(−3)0 | 3.22 | 2.27 | V, | H-like | 1s–2p | 2.29 | 45.4 |
| 13(−4)0 | 2.36 | 1.67 | Co, | H-like | 1s–2p | 1.66 | 44.7 |

The following developments are planned (1) to expand the present principle to the transmission scheme, and (2) to use bent crystals.

(E.O. Baronova, M.M. Stepanenko)

# References

1. T. Fujimoto: in Proc. US-Japan Plasma Polarization Spectroscopy Workshop, Report UCRL-ID-146907 (2001) p.1
2. A. Compton, S. Alison: in *X-rays in theory and experiment* (D. Van Nostrand 1960)
3. C.G. Barkla: Proc. Roy. Soc., London **77**, 247 (1906)
4. Qun Shen, K.D. Finkelshtein: Phys. Rev. B **45**, 5075 (1992)
5. P. Beiersdorfer, M. Slater: in Proc. US-Japan Plasma Polarization Spectroscopy Workshop, Report UCRL-ID-146907 (2001) p. 329
6. K.D. Finkelstein, C. Staffa, Qun Shen: Nucl. Instr. Meth. A **347**, 124 (1994)
7. L. Jakubowski, M. Sadowski, E. Baronova: Nucl. Fusion **44**, 395 (2004)
8. M.M. Stepanenko, E.O. Baronova: Instr. Exper. Tech. **42**, 1 (1999)
9. E.O. Baronova, M.M. Stepanenko, N.R. Pereira: Rev. Sci. Instrum. **72**, 1416 (2001)
10. E.O. Baronova, M.M. Stepanenko: Nucleonika **46**, Suppl. (2001)
11. L. Jakubowski, M. Sadowski, E.O. Baronova: Cechoslovac. J. Phys: Suppl. S3, **50**, 173 (2000)
12. V.V. Lider, E.O. Baronova, M.M. Stepanenko: Crystallography Reports, **46**, 341 (2001); Translated from Krystallografiya **46**, 391 (2001)
13. E.O. Baronova, M.M. Stepanenko: Plasma Phys. Control. Fusion, **45**, 1113 (2003)
14. N.R. Pereira, E.O. Baronova, M.M. Stepanenko: presented at APS Meeting, October, (2005).
15. N.R. Pereira: submitted to J. Modern Opt.

# Appendix A

# Light Polarization and Stokes Parameters

In this appendix, we introduce the polarization of light emitted by atoms both in the classical and quantum pictures. We then introduce the standard method to quantify the polarization characteristics of a light beam.

## A.1 Electric Dipole Radiation

For the purpose of showing why an atomic transition produces polarized light, we begin by adopting the classical picture. We assume a classical atom, i.e., an oscillating electric dipole; this dipole emits electromagnetic waves, i.e., light. This dipole is located at the origin of the coordinate system. First, we assume that the oscillation is on the $z$-axis, the polar axis, then this atom emits a characteristic radiation. This particular situation is treated in standard textbooks of classical electromagnetism. See Fig. A.1a. It is well known that the observed intensity has its characteristic angular dependence (Fig. A.1b), and when observed from a direction that has polar angle $\theta$ the observed intensity is given by

$$I_\pi(\theta) = \frac{1}{2} I_0 \sin^2 \theta \,, \tag{A.1}$$

where $I_0$ is a constant intensity, which is proportional to the square of the amplitude of the oscillator. In the present discussion, we assume that our detector has no polarization selectivity. The light is linearly polarized, irrespective of $\theta$ (Fig. A.1(a)), and no light is observed from the $\pm z$ directions (Fig. A.1a,b). Since it will turn out soon that this light is nothing but the $\pi$ light, which was introduced in Chap. 4, we adopt this notation in (A.1).

If the dipole is now rotating around the polar axis on the $x$–$y$ plane, and the sense of rotation is right-handed (Fig. A.1c), the observed intensity (Fig. A.1d) is given as

$$I_{\sigma+}(\theta) = \frac{1}{4} I_0 (1 + \cos^2 \theta) \,, \tag{A.2}$$

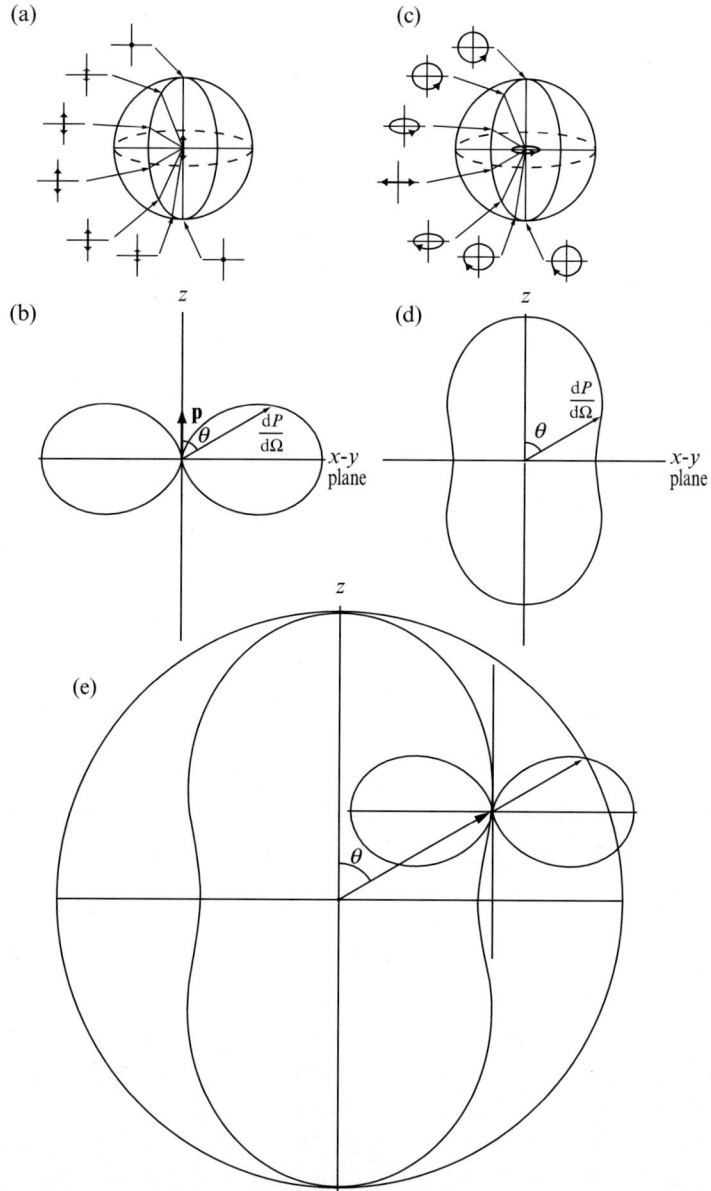

**Fig. A.1.** Polarization of the electric field seen by looking from different directions at a dipole oscillator executing (**a**) linear oscillations ($\pi$ light) and (**c**) circular oscillations ($\sigma+$ light). The observer is at a large fixed distance. Angular distribution of electric dipole radiation (**b**) for (**a**), and (**d**) of (**c**). The length of the radius vector is proportional to the intensity. (**e**) When the light is emitted by equal numbers of oscillators: $\pi$, $\sigma+$, and $\sigma-$, the total intensity of the light is independent of the observation direction (Reconstructed after A. Corney, *Atomic and Laser Spectroscopy.*)

where the amplitude, or more exactly, the magnitude, of the dipole is assumed the same as the first case. The state of polarization of the emitted light is a little complicated (Fig. A.1c): If observed from the positive $z$ direction the light is left-circularly polarized and from the negative $z$ direction, it is right-circularly polarized. When observed from an oblique direction, the light is left- or right-elliptically polarized. On the equatorial plane, the light is observed as linearly polarized with the polarization direction perpendicular to the polar axis. Thus, this light may be regarded as the $\sigma$ light. This is the reason why we have adopted the notation $\sigma+$ in (A.2). For a rotating dipole in the left-handed direction, the above argument is valid except for the sense of circular or elliptic polarization of the emitted light. The intensity for this latter case may be denoted as $I_{\sigma-}(\theta)$.

Suppose, we have a large number of oscillators, the frequencies of which are slightly different from each other so that the electromagnetic waves emitted by these oscillators do not interfere each other. In other words, we have incoherent superposition of the waves. We further suppose that we have equal numbers ($N$) of oscillators oscillating in the $z$-direction, those rotating in the right-handed sense and those in the left-handed sense. Then, the total intensity of the light emitted by these oscillators is

$$NI_\pi(\theta) + NI_{\sigma+}(\theta) + NI_{\sigma-}(\theta) = \frac{3}{2}NI_0, \tag{A.3}$$

being independent of the observation direction (Fig. A.1e). It can also be shown that the light is unpolarized.

In Chap. 4, in the quantum mechanical picture we discussed the presence of alignment in the ensemble of emitting atoms and the characteristics of the light emitted by this ensemble. Equation (4.5) expresses the intensity observed with a linear polarizer, the transmission axis of which is given by $e$. As noted in several places in the text, the polarization characteristics of the light for the transition of the 0–1 angular momentum pair are the same as those of the classical oscillator, which was discussed above. In (4.5), we assume $J_s = 0$ so that $J_p = 1$. Then, the second term is $-\frac{\sqrt{6}}{2}(1 - 3\cos^2\eta)\frac{a(p)}{n(p)}$ (see also the discussion after (7.9)), where $\eta$ is the angle of $e$ with respect to the polar axis. First, we assume that, in the upper level with $J_p = 1$, only the $M = 0$ state is populated with the $M = \pm 1$ states unpopulated (see Fig. 7.1a). In this case, $a(p)/n(p) = -2/\sqrt{6}$ (see Table 4.1). Suppose we observe this light with $e$ lying in the plane containing the polar axis, so that $\eta = \frac{\pi}{2} - \theta$. It is readily seen that the intensity is proportional to $3\sin^2\theta$, in accordance with (A.1). This is the $\pi$ light and when observed on the equatorial plane its intensity is proportional to 3. If only the $M = +1$ state is populated (Fig. 7.1a), on the other hand, the square brackets of (4.5) reduces to $\frac{3}{2}\sin^2\eta$. Since we assume here that we are observing a linearly polarized component, this situation is different from (A.2), in which we assumed no polarization selectivity. If $\eta = \pi/2$, the observed intensity is proportional to $3/2$; this is also the case for observation on the equatorial plane. This light is identified with the $\sigma$ light. This conclusion is the same for the case of the $M = -1$ population. It is thus concluded that if the three magnetic substates are equally populated, the $\pi$ light intensity and the $\sigma$ light intensity as observed on the $x$–$y$ plane are equal, or the observed light is unpolarized. It is also readily shown that the light is isotropically emitted and it is unpolarized.

We have thus shown that the classical electric dipole radiation is equivalent to the radiation emitted by atoms in the transition between the 0–1 angular momentum pair.

## A.2 Stokes Parameters

A standard way of specifying the state of polarization of a light beam is to use the Stokes parameters. We suppose that we have a beam of light, i.e., plane parallel and propagating, which is quasi-monochromatic, or the incoherent superposition of sinusoidal electromagnetic waves. We may imagine a situation in which we observe the light emitted by an ensemble of atoms in Fig. A.1 at a large distance on the $z$-axis. We rotate our ensemble of atoms by a certain angle; the emitted light is virtually a plane parallel light beam propagating in the positive $z$-direction, and we observe this beam facing the beam. Suppose we have a linear polarizer and two circular polarizers, each transmitting right-circularly and left-circularly polarized light, respectively. We measure the intensity of the beam with one of the polarizers placed in front of our detector. We denote the observed intensity as $I_\theta$ or $I_r$ or $I_l$, where $I_\theta$ is the intensity with the linear polarizer with its transmission axis making angle $\theta$ with respect to the $x$-axis and r and l denote the right-circularly and left-circularly polarized component, respectively. The Stokes parameters are defined as

$$\begin{aligned}
I &= I_0 + I_{\pi/2} = I_{\pi/4} + I_{3\pi/4} = I_r + I_l \\
Q &= I_0 - I_{\pi/2} \\
U &= I_{\pi/4} - I_{3\pi/4} \\
V &= I_r - I_l .
\end{aligned} \tag{A.4}$$

It may be shown that

$$Q^2 + U^2 + V^2 \leqslant I^2 . \tag{A.5}$$

The degree of polarization of the light beam is defined as

$$P = \frac{\sqrt{Q^2 + U^2 + V^2}}{I} . \tag{A.6}$$

Any light beam can be expressed as a superposition of completely polarized beam $(P = 1)$ and unpolarized (natural) beam $(P = 0)$,

$$I = I_{\text{pol}} + I_{\text{unpol}} . \tag{A.7}$$

# Appendix B

# Angular Momentum and Rotation Matrix

In the text, we have used some elements of angular momentum theory and rotation matrix without explanations. These elements are introduced and discussed in this Appendix. The first is the coupling coefficients of angular momenta and then the rotation matrix follows which describes the transformation of atomic states under spatial rotation.

## B.1 Angular Momentum Coupling

### B.1.1 3-$j$ Symbol

Suppose we have two quantum systems, each of which has angular momentum $\boldsymbol{j}_1$ and $\boldsymbol{j}_2$, respectively. They are coupled to form a single system which has its angular momentum $\boldsymbol{J}$. The angular momentum state of the latter system is expressed in terms of those of the former systems

$$|j_1 j_2 JM\rangle = \sum_{m_1 m_2} \langle j_1 j_2 m_1 m_2 | JM\rangle |j_1 m_1\rangle |j_2 m_2\rangle, \qquad (\text{B.1})$$

where $\langle j_1 j_2 m_1 m_2 | JM\rangle$ is called the Clebsch–Gordan coefficient. There is a constraint among $j_1$, $j_2$, and $J$, and $m_1 + m_2 = M$. This coefficient is sometimes expressed in terms of the 3-$j$ symbol

$$\langle j_1 j_2 m_1 m_2 | JM\rangle = (-)^{j_1 - j_2 + M} \sqrt{2J + 1} \begin{pmatrix} j_1 & j_2 & J \\ m_1 & m_2 & -M \end{pmatrix}, \qquad (\text{B.2})$$

where $(-)$ stands for $(-1)$ as in Chap. 4. The 3-$j$ symbol has symmetry properties:

$$\begin{pmatrix} j_1 & j_2 & j_3 \\ m_1 & m_2 & m_3 \end{pmatrix}$$

is

1. Invariant in a circular permutation of the three columns
2. Multiplied by $(-1)^{j_1+j_2+j_3}$ in a permutation of two columns
3. Multiplied by $(-1)^{j_1+j_2+j_3}$ when we simultaneously change the signs of $m_1$, $m_2$ and $m_3$

In literature, various formulas to calculate the 3-$j$ symbols and some special cases are given. We here note only two special cases which are sometime useful:

$$\begin{pmatrix} j & j & 0 \\ m & -m & 0 \end{pmatrix} = \frac{(-)^{j-m}}{\sqrt{2j+1}},$$

$$\begin{pmatrix} j & j & 1 \\ m & -m & 0 \end{pmatrix} = \frac{(-)^{j-m}m}{\sqrt{(2j+1)(j+1)j}}.$$

(B.3)

Extensive tables are published for numerical values of the 3-$j$ symbols, and we reproduce a part of them as Table B.1 [1]. This table gives values of

$$\begin{pmatrix} j_1 & j_2 & j_3 \\ m_1 & m_2 & m_3 \end{pmatrix}^2$$

in the following notation: given are the exponents of the prime numbers in the order 2, 3, 5, 7, ... Negative exponents are underscored. For example

$$2\underline{1}0\underline{2} = \frac{2^2 \times 3^1 \times 5^0}{7^2} = \frac{12}{49}.$$

An asterisk indicates a negative radical. To find the value of a 3-$j$ symbol, use the symmetry properties (1)–(3) to (a) interchange the columns so that $j_1 \geqslant j_2 \geqslant j_3$ and (b) change the signs (if necessary) of $m_1$, $m_2$, and $m_3$ so that $m_2 \leqslant 0$.

### B.1.2 6-$j$ Symbol

We have three angular momenta $j_1$, $j_2$, and $j_3$, which are coupled to form a single system with angular momentum $J$. The order of coupling can be either of the following: (1) $j_1+j_2 = g$, $g+j_3 = J$. The resultant angular momentum state may be expressed as

$$|(j_1 j_2)g, j_3; JM\rangle = \sum_{m_1 m_2 m_3 \mu} \langle j_1 j_2 m_1 m_2 | g\mu\rangle \langle g j_3 \mu m_3 | JM\rangle |j_1 m_1\rangle |j_2 m_2\rangle |j_3 m_3\rangle.$$

(B.4)

(2) $j_1 + j_3 = g'$, $g' + j_2 = J$

$$|(j_1 j_3)g', j_2; JM\rangle = \sum_{m_1 m_2 m_3 \mu'} \langle j_1 j_3 m_1 m_3 | g'\mu'\rangle \langle g' j_2 \mu' m_2 | JM\rangle |j_1 m_1\rangle |j_2 m_2\rangle |j_3 m_3\rangle.$$

(B.5)

**Table B.1.** Numerical values of 3-$j$ symbols. For explanation, see the text. "0" means null

| $j_1$ | $j_2$ | $j_3$ | $m_1$ | $m_2$ | $m_3$ | | $j_1$ | $j_2$ | $j_3$ | $m_1$ | $m_2$ | $m_3$ | |
|---|---|---|---|---|---|---|---|---|---|---|---|---|---|
| 1 | 1 | 0 | 0 | 0 | 0 | *01 | 2 | 1 | 1 | 1 | −1 | 0 | *101 |
| 2 | 1 | 1 | 0 | 0 | 0 | 111 | 2 | 1 | 1 | 2 | −1 | −1 | 001 |
| 2 | 2 | 0 | 0 | 0 | 0 | 001 | 2 | 3/2 | 1/2 | 0 | −1/2 | 1/2 | *101 |
| 3 | 2 | 1 | 0 | 0 | 0 | *0111 | 2 | 3/2 | 1/2 | 1 | −3/2 | 1/2 | 201 |
| 3 | 3 | 0 | 0 | 0 | 0 | *0001 | 2 | 3/2 | 1/2 | 1 | −1/2 | −1/2 | 211 |
| 3 | 3 | 2 | 0 | 0 | 0 | 2111 | 2 | 3/2 | 1/2 | 2 | −3/2 | −1/2 | *001 |
| 4 | 2 | 2 | 0 | 0 | 0 | 1011 | 2 | 3/2 | 3/2 | −1 | −1/2 | 3/2 | 101 |
| 4 | 3 | 1 | 0 | 0 | 0 | 2201 | 2 | 3/2 | 3/2 | 0 | −3/2 | 3/2 | *201 |
| 4 | 3 | 3 | 0 | 0 | 0 | *10011 | 2 | 3/2 | 3/2 | 0 | −1/2 | 1/2 | *201 |
| 4 | 4 | 0 | 0 | 0 | 0 | 02 | 2 | 3/2 | 3/2 | 1 | −3/2 | 1/2 | 101 |
| 4 | 4 | 2 | 0 | 0 | 0 | *22111 | 2 | 3/2 | 3/2 | 1 | −1/2 | −1/2 | 0 |
| 4 | 4 | 4 | 0 | 0 | 0 | 120111 | 2 | 3/2 | 3/2 | 2 | −3/2 | −1/2 | *101 |
| 1/2 | 1/2 | 0 | 1/2 | −1/2 | 0 | 1 | 2 | 3/2 | 3/2 | 2 | −1/2 | −3/2 | 101 |
| 1 | 1/2 | 1/2 | 0 | −1/2 | 1/2 | 11 | 2 | 2 | 0 | 0 | 0 | 0 | 001 |
| 1 | 1/2 | 1/2 | 1 | −1/2 | −1/2 | *01 | 2 | 2 | 0 | 1 | −1 | 0 | *001 |
| 1 | 1 | 0 | 0 | 0 | 0 | *01 | 2 | 2 | 0 | 2 | −2 | 0 | 001 |
| 1 | 1 | 0 | 1 | −1 | 0 | 01 | 2 | 2 | 1 | −1 | 0 | 1 | *101 |
| 1 | 1 | 1 | −1 | 0 | 1 | 11 | 2 | 2 | 1 | 0 | −1 | 1 | 101 |
| 1 | 1 | 1 | 0 | −1 | 1 | *11 | 2 | 2 | 1 | 0 | 0 | 0 | 0 |
| 1 | 1 | 1 | 0 | 0 | 0 | 0 | 2 | 2 | 1 | 1 | −2 | 1 | *011 |
| 1 | 1 | 1 | 1 | −1 | 0 | 11 | 2 | 2 | 1 | 1 | −1 | 0 | *111 |
| 1 | 1 | 1 | 1 | 0 | −1 | *11 | 2 | 2 | 1 | 1 | 0 | −1 | 101 |
| 3/2 | 1 | 1/2 | −1/2 | 0 | 1/2 | *11 | 2 | 2 | 1 | 2 | −2 | 0 | 111 |
| 3/2 | 1 | 1/2 | 1/2 | −1 | 1/2 | 21 | 2 | 2 | 1 | 2 | −1 | −1 | *011 |
| 3/2 | 1 | 1/2 | 1/2 | 0 | −1/2 | 11 | 2 | 2 | 2 | −2 | 0 | 2 | 1011 |
| 3/2 | 1 | 1/2 | 3/2 | −1 | −1/2 | *2 | 2 | 2 | 2 | −1 | −1 | 2 | *0111 |
| 3/2 | 3/2 | 0 | 1/2 | −1/2 | 0 | *2 | 2 | 2 | 2 | −1 | 0 | 1 | 1011 |
| 3/2 | 3/2 | 0 | 3/2 | −3/2 | 0 | 2 | 2 | 2 | 2 | 0 | −2 | 2 | 1011 |
| 3/2 | 3/2 | 1 | −1/2 | −1/2 | 1 | 111 | 2 | 2 | 2 | 0 | −1 | 1 | 1011 |
| 3/2 | 3/2 | 1 | 1/2 | −3/2 | 1 | *101 | 2 | 2 | 2 | 0 | 0 | 0 | *1011 |
| 3/2 | 3/2 | 1 | 1/2 | −1/2 | 0 | *211 | 2 | 2 | 2 | 1 | −2 | 1 | *0111 |
| 3/2 | 3/2 | 1 | 3/2 | −3/2 | 0 | 211 | 2 | 2 | 2 | 1 | −1 | 0 | 1011 |
| 3/2 | 3/2 | 1 | 3/2 | −1/2 | −1 | *101 | 2 | 2 | 2 | 1 | 0 | −1 | 1011 |
| 2 | 1 | 1 | −1 | 0 | 1 | *101 | 2 | 2 | 2 | 2 | −2 | 0 | 1011 |
| 2 | 1 | 1 | 0 | −1 | 1 | 111 | 2 | 2 | 2 | 2 | −1 | −1 | *0111 |
| 2 | 1 | 1 | 0 | 0 | 0 | 111 | 2 | 2 | 2 | 2 | 0 | −2 | 1011 |

These two expressions are related to each other by

$$|(j_1 j_2)g, j_3; JM\rangle = \sum_{g'} |(j_1 j_3)g', j_2; JM\rangle$$

$$\times \sqrt{(2g+1)(2g'+1)}(-)^{j_1+j_2+j_3+J} \left\{ \begin{matrix} j_1 & j_2 & g \\ j_3 & J & g' \end{matrix} \right\}, \quad (B.6)$$

where $\{:::\}$ is called the 6-$j$ symbol. It has symmetry relationships; it is invariant under

1. An interchange of its columns
2. An interchange of any two numbers in the bottom row with the corresponding two numbers in the top row

Thus, we have

$$\left\{ \begin{matrix} j_1 & j_2 & j_3 \\ g_1 & g_2 & g_3 \end{matrix} \right\} = \left\{ \begin{matrix} j_2 & j_1 & j_3 \\ g_2 & g_1 & g_3 \end{matrix} \right\} = \left\{ \begin{matrix} g_1 & g_2 & j_3 \\ j_1 & j_2 & g_3 \end{matrix} \right\} = \cdots$$

A useful special case is that one of the $j$s is null

$$\left\{ \begin{matrix} j_1 & j_2 & 0 \\ g_1 & g_2 & g_3 \end{matrix} \right\} = (-)^{j_1+g_1+g_3} \frac{\delta_{j_1 j_2} \delta_{g_1 g_2}}{\sqrt{(2j_1+1)(2g_1+1)}}, \quad (B.7)$$

where $|j_1 - g_1| \leqslant g_3 \leqslant |j_1 + g_1|$ must be met. Extensive tables are published, and we reproduce a part of them as Table B.2. The meaning of the notation is the same as in Table B.1. To find the value of a 6-$j$ symbol use the above symmetry properties to (a) place the largest of the six parameters in the upper left-hand corner ($j_1$ position), (b) place the largest of the remaining four parameters in the middle of the top row ($j_2$ position), and (c) make $g_1 > g_2$ if $j_1 = j_2$.

## B.2 Rotation Matrix

Suppose we rotate an angular momentum state in space. Let $(\phi, \theta, \gamma)$ be the Euler angle of this rotation (Fig. B.1).

We define the rotation operator $R$ so that $R(\phi\theta\gamma)|JM'\rangle$ is the result of this rotation. The matrix element is expressed as

$$R^{(J)}_{MM'}(\phi\theta\gamma) \equiv \langle JM|R(\phi\theta\gamma)|JM'\rangle$$

$$\equiv \langle JM|e^{-i\phi J_z} e^{-i\theta J_y} e^{-i\gamma J_z}|JM'\rangle, \quad (B.8)$$

where $J_y$ and $J_z$ are the operators of projection of the total angular momentum onto the $y$-axis and the $z$-axis, respectively. The matrix element can be decomposed into

**Table B.2.** Numerical values of 6-$j$ symbols. For explanation, see the text

| $j_1$ | $j_2$ | $j_3$ | $g_1$ | $g_2$ | $g_3$ | | $j_1$ | $j_2$ | $j_3$ | $g_1$ | $g_2$ | $g_3$ | |
|---|---|---|---|---|---|---|---|---|---|---|---|---|---|
| 1/2 | 1/2 | 0 | 0 | 0 | 1/2 | *1 | 2 | 3/2 | 1/2 | 3/2 | 1 | 1 | 311 |
| 1/2 | 1/2 | 0 | 1/2 | 1/2 | 0 | *2 | 2 | 3/2 | 1/2 | 2 | 3/2 | 1/2 | 402 |
| 1 | 1/2 | 1/2 | 0 | 1/2 | 1/2 | 2 | 2 | 3/2 | 1/2 | 2 | 3/2 | 3/2 | *202 |
| 1 | 1/2 | 1/2 | 1 | 1/2 | 1/2 | 22 | 2 | 3/2 | 3/2 | 0 | 3/2 | 3/2 | *4 |
| 1 | 1 | 0 | 0 | 0 | 1 | 01 | 2 | 3/2 | 3/2 | 1/2 | 1 | 1 | *31 |
| 1 | 1 | 0 | 1/2 | 1/2 | 1/2 | 11 | 2 | 3/2 | 3/2 | 1 | 1/2 | 3/2 | *201 |
| 1 | 1 | 0 | 1 | 1 | 0 | 02 | 2 | 3/2 | 3/2 | 1 | 3/2 | 1/2 | *201 |
| 1 | 1 | 0 | 1 | 1 | 1 | *02 | 2 | 3/2 | 3/2 | 1 | 3/2 | 3/2 | 402 |
| 1 | 1 | 1 | 1/2 | 1/2 | 1/2 | *02 | 2 | 3/2 | 3/2 | 3/2 | 1 | 1 | *112 |
| 1 | 1 | 1 | 1 | 0 | 1 | *02 | 2 | 3/2 | 3/2 | 2 | 1/2 | 3/2 | *202 |
| 1 | 1 | 1 | 1 | 1 | 0 | *02 | 2 | 3/2 | 3/2 | 2 | 3/2 | 1/2 | *202 |
| 1 | 1 | 1 | 1 | 1 | 1 | 22 | 2 | 3/2 | 3/2 | 2 | 3/2 | 3/2 | 422 |
| 3/2 | 1 | 1/2 | 0 | 1/2 | 1 | *11 | 2 | 2 | 0 | 0 | 0 | 2 | 001 |
| 3/2 | 1 | 1/2 | 1/2 | 1 | 1/2 | *02 | 2 | 2 | 0 | 1/2 | 1/2 | 3/2 | 101 |
| 3/2 | 1 | 1/2 | 1 | 1/2 | 1 | *22 | 2 | 2 | 0 | 1 | 1 | 1 | 011 |
| 3/2 | 1 | 1/2 | 3/2 | 1 | 1/2 | *42 | 2 | 2 | 0 | 1 | 1 | 2 | *011 |
| 3/2 | 3/2 | 0 | 0 | 0 | 3/2 | *2 | 2 | 2 | 0 | 3/2 | 3/2 | 1/2 | 201 |
| 3/2 | 3/2 | 0 | 1/2 | 1/2 | 1 | *3 | 2 | 2 | 0 | 3/2 | 3/2 | 3/2 | *201 |
| 3/2 | 3/2 | 0 | 1 | 1 | 1/2 | *21 | 2 | 2 | 0 | 2 | 2 | 0 | 002 |
| 3/2 | 3/2 | 0 | 1 | 1 | 3/2 | 21 | 2 | 2 | 0 | 2 | 2 | 1 | *002 |
| 3/2 | 3/2 | 0 | 3/2 | 3/2 | 0 | *4 | 2 | 2 | 0 | 2 | 2 | 2 | 002 |
| 3/2 | 3/2 | 0 | 3/2 | 3/2 | 1 | 4 | 2 | 2 | 1 | 1/2 | 1/2 | 3/2 | *201 |
| 3/2 | 3/2 | 1 | 1/2 | 1/2 | 1 | 321 | 2 | 2 | 1 | 1 | 0 | 2 | *011 |
| 3/2 | 3/2 | 1 | 1 | 0 | 3/2 | 21 | 2 | 2 | 1 | 1 | 1 | 1 | *201 |
| 3/2 | 3/2 | 1 | 1 | 1 | 1/2 | 321 | 2 | 2 | 1 | 1 | 1 | 2 | 221 |
| 3/2 | 3/2 | 1 | 1 | 1 | 3/2 | *121 | 2 | 2 | 1 | 3/2 | 1/2 | 3/2 | *301 |
| 3/2 | 3/2 | 1 | 3/2 | 1/2 | 1 | 22 | 2 | 2 | 1 | 3/2 | 3/2 | 1/2 | *322 |
| 3/2 | 3/2 | 1 | 3/2 | 3/2 | 0 | 4 | 2 | 2 | 1 | 3/2 | 3/2 | 3/2 | 102 |
| 3/2 | 3/2 | 1 | 3/2 | 3/2 | 1 | *42202 | 2 | 2 | 1 | 2 | 1 | 1 | *202 |
| 2 | 1 | 1 | 0 | 1 | 1 | 02 | 2 | 2 | 1 | 2 | 1 | 2 | 2121 |
| 2 | 1 | 1 | 1 | 1 | 1 | 22 | 2 | 2 | 1 | 2 | 2 | 0 | *002 |
| 2 | 1 | 1 | 2 | 1 | 1 | 222 | 2 | 2 | 1 | 2 | 2 | 1 | 22 |
| 2 | 3/2 | 1/2 | 0 | 1/2 | 3/2 | 3 | 2 | 2 | 1 | 2 | 2 | 2 | *202 |
| 2 | 3/2 | 1/2 | 1/2 | 1 | 1 | 21 | 2 | 2 | 2 | 1 | 1 | 1 | 2121 |
| 2 | 3/2 | 1/2 | 1 | 1/2 | 3/2 | 301 | 2 | 2 | 2 | 1 | 1 | 2 | 2121 |
| 2 | 3/2 | 1/2 | 1 | 3/2 | 1/2 | 4 | 2 | 2 | 2 | 3/2 | 1/2 | 3/2 | 3021 |
| 2 | 3/2 | 1/2 | 1 | 3/2 | 3/2 | *201 | 2 | 2 | 2 | 3/2 | 3/2 | 1/2 | 3021 |

$$R(\phi,\theta,\gamma)$$

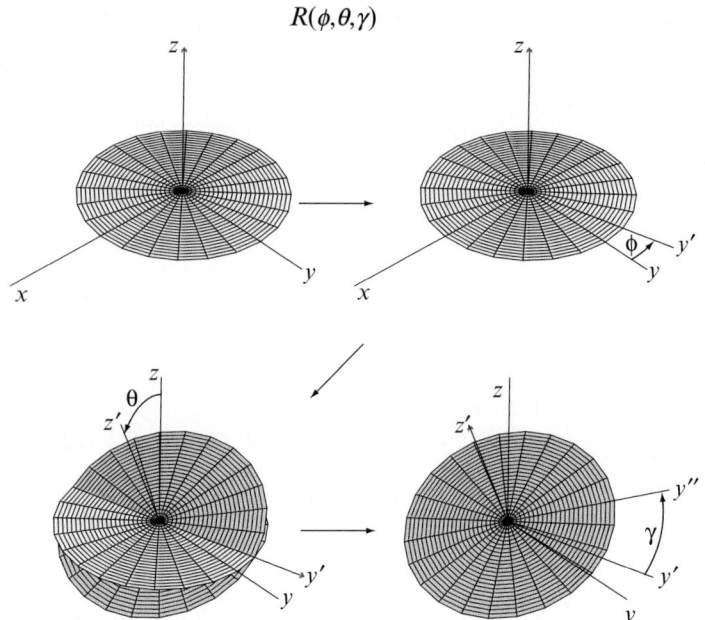

**Fig. B.1.** Three steps of Euler rotations. Fist, we rotate the rigid body about the $z$-axis by angle $\phi$. The second rotation is performed about the $y'$-axis, which is the body-fixed $y$-axis after the first rotation, by angle $\theta$. The third rotation is about the $z'$-axis by angle $\gamma$. The body $y$-axis now becomes the $y''$-axis [2]

$$R^{(J)}_{MM'}(\phi\theta\gamma) = \mathrm{e}^{-\mathrm{i}\phi M} r^{(J)}_{MM'}(\theta)\mathrm{e}^{-\mathrm{i}\gamma M'} \tag{B.9}$$

with

$$r^{(J)}_{MM'}(\theta) \equiv \langle JM|\mathrm{e}^{-\mathrm{i}\theta J_y}|JM'\rangle. \tag{B.10}$$

When $r^{(J)}_{MM'}(\theta)$ is known, the calculation of a rotation matrix is straightforward. We give a special case for $J = 1$:

$$r^{(1)}(\theta) = \begin{pmatrix} (1+\cos\theta)/2 & -\sin\theta/\sqrt{2} & (1-\cos\theta)/2 \\ \sin\theta/\sqrt{2} & \cos\theta & -\sin\theta/\sqrt{2} \\ (1-\cos\theta)/2 & \sin\theta/\sqrt{2} & (1+\cos\theta)/2 \end{pmatrix}. \tag{B.11}$$

The matrix for other $J$-values can be constructed from (B.11). It can also be directly obtained from Wigner's formula:

$$r^{(J)}_{MM'}(\theta) = \sum_K (-)^{K+M-M'} \frac{\sqrt{(J+M)!(J-M)!(J+M')!(J-M')!}}{(J-K-M)!K!(J+M'-K)!(K+M-M')!}$$

$$\times \left(\cos\frac{\theta}{2}\right)^{2J-2K-M+M'} \left(\sin\frac{\theta}{2}\right)^{2K+M-M'}. \tag{B.12}$$

# References

1. M. Weissbluth, *Atoms and Molecules* (Academic, New York, 1978)
2. J.J. Sakurai, *Modern Quantum Mechanics* (Addison-Wesley, Redwood City, 1985)

# Appendix C

# Density Matrix: Light Observation and Relaxation

In this Appendix, we review briefly the density matrix of an atomic ensemble, its time development, observation of light emitted by this ensemble, and its isotropic relaxation. We follow the formulation by Omont [1]. The standard textbook [2] is also referred to.

## C.1 Density Matrix

Suppose we have an ensemble of quantum systems, not necessarily atoms. The basic definition of the density operator for this ensemble is

$$\rho = \sum_m p_m |m\rangle\langle m|, \tag{C.1}$$

where $p_m$ is the number of systems to be found in state $m$. In many textbooks, $p_m$ is normalized to unity, so that $p_m$ is called "the probability of finding the system" or "the fractional population." In the present book, $p_m$ is not normalized so that it corresponds to the number of atoms in state $m$ in the case of an atomic ensemble. In any orthogonal and normalized basis, this operator can be written as

$$\rho = \sum_{ij} \rho_{ij} |i\rangle\langle j|. \tag{C.2}$$

The matrix $\rho_{ij}$ is called the density matrix. We now confine ourselves to an ensemble of atoms. The density operator expressed in terms of atomic states is

$$\rho = \sum_{\alpha JM\beta GN} \rho_{\alpha JM,\beta GN} |\alpha JM\rangle\langle \beta GN| \tag{C.3}$$

with the density matrix components

$$\rho_{\alpha JM,\beta GN} = \langle \alpha JM|\rho|\beta GN\rangle. \tag{C.4}$$

Here $J$ and $G$ are the total angular momentum quantum numbers, $M$ and $N$ are their projection onto the quantization axis and $\alpha$ and $\beta$ are other parameters necessary to specify an atomic state. The trace of this operator is the total population of the atoms:

$$\operatorname{Tr}\rho = n\,. \tag{C.5}$$

As noted above, in many cases, the trace of the density operator is normalized, i.e., $n = 1$. The observed value of an operator $A$ is given by

$$\operatorname{Tr}(\rho A)\,. \tag{C.6}$$

We now rotate the atomic system. According to Appendix B, transformation of a standard basis for atomic states may be expressed as

$$R|\alpha JM\rangle = \sum_{M'} |\alpha JM'\rangle R^{(J)}_{M'M}\,. \tag{C.7}$$

Since the set $\langle\beta GN|$ is contragredient to $|\beta GN\rangle$, the set $(-)^{G-N}\langle\beta GN|$ transforms like $|\beta G-N\rangle$. Accordingly, the set $(-)^{G-N}|\alpha JM\rangle\langle\beta GN|$ transforms like the product of two angular momentum states $|\alpha JM\rangle|\beta G-N\rangle$. We can construct irreducible sets in analogy with the coupling of two angular momenta:

$$T^{(k)}_q(\alpha J, \beta G) = \sum_{M} (-)^{G-N}\langle JGM-N|kq\rangle|\alpha JM\rangle\langle\beta GN|\,, \tag{C.8}$$

where $\langle JGM-N|kq\rangle$ is the Clebsch–Gordan coefficient (B.2). From the symmetry property (3) of the 3-$j$ symbols, for a given rank $k$  ($\geqslant 0$) the independent elements of $T^{(k)}_q$ are those with $q$ values that are nonnegative, i.e., $0 \leqslant q \leqslant k$. It is obvious from the definition (C.8) that $T^{(k)}_q$ is transformed under rotation like a state vector $|kq\rangle$:

$$RT^{(k)}_q(\alpha J, \beta G)R^{-1} = \sum_{q'=-k}^{k} T^{(k)}_{q'}(\alpha J, \beta G)R^{(k)}_{q'q}\,, \tag{C.9}$$

where the operator $R^{-1}$ on the left-hand side is meant to apply to the contragredient bra. As (C.9) indicates, the set $T^{(k)}_q$ is irreducible under rotation, i.e., the rank $k$ does not change.

The density operator can be expanded in terms of the irreducible set thus defined:

$$\rho = \sum_{\substack{\alpha J\,\beta G \\ kq}} \rho^k_q(\alpha J, \beta G)T^{(k)}_q(\alpha J, \beta G)\,, \tag{C.10}$$

where $q$ runs from $-k$ to $k$. It is obvious that, under rotation, the expansion coefficients $\rho^k_q$ transform within the same rank $k$. So, they are called the

**Table C.1.** $\rho_q^k$ and $\rho_{MN}$

| $J$ | $\rho_q^k$ | $\rho_{MN}$ |
|-----|------------|-------------|
| 1/2 | $\rho_0^0$ | $\frac{1}{\sqrt{2}}\left(\rho_{\frac{1}{2}\,\frac{1}{2}} + \rho_{-\frac{1}{2}\,-\frac{1}{2}}\right)$ |
|     | $\rho_0^1$ | $\frac{1}{\sqrt{2}}\left(\rho_{\frac{1}{2}\,\frac{1}{2}} - \rho_{-\frac{1}{2}\,-\frac{1}{2}}\right)$ |
|     | $\rho_1^1$ | $-\rho_{\frac{1}{2}\,-\frac{1}{2}}$ |
| 1   | $\rho_0^0$ | $\frac{1}{\sqrt{3}}(\rho_{11} + \rho_{00} + \rho_{-1\,-1})$ |
|     | $\rho_0^1$ | $\frac{1}{\sqrt{2}}(\rho_{11} - \rho_{-1\,-1})$ |
|     | $\rho_1^1$ | $-\frac{1}{\sqrt{2}}(\rho_{10} + \rho_{0\,-1})$ |
|     | $\rho_0^2$ | $\frac{1}{\sqrt{6}}(\rho_{11} - 2\rho_{00} + \rho_{-1\,-1})$ |
|     | $\rho_1^2$ | $-\frac{1}{\sqrt{2}}(\rho_{10} - \rho_{0\,-1})$ |
|     | $\rho_2^2$ | $\rho_{-1\,-1}$ |

irreducible components of the density matrix. The relationships between the irreducible and ordinary components are derived from (C.8):

$$\rho_{\alpha JM,\beta GN} = (-)^{G-N}\sum_{kq}\langle JGM -N|kq\rangle\rho_q^k(\alpha J,\beta G)\,,\tag{C.11}$$

$$\rho_q^k(\alpha J,\beta G) = \sum_{M}(-)^{G-N}\langle JGM -N|kq\rangle\rho_{\alpha JM,\beta GN}\,.\tag{C.12}$$

Until this point we have followed our formulation on the assumption that there can be a coherence, or a phase correlation, between different levels $\alpha J$ and $\beta G$, e.g., $1^1$S and $2^1$P levels of neutral helium atoms. However, this situation is unlikely to occur in practical PPS experiments. At this point we remove this assumption. Then if $\alpha J \neq \beta G$, quantities with $(\alpha J,\beta G)$ in the above equations disappear. Henceforth, we assume $\beta G = \alpha J$. Thus, $\rho_q^k(\alpha J,\alpha J)$ may be written simply as $\rho_q^k(p)$, where $p$ stands for $\alpha J$. As mentioned in Chap. 4, $\rho_0^0(p)$ is the "population" of level $p$, which gives the conventional population by $n(p) = \sqrt{2J+1}\rho_0^0(p)$. $\rho_q^1(p)$ with $q = 0$ and 1, and $\rho_q^2(p)$ with $q = 0, 1$, and 2 represent, respectively, the orientation and the alignment of atoms in level $p$. Table C.1 shows examples of explicit expressions of (C.12).

## C.2 Temporal Development

An atomic state follows the Schrödinger equation

$$i\hbar\frac{\partial}{\partial t}|\psi(t)\rangle = H(t)|\psi(t)\rangle\,,\tag{C.13}$$

where $H$ denotes the Hamiltonian of the system. It can be shown that the density matrix follows the equation of motion

$$i\hbar\frac{\partial}{\partial t}\rho(p;t) = [H(t),\rho(p;t)]\tag{C.14}$$

with the commutator

$$[H, \rho] \equiv H\rho - \rho H \,. \tag{C.15}$$

In the present discussion, we restrict ourselves to the atomic ensemble in level $p$. It is sometimes convenient to express (C.14) as

$$i\hbar \frac{\partial}{\partial t}\rho = [H_{\mathrm{a}}, \rho] + [H_{\mathrm{F}}, \rho] + [i\hbar \frac{\partial}{\partial t}\rho]_{\mathrm{rel}} + [i\hbar \frac{\partial}{\partial t}\rho]_{\mathrm{p}} \,, \tag{C.16}$$

where, on the right-hand side, the first term represents the time development of the atomic system in the absence of any external perturbation, the second term the effect of applied fields, the third term the relaxation, or, in the case of isotropic relaxation, the decay, and the last term the creation or pumping of this atomic ensemble.

## C.3 Observation

We observe the intensity of radiation emitted by our ensemble of atoms in level $|\alpha J\rangle$ in making a transition to $|\beta G\rangle$. We are at distance $l$ from the atoms, and a polarizer which transmits light with the polarization vector $\boldsymbol{e}$ is placed in front of our detector. The observed intensity (the energy flux per unit area) is given according to (C.6):

$$I(\alpha J, \beta G; \boldsymbol{e}) = C_{\mathrm{D}} \operatorname{Tr}\left[\mathfrak{D}(\beta G; \boldsymbol{e})\rho(\alpha J)\right] \,. \tag{C.17}$$

The order of the two operators is opposite to that in (C.6). The order is trivial in the present case, but the order in (C.17) was adopted from consistency with quantum electrodynamics [1]. Here the geometrical factor is given by

$$C_{\mathrm{D}} = \omega_{\alpha J,\beta G}^{4}/2\pi c^{3}l^{2}, \tag{C.18}$$

where $\omega_{\alpha J,\beta G}$ is the angular frequency of the transition line and $c$ is the speed of light. The detection operator is given in the present case as

$$\mathfrak{D}(\beta G; \boldsymbol{e}) = \boldsymbol{e} \cdot \boldsymbol{d} \sum_{N} |\beta G N\rangle\langle\beta G N| \boldsymbol{e}^{*} \cdot \boldsymbol{d}, \tag{C.19}$$

where $\boldsymbol{d} = -\sum e\boldsymbol{r}$ is the electric dipole operator. The interaction of the electric dipole and the radiation (angular part) is expressed as (see the beginning of Chap. 7)

$$\boldsymbol{e} \cdot \boldsymbol{d} = \sum_{s=-1}^{1} (-)^{s} e_{-s} d_{s} \tag{C.20}$$

with

$$\begin{aligned} e_0 &= \gamma, \\ e_{\pm 1} &= \mp \frac{1}{\sqrt{2}}\left(\alpha \pm i\beta\right), \end{aligned} \tag{C.21}$$

where $\alpha$, $\beta$, and $\gamma$ are the Cartesian components of $\boldsymbol{e}$ onto the $x$-, $y$-, and $z$-axes, respectively. These equations have appeared as (7.4) and (7.5) already. The trace on the right-hand side of (C.17) is explicitly written as

$$\mathrm{Tr}\left[\mathfrak{D}(\beta G; \boldsymbol{e})\rho(\alpha J)\right] =$$
$$\sum_{MM'} \langle \alpha JM | \boldsymbol{e} \cdot \boldsymbol{d} \sum_N |\beta GN\rangle\langle\beta GN| \boldsymbol{e}^* \cdot \boldsymbol{d} |\alpha JM'\rangle\langle\alpha JM'|\rho|\alpha JM\rangle .$$

$$(\text{C.22})$$

## C.4 Examples

In this section, we examine a couple of examples for the purpose of showing how the preceding formulations are applied in practical situations. Throughout this section, we take a transition of the 0–1 angular momentum pair in Fig. 7.1a, as an example. These levels are $|\beta 00\rangle$ and $|\alpha 1M\rangle$, respectively. The summation over the lower states in (C.19) and (C.22) has just one term, $\beta 00$. In spite of (C.5), we assume in this section that the density matrix is normalized, i.e., $n = 1$.

### C.4.1 $\pi$-Light Excitation

We excite our ensemble of atoms from the ground state $\beta$ to the excited level $\alpha$ with an electron beam traveling in the $z$ or $-z$ direction at the energy just above the excitation threshold. As discussed at the beginning of Chap. 1, only the magnetic sublevel $M = 0$ is populated and the $M = \pm 1$ sublevels are unpopulated. Or, we excite these atoms with a laser light beam which is linearly polarized in the $z$ direction, the $\pi$ light. This is the origin of the title of this section. It is obvious that the density matrix of the upper level is given as

$$\rho_z(\alpha) = \begin{pmatrix} 0 & 0 & 0 \\ 0 & 1 & 0 \\ 0 & 0 & 0 \end{pmatrix} . \qquad (\text{C.23})$$

We now observe the light this system emits. First, the polarization axis of our polarizer is in the $z$ direction, i.e., we are observing the $\pi$ light. Then, $\boldsymbol{e}$ has $\alpha = \beta = 0$ and $\gamma = 1$, so that $\boldsymbol{e} \cdot \boldsymbol{d} = \boldsymbol{e}^* \cdot \boldsymbol{d} = d_0$, which is proportional to the $z$-component of $\sum \boldsymbol{r}$. Since the dipole matrix element $d_0$ involves no change in the projection of angular momenta, (C.22) reduces to

$$\mathrm{Tr}[\mathfrak{D}(\beta 0; z)\rho_z(\alpha 1)] = \langle \alpha 10|d_0|\beta 00\rangle\langle\beta 00|d_0|\alpha 10\rangle\langle\alpha 10|\rho_z|\alpha 10\rangle$$
$$= \frac{1}{3}|\langle\beta 0||d||\alpha 1\rangle|^2 , \qquad (\text{C.24})$$

where $\langle\beta0||d||\alpha1\rangle$ is the reduced matrix element of the electric dipole. In the above derivation the Wigner–Eckart theorem

$$\langle\alpha JM|d_q^{(1)}|\beta GN\rangle = \langle G1Nq|JM\rangle\frac{\langle\alpha J||d^{(1)}||\beta G\rangle}{\sqrt{2J+1}}$$

$$= (-)^{J-M}\begin{pmatrix} J & 1 & G \\ -M & q & N \end{pmatrix}\langle\alpha J||d^{(1)}||\beta G\rangle \tag{C.25}$$

and the relationship

$$\langle\alpha J||d^{(1)}||\beta G\rangle^* = (-)^{G-J}\langle\beta G||d^{(1)\dagger}||\alpha J\rangle \tag{C.26}$$

have been used. To the dipole operator, superscript "(1)" has been added so as to show explicitly its rank as an irreducible tensor (a vector in this case), and $d^{(1)\dagger}$ is the same as $d^{(1)}$ in the present case.

The intensity of the polarized component in the $y$ direction is calculated from $\beta = 1, \alpha = \gamma = 0$, for $e$ so that $e\cdot d = e^*\cdot d = \frac{i}{\sqrt{2}}(d_{+1}+d_{-1})$. The trace is given as

$$\mathrm{Tr}\left[\mathfrak{D}(\beta0;y)\rho_z(\alpha1)\right] = -\frac{1}{2}\langle\alpha10|d_{+1}+d_{-1}|\beta00\rangle$$

$$\times \langle\beta00|d_{+1}+d_{-1}|\alpha10\rangle\langle\alpha10|\rho_z|\alpha10\rangle$$

$$= 0. \tag{C.27}$$

The vanishing matrix elements are understood from

$$d_{\pm1} = \mp\frac{1}{\sqrt{2}}(d_x \pm id_y) = \mp\frac{1}{\sqrt{2}}d_\perp e^{\pm i\phi} \tag{C.28}$$

and the atomic states in (C.27) which have no $\phi$-dependence. Here $d_\perp$ means the component of the dipole perpendicular to the quantization axis.

### C.4.2  σ-Light Excitation

Suppose our laser light is polarized in the $y$ direction. Starting from the $\pi$-light excitation, (C.23), we rotate the system with the Euler angle $(\phi, \theta, \gamma) = (\pi/2, \pi/2, 0)$. The rotation matrix is

$$r^{(1)}\left(\frac{\pi}{2}\right) = \begin{pmatrix} 1/2 & -1/\sqrt{2} & 1/2 \\ 1/\sqrt{2} & 0 & -1/\sqrt{2} \\ 1/2 & 1/\sqrt{2} & 1/2 \end{pmatrix}, \tag{C.29}$$

and $e^{-i(\pi/2)M}$ is trivial. According to (B.8), the resulting density matrix is

$$\rho_y(p) = \begin{pmatrix} 1/2 & 0 & 1/2 \\ 0 & 0 & 0 \\ 1/2 & 0 & 1/2 \end{pmatrix}. \tag{C.30}$$

Besides the equal populations in the $M = \pm 1$ sublevels, there is a coherence between these states.

By following the procedure similar to the above, i.e., (C.22) with (C.25), we can readily show the polarization characteristics of the observed radiation, i.e.,

$$\mathrm{Tr}\left[\mathfrak{D}(\beta 0; y)\rho_y(\alpha 1)\right] = \frac{1}{3}|\langle\beta 0||d||\alpha 1\rangle|^2 \tag{C.31a}$$

and

$$\mathrm{Tr}\left[\mathfrak{D}(\beta 0; x)\rho_y(\alpha 1)\right] = \mathrm{Tr}\left[\mathfrak{D}(\beta 0; z)\rho_y(\alpha 1)\right] = 0. \tag{C.31b}$$

### C.4.3 Magic-Angle Excitation

The magic angle $\theta_m$ is defined such that the direction of a vector starting from the coordinate origin makes equal angles with respect to the $x$-, $y$-, and $z$-axes. This angle $\theta_m$ is called the magic angle and satisfies $\cos^2\theta_m = 1/3$ (see (4.5)). We now start with the $\pi$-light excitation, (C.23), and rotate the system by $(\phi, \theta, \gamma) = (\pi/4, \theta_m, 0)$. Then, we have the initial state of (6.8). Derivation is left with the reader. From Table C.1, the irreducible components of the original density matrix for the $\pi$-light excitation, (C.23), are

$$\rho_0^2 = -2/\sqrt{6}, \quad \rho_1^2 = 0, \quad \rho_2^2 = 0, \tag{C.32a}$$

while those for the magic-angle excitation are

$$\rho_0^2 = 0, \quad \rho_1^2 = -i/3, \quad \rho_2^2 = i/3. \tag{C.32b}$$

The population imbalance in the former case is transformed to coherences in the latter. This transformation can be practiced directly from Wigner's formula, (B.12) for $r_{MM'}^{(2)}(\theta_m)$. This problem is also left with the reader.

### C.4.4 Isotropic Excitation

Among the magnetic sublevels, we have equal populations and no coherence:

$$\rho_\circ = \begin{pmatrix} 1/3 & 0 & 0 \\ 0 & 1/3 & 0 \\ 0 & 0 & 1/3 \end{pmatrix}, \tag{C.33}$$

where circle stands for "isotropic" (see also (6.7) and (6.8)). We now calculate the trace of (C.22) for this case for $e$ in the $z$ direction, or the $\pi$ light, and also in the $x$ or $y$ direction. It is readily seen that

$$\mathrm{Tr}[\mathfrak{D}(\beta 0; z)\rho_\circ(\alpha 1)] = \mathrm{Tr}[\mathfrak{D}(\beta 0; x)\rho_\circ(\alpha 1)] = \mathrm{Tr}[\mathfrak{D}(\beta 0; y)\rho_\circ(\alpha 1)]$$

$$= \frac{1}{9}|\langle\beta 0||d^{(1)}||\alpha 1\rangle|^2. \tag{C.34}$$

The emitted radiation is unpolarized and its intensity distribution is isotropic. In this case, the total energy emitted in unit time by this ensemble of atoms is

$$4\pi l^2 C_{\mathrm{D}} \mathrm{Tr}[\mathfrak{D}(\beta 0; e)\rho_\circ(\alpha 1)] \times 2 = \frac{4}{9}\frac{\omega^4_{\alpha 1,\beta 0}}{c^3}|\langle\beta 0||d^{(1)}||\alpha 1\rangle|^2$$
$$= A(p, q)\hbar\omega_{p,q}. \tag{C.35}$$

Remember that our density matrix is normalized here. In (C.35), we have taken into account two polarized components, and the Einstein $A$ coefficient has been defined as

$$A(\alpha J, \beta G) = \frac{1}{(2J+1)}\frac{4}{3}\frac{\omega^3_{\alpha J,\beta G}}{\hbar c^3}|\langle\beta G||d^{(1)}||\alpha J\rangle|^2. \tag{C.36}$$

### C.4.5 Magnetic Field

We adopt here the density matrix in the ordinary representation rather than the irreducible one. A magnetic field $\boldsymbol{B}$ is applied in the $z$ direction. The perturbation Hamiltonian in (C.16) is

$$H_{\mathrm{F}} = -\boldsymbol{\mu}\cdot\boldsymbol{B}$$
$$= -\hbar\omega J_z, \tag{C.37}$$

where the Larmor (angular) frequency is given as $\omega = \mu_{\mathrm{B}}g_J B$. Here $\mu_{\mathrm{B}} \equiv \hbar e/2m$ is the Bohr magneton and $g_J$ is the Landé $g$-factor. In the matrix form, (C.37) is written as

$$H_{\mathrm{F}} = -\hbar\omega\begin{pmatrix} 1 & 0 & 0 \\ 0 & 0 & 0 \\ 0 & 0 & -1 \end{pmatrix}. \tag{C.38}$$

In (C.16), the first term reduces to zero, and we consider only the perturbation term here. Equation (C.16) leads to

$$i\frac{\partial}{\partial t}\begin{pmatrix} \rho_{11} & \rho_{10} & \rho_{1-1} \\ \rho_{01} & \rho_{00} & \rho_{0-1} \\ \rho_{-11} & \rho_{-10} & \rho_{1-1} \end{pmatrix} = i\omega\begin{pmatrix} 0 & \rho_{10} & 2\rho_{1-1} \\ -\rho_{01} & 0 & \rho_{0-1} \\ -2\rho_{-11} & -\rho_{-10} & 0 \end{pmatrix}. \tag{C.39}$$

The solution is obvious. For example,

$$\rho_{1-1}(t) = \rho_{1-1}(0)\,e^{2i\omega t}, \qquad \rho_{-11}(t) = \rho_{-11}(0)\,e^{-2i\omega t}. \tag{C.40}$$

At time zero, we excite our atomic system with the $y$-polarized light, (C.30). The time development of our ensemble is given as

$$\rho_y(t) = \frac{1}{2}\begin{pmatrix} 1 & 0 & e^{2i\omega t} \\ 0 & 0 & 0 \\ e^{-2i\omega t} & 0 & 1 \end{pmatrix}. \tag{C.41}$$

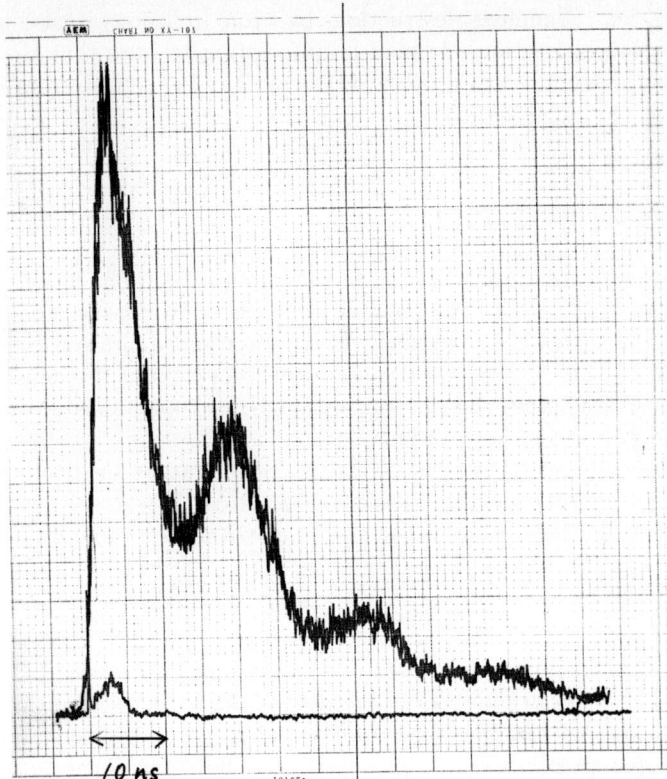

**Fig. C.1.** An example of observation of (C.42) in a LIFS experiment: pulsed laser light, which is polarized in the $y$ direction, is incident on a neon discharge plasma, a magnetic field is applied in the $z$ direction and the $y$ component of the fluorescence is observed from the $z$ direction. The Ne I $\lambda616.3\,\text{nm}$ $(1\text{s}_3(J=0)-2\text{p}_2(J=1))$ line both for excitation and observation (Fig. 7.1a). The time resolution of the system is indicated by the small peak of the scattered laser light signal in the background, i.e. the signal with the discharge current off. The observed sinusoidal-decay signal consists of the Larmor precession, (C.42), the population decay, (C.46) later, and disalignment

From (C.20)–(C.22) it is obvious that the observed intensity of the $z$-polarized component vanishes, (see (C.31b)). The trace for observation of the $y$-polarized component is readily calculated to be

$$\text{Tr}[\mathfrak{D}(\beta0;y)\rho_y(\alpha1;t)] = \frac{1}{6}|\langle\beta0||d||\alpha1\rangle|^2(1+\cos 2\omega t). \qquad (\text{C.42})$$

The real observed signal contains the decay factor, which we ignored in deriving (C.42). Figure C.1 shows an example of the results obtained in such an excitation-observation scheme. The excitation laser pulse is indicated in the scattered-light signal in the background. Under this experimental condition, relaxation, i.e., alignment destruction, by atom collisions is also significant (see the following section). It may be interesting to note that the intensity of the $x$-component is given by (C.42) with the plus sign in front of $\cos 2\omega t$ replaced by the minus sign.


## C.5 Relaxation

In the text, we were concerned with the decay, or relaxation, of the population and alignment of our atomic ensemble. The former was called the depopulation, and the latter the alignment destruction and expressed by $C^{2,2}(p,p)$ in Fig. 4.2. In this section, we discuss these processes in the density matrix formalism. We assume that the relaxation is isotropic. This is the case for radiative decay and atom collisions.

Following Omont [1], we denote the relaxation rate of the irreducible component of the density matrix of rank $k$ as $g_k$. It can be shown that the rate is dependent on $k$ but is common for all the $q$s within the same $k$:

$$\frac{\mathrm{d}}{\mathrm{d}t}\rho_q^k = -g_k\rho_q^k.$$  (C.43)

We call $g_0$ the depopulation rate, $g_1$ the orientation destruction rate, and $g_2$ the *alignment destruction* rate. We now examine the relationship between these rates and the transition rates for the elements of the density matrix in the ordinary representation $\rho_{MN} \rightarrow \rho_{M'N'}$. The transition rate within the $|\alpha JM\rangle$ multiplet is given as

$$G_{MN,M'N'} = \sum_{kq}(-)^{2J+M+M'}\langle JJM-N|kq\rangle\langle JJM'-N'|kq\rangle g_k(J),$$  (C.44)

where $q$ runs from $-k$ to $k$. We may define the rates

$$h_k = g_k - g_0$$  (C.45)

and call $h_1$ the disorientation rate and $h_2$ the *disalignment* rate. It may be readily shown that the depopulation process is expressed as

$$\frac{\mathrm{d}}{\mathrm{d}t}\rho_0^0(J) = -g_0\rho_0^0(J).$$  (C.46)

For the disalignment process, i.e., transitions within the $|\alpha JM\rangle$ multiplet, we take the case of $J=1$ and examine it in detail. From (C.44) with (C.45) we have

$$\frac{d}{dt} \begin{pmatrix} \rho_{11} & \rho_{10} & \rho_{1-1} \\ \rho_{01} & \rho_{00} & \rho_{0-1} \\ \rho_{-11} & \rho_{-10} & \rho_{-1-1} \end{pmatrix}$$

$$= -\frac{h_2}{6} \begin{pmatrix} \rho_{11} - 2\rho_{00} + \rho_{-1-1} & 3(\rho_{10} - \rho_{0-1}) & 6\rho_{1-1} \\ 3(\rho_{01} - \rho_{-10}) & -2(\rho_{11} - 2\rho_{00} + \rho_{-1-1}) & 3(\rho_{0-1} - \rho_{10}) \\ 6\rho_{-11} & 3(\rho_{-10} - \rho_{01}) & \rho_{11} - 2\rho_{00} + \rho_{-1-1} \end{pmatrix}.$$

$$(C.47)$$

The reader may verify that the three elements of the alignment, $\rho_0^2$, $\rho_1^2$, and $\rho_2^2$, in Table C.1 decay with the same rate $(g_0 + h_2 = g_2)$. This is the property on which the experiment in Sect. 6.1.4 was based (see (C.32a) and (C.32b)).

# References

1. A. Omont, Prog. Quant. Electron. **5**, 69 (1977)
2. K. Blum, *Density Matrix Theory and Applications*, 2nd edn. (Plenum, New York, 1996)

# Appendix D

# Hanle Effect

In this appendix, we introduce the Hanle effect, which played an important role in developing PPS.

## D.1 Classical Picture

The classical atom consists of an electron which is attracted by the ion core with the harmonic force. Suppose the atom is located at the origin. A pulsed beam of light propagating in the $z$ direction with its polarization directed in the $y$ direction illuminates the atom and the atomic electron is accelerated by the electric field. The electron begins to oscillate in the $y$ direction, and the oscillation decays with the natural lifetime, i.e., the spontaneous decay. The emitted light, the resonance fluorescence, is polarized in the $y$ direction and it also decays with the same rate. Suppose, a magnetic field is applied in the $z$ direction. In the oscillation motion, the electron is exerted by the Lorentz force, and its trajectory is modified; the oscillation direction rotates around the $z$-axis, i.e., the Larmor precession. The direction of the polarization of the emitted light accordingly rotates. Suppose we observe the $y$-polarized component and the $x$-polarized component of the emitted light separately, and integrate the signals over the whole decay time. Owing to the presence of the magnetic field, the averaged polarization degree becomes smaller. The rate of decrease is determined by the balance between the decay time of the atom and the Larmor precession frequency, which is proportional to the strength of the applied magnetic field. Therefore, the observed polarization degree against the magnetic field has a Lorentzian profile with its FWHM (full width at half maximum) given by the natural lifetime. Here we have assumed that the Landé $g$-factor is known. This effect is known as the Hanle effect, and the plot of the polarization degree against the magnetic field strength is called the Hanle signal.

If the excitation-observation geometry is different, the observed profile can be of a dispersion shape. For example, when our polarizer is directed in the $\pm 45°$ with respect to the $y$ direction, we obtain a dispersion shape profile.

When our atom suffers collisional relaxation, the effective lifetime shortens, and the FWHM increases. Thus, FWHM against, for example, the atom density gives a straight line. From this plot, we can determine the natural lifetime from the intercept in the zero-density limit and the collisional relaxation rate coefficient from the slope of this line.

## D.2 Quantum Picture

We have an atomic ensemble at the origin, and this system is illuminated by the beam of light in the same way as the above. We assume the 0–1 angular momentum pair for our transition. The initial density matrix is given by (C.30), and its time development is given by (C.41). To be realistic, we include the spontaneous decay. Then, the density matrix is multiplied by the decay factor $e^{-g_0 t}$, and the intensity of the emitted radiation with our polarizer directed in the $y$ direction is given by (C.42) with the same decay factor multiplied. The time-integrated signal is

$$S_y \propto \frac{1}{g_0} \left[ 1 + \frac{1}{1 + (2\omega/g_0)^2} \right] , \tag{D.1}$$

where we have dropped the constant factor. Here, the Larmor (angular) frequency is $\omega = \mu_B g_J B$. See (C.37) and below. The observed intensity of the $x$ polarized component is, as noted after (C.42), given by (D.1) with the plus sign between the two terms in the brackets in (D.1) replaced by the minus sign. Therefore, the polarization degree is given as

$$P = \frac{S_y - S_x}{S_y + S_x} = \frac{1}{1 + (2\omega/g_0)^2} , \tag{D.2}$$

From the FWHM of the Lorentzian profile, we can obtain $g_0$, the spontaneous decay rate.

Now, we include collisional relaxation as expressed by $h_1$ and $h_2$ (see Sect. C.5). $g_0$ may include collisional depopulation (see Fig. E.2). It is interesting to note that the disorientation, $h_1$, has no influence on the Hanle signal. The disalignment, $h_2$, affects the Hanle signal exactly in the same way as the spontaneous decay. That is, the observed polarization degree is given by (D.2) with $g_0$ replaced by $g_2$ (see (C.45)). The proofs are left with the reader.

# Appendix E

# Method to Determine the Population

In this Appendix, we consider one of the practical, but basic, problems: i.e., by observing an emission line from excited atoms, how we can determine the population of the upper-level atoms. This problem may seem rather trivial, but, if these atoms are anisotropically excited, it is less straightforward. A typical example is the beam excitation experiment as mentioned in the beginning of Chap. 1 (see (6.2)). Suppose we want to determine $n(p)$ from the observed line intensity of the transition $p \to s$ from a certain direction with respect to the $z$-axis, the symmetry axis. This may be for the purpose of, e.g., in the beam excitation experiment, determining the excitation cross section $Q_0^{0,0}(r,p)$, or some other purposes in which $n(p)$ should be known without ambiguities. A typical example is determination of natural lifetime of the upper-level atoms. Since $f_2(v)$ is present in this example, an alignment is also created in level $p$ (see (6.4)). As (4.5) shows, the observed intensity, in general, is given not only by the population but also by the alignment, the second term in the brackets, so that we have to eliminate the contribution from the term proportional to $a(p)/n(p)$. One configuration which immediately meets this requirement is to have a polarizer, the transmission axis of which is directed in $\eta = 54.7°$, the magic angle, so that $\cos^2 \eta = 1/3$. In many cases, however, the light source itself is already weak and adding a polarizer further reduces the signal-to-noise ratio in the observation. In the vuv (vacuum ultraviolet) region, we even do not have a polarizer with sufficient efficiency. So, this configuration is rather undesirable.

Instead, we could observe the radiation emitted by the atoms from the direction of 54.7° with respect to the $z$-axis. We may call this the magic angle observation. In this case, for the $\pi$ light $\eta = 35.3°$ so that $(1 - 3\cos^2 \eta) = -1$ and for the $\sigma$ light $\eta = \pi/2$ so that $(1 - 3\cos^2 \eta) = 1$. By adding these two component intensities, we eliminate the contribution from the alignment. When we use a grating spectrometer to resolve this line, however, it usually has different efficiencies for different polarized components with respect to the line of sight of the spectrometer. When the angle of incidence of the light incident on the grating surface is close to the blaze angle, at which the reflection efficiency is about the maximum, both the components, i.e., the polarized light whose electric vector oscillates in the parallel direction or the perpendicular direction to the grooves of the grating, have approximately the same efficiency. At shorter wavelengths, the former component tends to have higher efficiencies and at longer wavelengths, the latter component has higher efficiencies. This tendency is seen in Fig. 14.8: the $\pi$-light is the light polarized in the perpendicular direction to the grooves of the grating, i.e., the p-light. For the purpose of making the efficiencies of the grating virtually equal for both the $\pi$ and $\sigma$ components of the emission line, the spectrometer may be tilted by 45° around its line of sight. This method is, however, sometimes inconvenient, since the entrance slit is not parallel or perpendicular to the symmetry axis of the experimental geometry any more.

Another method almost equivalent to the magic-angle observation is that, by applying a magnetic field in the direction of the line of sight, we rotate the aligned atom system in time around the line of sight. Figure E.1 shows an example [1]. In this experiment neon atoms in the metastable state ($1s_3$: Paschen notation; see Figs. 6.19 and 7.1.) in a glow discharge plasma is excited by a linearly polarized laser pulse to the $2p_2$ level, and the subsequent fluorescence of the $1s_2$–$2p_2$ transition line is observed. The polarization direction of the laser light with respect to the line of sight is 54.7°, the magic angle. This excitation is virtually equal to the beam excitation at the excitation threshold (within the classical arguments), as discussed in the beginning of Chap. 1. The direction of polarization of the excitation laser light corresponds to the beam direction in the beam excitation. By this excitation an alignment is created in the upper level atoms along with the population. The alignment may further be relaxed by atom collisions and/or radiation reabsorption during the lifetime. These relaxation phenomena are discussed in Sects. 6.3 and 7.2, respectively. In this demonstration experiment, for the purpose of illustration, the polarization effect is made pronounced: A plastic polarizer is placed in front of the entrance slit of our spectrometer, and our system is made sensitive only to the component of the fluorescence polarized in the direction of the laser beam. In the absence of the magnetic field (0 G), the observed intensity decays with a certain apparent decay time constant. When a magnetic field is applied in the direction of the line of sight, an oscillation appears in the decay curve. This is due to the Larmor precession of the produced alignment, or the magnetic dipoles, around the magnetic field (see Appendix C). With the increase in the magnetic field strength, the oscillation frequency increases, and finally it becomes too high to be resolved by our observation system. At the magnetic field of 30 G, the oscillation virtually disappears and the effect of the alignment is smeared out. In this limit, the decay curve represents correctly the decay of the population, being independent of the alignment and its relaxation.

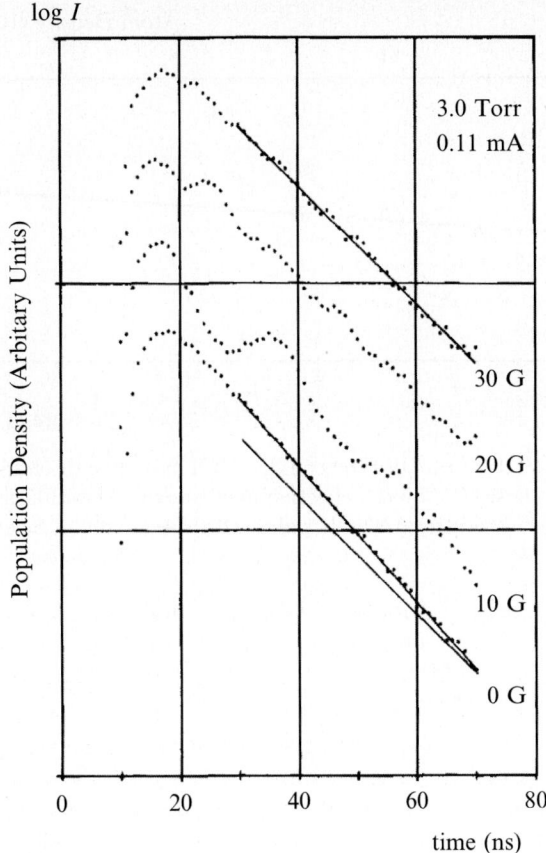

**Fig. E.1.** Temporal decay of the observed emission line intensity subsequent to excitation by a laser pulse, which is polarized in the magic angle. A magnetic field is applied in the direction of the line of sight. "0 G" means the absence of the magnetic field. The apparent decay is substantially different from the population decay as determined from the intensity decay at "30 G". (Quoted from [1], with permission from The Royal Swedish Academy of Science.)

It is seen that the apparent decay of the observed intensity without a magnetic field is substantially different from the true decay of the population. This difference is due to the disalignment, i.e., a decay of $a(p)/n(p)$ with time, which is due to atom collisions and/or radiation reabsorption. As noted above these processes are discussed in the text. It may be noted that when disalignment is absent, the apparent decay coincides with the population decay in any excitation-observation geometry. In this case, there should be no magnetic field, of course.

If the situation permits, (4.6) could be directly adopted for the present purpose; $I_\pi$ and $I_\sigma$ are measured separately and the decay of $(I_\pi + 2I_\sigma)$ is

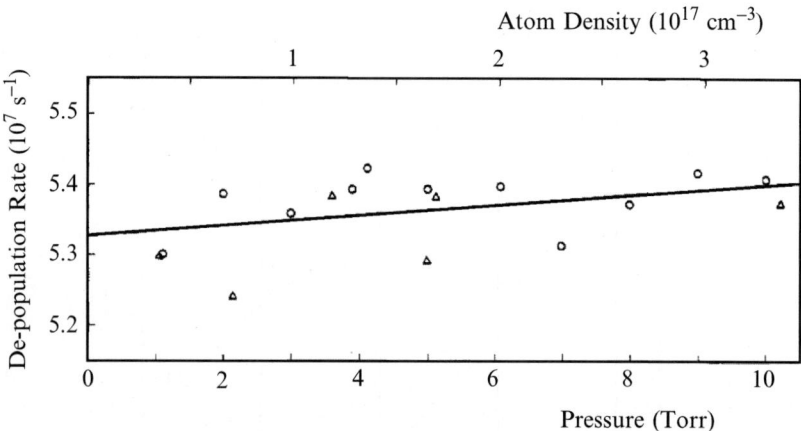

**Fig. E.2.** Depopulation rate of the neon 2p$_2$ level in a neon discharge plasma. (*circle*): from (4.6); (*triangle*): by the effective magic-angle excitation method (Quoted from [1], with permission from The Royal Swedish Academy of Science.)

determined. Figure E.2 compares the depopulation rates determined by these two methods [1]. It is seen that both the methods give virtually identical results.

## Reference

1. T. Fujimoto, C. Goto, K. Fukuda, Phys. Scripta **26**, 443 (1982)

# Index

Springer Series on
# ATOMIC, OPTICAL, AND PLASMA PHYSICS

Springer Series on
# ATOMIC, OPTICAL, AND PLASMA PHYSICS

Printing: Krips bv, Meppel, The Netherlands
Binding: Stürtz, Würzburg, Germany